Springer Undergraduate Mathematics Series

The Springer Undergraduate Mathematics Series (SUMS) is a series designed for undergraduates in mathematics and the sciences worldwide. From core foundational material to final year topics, SUMS books take a fresh and modern approach. Textual explanations are supported by a wealth of examples, problems and fully-worked solutions, with particular attention paid to universal areas of difficulty. These practical and concise texts are designed for a one- or two-semester course but the self-study approach makes them ideal for independent use.

More information about this series at http://www.springer.com/series/3423

Franz Lemmermeyer

Quadratic Number Fields

 Springer

Franz Lemmermeyer (iD)
Jagstzell, Baden-Württemberg, Germany

ISSN 1615-2085 ISSN 2197-4144 (electronic)
Springer Undergraduate Mathematics Series
ISBN 978-3-030-78651-9 ISBN 978-3-030-78652-6 (eBook)
https://doi.org/10.1007/978-3-030-78652-6

Mathematics Subject Classification: 11R11, 11D09, 11D25

Translation from the German language edition: Quadratische Zahlkörper by Franz Lemmermeyer,
© Springer-Verlag GmbH Deutschland 2017. Published by Springer Spektrum. All Rights Reserved.

This Springer imprint is published by the registered company Springer Nature Switzerland AG.
The registered company address is: Gewerbestrasse 11, 6330 Cham, Switzerland

Preface

This book evolved from a manuscript for an introductory lecture series at the University of the Saarland in Saarbrücken in 1999. The goal was to present the arithmetic of quadratic number fields and to explain how to apply the results to problems in elementary number theory.

I expect the readers to be familiar with notions such as prime numbers and residue class rings from elementary number theory, and with fundamental theorems such as unique factorization, Fermat's Little Theorem, and the quadratic reciprocity law.[1] The theory of quadratic number fields deals with similar theorems in bigger rings of integers, for example the ring of Gaussian integers, which consists of all numbers of the form $a + bi$, where a and b are ordinary integers and where $i^2 = -1$. In this ring, 5 is not a prime anymore because $5 = (1 + 2i)(1 - 2i)$. Whether an ordinary prime number p remains prime in this ring depends on the Legendre symbol $(\frac{-1}{p})$; in general quadratic number rings, the behaviour of prime numbers also depends on Legendre symbols. In this connection, we will learn that quadratic reciprocity, which is perhaps perceived as a curiosity by someone who has never looked beyond the horizon of elementary number theory, is a fundamental result that governs the behaviour of prime numbers in quadratic number rings.

Finally, the theory of quadratic number fields has numerous applications to elementary number theory. It puts several results such as the Two-Squares Theorem, which asserts that primes of the form $p = 4n + 1$ can be written as the sum of two squares, into a bigger perspective, and it allows you to solve Diophantine equations, in particular certain Bachet–Mordell equations $y^2 = x^3 + k$, special cases of the Catalan equation $x^p + y^q = 1$, or the Fermat equations $x^n + y^n = z^n$ for $n \leq 5$.

Any book on quadratic number fields has to cover a set of standard topics such as rings of integers, unique factorization into ideals, finiteness of the class group and the solvability of the Pell equation. It is the topics outside the standard curriculum that "define" this book, so I would like to say a few words about them here.

---

[1] This result will be proved in several ways in this book, but I assume that the readers know how to apply it.

Chapter 1 on the "prehistory" of algebraic number theory is a reflection of my occupation with the history of number theory. Mathematics is not an accumulation of true results, it is a development of ideas. One such idea in number theory is Dedekind's definition of prime numbers by their property that primes divide a factor whenever they divide some product; another one is his definition of ideals, which shaped the development of algebra. Similarly, the proof of the Unique Factorization Theorem is not difficult once it is formulated; a lot more important than the mere truth of this results is the *idea* that it may serve as the foundation of elementary number theory. Some of the concepts used in Chap. 1 are defined and studied properly only in subsequent chapters; readers should regard this chapter as a view on the promised land—all the concepts showing up there will be discussed properly in subsequent chapters.

Another unusual topic is the arithmetic of Pell conics presented in Chap. 2. I find the interpretation of the technique of Vieta jumping in terms of group laws on conics immensely pleasing; its ultimate goal is making the theory of elliptic curves more accessible to beginners.

Chapter 3 on the quadratic reciprocity law is a new addition. As in the last chapter on quadratic Gauss sums, the presentation I have given differs from the usual approach. I am convinced that importing the idea of modularity, which is central in class field theory and in the arithmetic of elliptic curves, to elementary number theory is a step that is long overdue. The notion of modularity is taken up again in the last chapter, where we formulate several results whose proofs are beyond the scope of the present book.

The topics covered in the other chapters up to the ambiguous class number formula are classical, and I am sure that much of it comes from the books on algebraic number theory by Harvey Cohn (in particular [22]). These chapters contain the basic arithmetic of quadratic number fields: rings of integers, units, ideals and ideal class groups.

Prerequisites for reading this book are basic notions of linear algebra (vector spaces, linear maps, matrices) and a familiarity with elementary number theory. Quadratic reciprocity will be proved from scratch in Chap. 3, but I assume some familiarity with its use in theoretic problems. Only in the last two chapters some abstract algebra (the basic isomorphism theorems, or the irreducibility of cyclotomic polynomials) is required.

I have profited from a number of useful comments by Dirk Bachmann and Heiko Hellwig. Chip Snyder deserves a huge thank you for reading a first draft of the English translation and for correcting many errors and pointing out gaps. The two anonymous referees and Remi Lodh provided me with lots of useful comments—I thank them all for their support.

Jagstzell, Germany   Franz Lemmermeyer
March 2021

Contents

Chapter 1
Prehistory

The idea of transferring the arithmetic of the ordinary integers to quadratic number rings appears to be so natural to those who are familiar with some abstract algebra that we tend to underestimate the achievement by Carl Friedrich Gauss, who paved the way by studying integers of the form $a + b\sqrt{-1}$ in the early nineteenth century. Before I discuss the contributions of Gauss and his successors Ernst Eduard Kummer and Richard Dedekind, I would like to show the immense difficulties that Leonhard Euler had to cope with when he used algebraic numbers for solving problems going back to Pierre Fermat, Claude Gaspard Bachet de Meriziac and ultimately even to Diophantus. Those who would like to familiarize themselves with the number theoretical work of Fermat are well advised to study André Weil's excellent book [132] (and, if they read German, [88]).

1.1 Pythagoras and Euclid

One of the oldest nontrivial theorems in geometry is the Theorem of Pythagoras, according to which the sides a, b, and c of a right triangle, where c is the side opposite to the right angle, satisfy the equation $a^2 + b^2 = c^2$. All ancient cultures (in particular, those living near the silk road (a trade route connecting the Mediterranean area with China and India), namely the Babylonians, the Chinese, the Hindus, and the Greeks) realized that this equation has integral solutions. The cuneiform tablet Plimpton 322 shows that already the Babylonians around 1800 BC knew how to generate arbitrarily many integral solutions of this equation. One possible way of discovering a method for generating triples (a, b, c) of integers satisfying $a^2 + b^2 = c^2$ is the following.

© The Author(s), under exclusive license to Springer Nature Switzerland AG 2021
F. Lemmermeyer, *Quadratic Number Fields*, Springer Undergraduate
Mathematics Series, https://doi.org/10.1007/978-3-030-78652-6_1

In some of the Babylonian problems[1] the task is computing a pair of rational reciprocal numbers m and $\frac{1}{m}$ from their sum $a = m + \frac{1}{m}$. The numbers m and $\frac{1}{m}$ are solutions of the quadratic equation $(x - m)(x - \frac{1}{m}) = x^2 - ax + 1 = 0$, and we know that this equation has a rational solution if and only if its discriminant $a^2 - 4$ is a rational square. Working backwards we see that this is easily achieved by setting $a = m + \frac{1}{m}$ for a rational number m, and in fact

$$\left(m + \frac{1}{m}\right)^2 - 4 = \frac{m^4 + 2m^2 + 1}{m^2} - 4 = \left(m - \frac{1}{m}\right)^2.$$

Clearing the denominators yields the solution

$$(m^2 + 1)^2 - (2m)^2 = (m^2 - 1)^2 \tag{1.1}$$

of the Pythagorean equation. If we set $m = \frac{t}{u}$ and clear denominators again we obtain

$$(t^2 - u^2)^2 + (2tu)^2 = (t^2 + u^2)^2. \tag{1.2}$$

With hindsight we can see that the basis of our derivation of (1.1) is the fact that the equation $x^2 + y^2 = z^2$ can be written as $x^2 = z^2 - y^2$, and that a difference of squares can be factored:

$$(m^2 + 1)^2 - (2m)^2 = (m^2 + 2m + 1)(m^2 - 2m + 1)$$

$$= (m + 1)^2(m - 1)^2 = (m^2 - 1)^2.$$

We can also find Pythagorean triples starting with the famous diagram of the "square in the middle" in Fig. 1.1. The area of the large square (Fig. 1.1) is $(a + b)^2$; since it is composed of the small square in the middle and four rectangles, this must be equal to $(a - b)^2 + 4ab$. Thus $(a - b)^2 + 4ab = (a + b)^2$, or, after dividing through by 4,

$$ab = \left(\frac{a + b}{2}\right)^2 - \left(\frac{a - b}{2}\right)^2, \tag{1.3}$$

which again shows that the difference of two squares is a product.

In order to find Pythagorean Triples we make ab equal to a square, for example, by setting $a = m^2$ and $b = n^2$. Then we obtain

$$(m^2 - n^2)^2 + (2mn)^2 = (m^2 + n^2)^2.$$

---

[1] For learning more about the methods used in "Babylonian algebra" see [63].

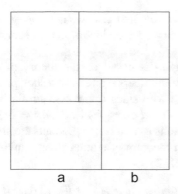 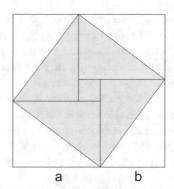

$$a \qquad b \qquad\qquad a \qquad b$$

Fig. 1.1 Geometric derivation of Pythagorean triples

If we draw four diagonals into the Babylonian square in Fig. 1.1, we get a proof of the Pythagorean Theorem for free. In fact, the area of the shaded square is c^2, where c denotes the hypotenuse of the right triangle with legs a and b; on the other hand, it is also equal to $4 \cdot \frac{ab}{2} + (a-b)^2 = 2ab + a^2 - 2ab + b^2 = a^2 + b^2$.

The verification that the triples $(t^2 - u^2, 2tu, t^2 + u^2)$ are solutions of the equation $x^2 + y^2 = z^2$ requires only *elementary algebra*, and it shows that Pythagorean triples exist in arbitrary commutative rings. The proof that there are no other solutions in ordinary integers, on the other hand, requires *arithmetic*; one such proof uses the Unique Factorization Theorem for integers. We regard unique factorization as the natural foundation of elementary number theory, and we do so to such an extent that we find it hard to believe that this is a modern insight. As a matter of fact you will look in vain for the concept of unique factorization in Euclid's *Elements*: Euclid based number theory on the *Four Numbers Theorem* (see [79, 102]), which is Proposition 19 in Book VII of his *Elements*:

Theorem 1.1 (Four Numbers Theorem) *Let a, b, c, d be natural numbers with $ab = cd$. Then there exist natural numbers x, y, z, w such that $a = xy$, $b = zw$, $c = xz$, and $d = yw$.*

The basic idea behind the Four Numbers Theorem is rather natural: Each decomposition of a number into different factors can be explained by a refined decomposition. The factorization $2 \cdot 12 = 4 \cdot 6$, for example, may be refined to $2 \cdot (2 \cdot 6) = (2 \cdot 2) \cdot 6$. The Four Numbers Theorem quickly implies Euclid's Lemma (Prop. VII.30):

Lemma 1.2 (Euclid's Lemma) *If a prime number p (in the classical sense, i.e., a number > 1 only divisible by 1 and itself) divides a product ab of natural numbers, then p divides one of the factors a or b (or both).*

In fact, since p is a divisor of ab we can write $ab = pc$ for some $c \in \mathbb{N}$. According to the Four Numbers Theorem 1.1 we must have $a = xy$, $b = zw$, $p = xz$, and $c = yw$ for integers x, y, z, and w. Since p does not have a nontrivial factorization, we must have $x = 1$ or $z = 1$. In the first case we get $a = y$, that is $c = aw$ and hence $ab = paw$; but then $b = pw$ and p divides b. In the second case we conclude similarly that p is a divisor of a.

Euclid's investigation of Pythagorean triples is based on the theory of plane numbers. These are products ab of two natural numbers a and b; two such products ab and cd are called *similar* if the rectangle with sides a and b is similar to the rectangle with sides c and d, that is, if $a : b = c : d$ or, equivalently, $ad = bc$. This implies the equation $ab \cdot d^2 = cd \cdot b^2$, which tells us that if two products are similar, then they differ by square factors. The main theorem of similar plane numbers is contained in Euclid's Propositions IX.1 and IX.2:

Theorem 1.3 *The product[2] of two natural numbers is a square if and only if these numbers are similar.*

As an example consider the product $6 \cdot 24 = 12^2$, which is a square; in accordance with Euclid the products $6 = 2 \cdot 3$ and $24 = 4 \cdot 6$ are similar plane numbers.

The connection with Pythagorean triples arises from an identity that Euclid presents in geometric clothing in Book II of his *Elements* and that already the Babylonians were aware of, namely Eq. (1.3), from which we already have derived a parametrization of Pythagorean triples.

From Theorem 1.3 we deduce an observation which is the basis of many applications to Diophantine problems, and which occurs in Euclid's *Elements* as a very special case of Proposition VIII.7:

Theorem 1.4 (Square Lemma) *If a and b are coprime natural numbers and if ab is a square, then a and b must be squares.*

According to Theorem 1.3 we have $a = rs$, $b = tu$, as well as $r : t = s : u$; the last relation yields $ru = st$. According to the Four Numbers Theorem 1.1 there exist numbers x, y, z, w such that $r = xy$, $u = zw$, $s = xz$, and $t = yw$. Since $a = rs = x^2yz$ and $b = tu = yzw^2$ are coprime, we must have $y = 1$ and $z = 1$, and thus $a = x^2$ and $b = w^2$ are squares.

Without invoking the Four Numbers Theorem we can conclude from $\gcd(r, t) = \gcd(s, u) = 1$ and $ru = st$ that r divides s and that s divides r, which in turn implies $r = s$. In the same way we deduce that $t = u$, and this shows again that a and b are squares.

We now would like to use these methods for showing that the Euclidean equation (1.2) contains all *primitive* Pythagorean triples; these are triples (a, b, c) with pairwise coprime natural numbers a, b, and c. Assume that (a, b, c) is such a triple. Then it is easily seen that c must be odd, and that a and b have different parity. We

---

[2]For Euclid, the product of a number is the representation of a number as a product, not the result. When Euclid wants the result of a product, he uses a clumsy phrase such as "if two numbers multiplied make a number."

may therefore assume that a is odd and $b = 2u$ is even and then obtain

$$(2u)^2 = b^2 = c^2 - a^2 = (c-a)(c+a), \quad \text{i.e.,} \quad u^2 = \frac{c-a}{2} \cdot \frac{c+a}{2}.$$

Observe that the last equation is just (1.3). Since a and c are coprime, so are the two factors on the right hand side (as a matter of fact, any number dividing both $\frac{c-a}{2}$ and $\frac{c+a}{2}$ divides their sum c and their difference a), and according to Theorem 1.4 both numbers must be squares. Thus $\frac{c-a}{2} = t^2$ and $\frac{c+a}{2} = s^2$, which immediately implies $c = s^2 + t^2$ and $a = s^2 - t^2$. Since $u = st$ we also have $b = 2u = 2st$, and we have shown that each primitive Pythagorean triple is contained in (1.2):

Proposition 1.5 *If (x, y, z) is a primitive Pythagorean triple and if x is even, then there exist numbers a and b such that $x = 2ab$, $y = a^2 - b^2$ and $z = a^2 + b^2$.*

We will meet the main idea behind Euclid's classification of Pythagorean triples in Prop. X.29a over and over again: The transformation of an additive into a multiplicative problem. This idea already occurred in the derivation of Pythagorean triples using the Babylonian "square in the middle."

Just as the Babylonians, the Chinese[3] solved numerous geometric problems in which they gave rational values to the side lengths of right triangles, and so did Hindu mathematicians such as Brahmagupta. Diophantus, and later also Fermat and his contemporaries, solved a huge number of problems in which the rational or integral sides of right triangles with additional properties had to be determined. This shows that one of the sources of number theory is metrical geometry.

1.2 Diophantus

During the European Middle Ages, sciences in Europe were almost non-existent. Even Euclid's *Elements* had been largely forgotten. Only the contact with the Muslim occupants of Spain and Sicily in the eleventh and twelfth century AD made the few European scientists (almost all of them were monks and bishops, since the monastery schools—the only schools where Latin was taught—were reserved for the future clergy) aware of the existence of classical works on medicine, astronomy, and mathematics. The invention of the printing press by Gutenberg made the *Elements* available to people outside of monasteries (in particular once they had been translated into other languages) and without access to the large libraries.

The most important event for the development of algebra and number theory was without doubt the discovery of the *Arithmetica* by Diophantus, who must have lived between 200 BC and 300 AD in Alexandria, a city near the Nile Delta founded by Alexander the Great in 331 BC shortly before his death. Alexander's

---

[3]Presentations of their work may be found in Vogel [126, 127] and in Chemla and Guo [19].

successor decided to build a library there, which attracted scientists from the whole Mediterranean world[4] and made it the scientific center of antiquity. Among the most famous scientists who have worked there are Euclid, the author of the *Elements*, Archimedes, one of the greatest mathematicians and, like Heron, also an exceptionally gifted engineer, the astronomer Ptolemy and Diophantus. Six of his thirteen books on arithmetic problems have survived in Greek, and another four (discovered only in the 1970s) in Arabic translation.

In this chapter we will discuss the following two problems solved by Diophantus in his *Arithmetica*: His construction of Pythagorean triples and Problem VI.19, which was to play an important role in the development of algebraic number theory.

We remark in advance that Diophantus already used some algebraic notation: He had symbols for *one* unknown and its powers up to the sixth.

Pythagorean Triples Diophantus treats the problem of finding Pythagorean triples in the following form:

II.10 To decompose a given square number into two squares.

Diophantus shows how to decompose 16 into a sum of two squares; he sets the first square equal to x^2, and then the second square is $16 - x^2$. Now he writes:

We form the square of an arbitrary multiple of x reduced by the side of the given square, say $2x - 4$.

The side of the square 16 is 4; subtracting 4 from a multiple of x we get $mx - 4$. Since Diophantus does not have symbols for more than one unknown, he picks $m = 2$, but it is clear that any other choice of m would work equally well. The square of $2x - 4$ then must equal the second number, that is, we have the equation

$$(2x - 4)^2 = 4x^2 - 16x + 16 = 16 - x^2.$$

Now we see the idea behind the choice of $2x - 4$: On both sides of the equation we have the constant term 16, which can be canceled; adding x^2 we obtain $5x^2 = 16x$, hence $x = \frac{16}{5}$. Thus one square is $\frac{256}{25}$, the other $\frac{144}{25}$.

Many centuries later it was observed that the substitution $y = 2x - 4$ may be interpreted as the equation of a line in the Euclidean plane. The equations of Diophantus then may be visualized geometrically as intersecting this line with the circle with radius 4 around the origin (see Fig. 1.2).

In order to decompose a^2 into a sum of two squares let us call the smaller square x^2; then $a^2 - x^2 = y^2$ must also be a square. The substitution $y = mx - a$ yields

$$a^2 - x^2 = m^2x^2 - 2amx + a^2,$$

hence $2amx = (m^2 + 1)x^2$ and thus $x = \frac{2am}{m^2+1}$.

---

[4]This included the Hellenistic world. Among the scientists believed to have studied in Alexandria are Archimedes from Syracuse in Sicily and Eratosthenes from Cyrene in North Africa. It is also conceivable that well-educated scribes from Mesopotamia preferred the boomtown Alexandria to the declining cities in Mesopotamia.

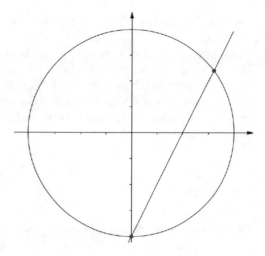

Fig. 1.2 Parametrization of Pythagorean triples

Now we find that the second square number is the square of $y = mx - a = a \cdot \frac{m^2-1}{m^2+1}$. This shows that

$$\left(\frac{2am}{m^2+1}\right)^2 + \left(a \cdot \frac{m^2-1}{m^2+1}\right)^2 = a^2,$$

and after canceling a^2 and getting rid of the denominator we recover (1.1).

Problem VI.19

To find a right-angled triangle in which the area increased by the hypotenuse is a square, and the perimeter is a cube.

Diophantus solves this problem[5] as follows. He denotes the area by x and the hypotenuse as a square minus x, say $c = 16 - x$. The product of the legs is $2x$; if one leg was equal to 2, the other would be x, and the perimeter $2+x+16-x = 18$, which is not a cube. Thus, says Diophantus, we need a square which increased by 2 makes a cube.

If one side of the square is $m + 1$ and the side of the cube is $m - 1$, then we must solve $m^3 - 3m^2 + 3m - 1 = m^2 + 2m + 3$, which gives $m = 4$. Thus the side of the square is 5 that of the cube 3.

If x denotes the area of the original triangle and $25 - x$ its hypotenuse, and if 2 and x are its legs, then the theorem of Pythagoras gives us $x^2 - 50x + 625 = x^2 + 4$, i.e., $x = \frac{621}{50}$, and the problem is solved.

Diophantus was forced to choose the substitution $c = 16 - x$; his calculations, however, may be transferred to the general substitution $c = k^2 - x$. At one point,

---

[5]This is problem VI.17 in Heath [59]; some problems are enumerated in a different way in different editions.

however, Diophantus "cheated": The solution $m = 4$ of the cubic equation $m^3 - 3m^2+3m-1 = m^2+2m+3$ can be found by writing it in the form $m^3+m = 4m^2+4$, which implies $m(m^2 + 1) = 4(m^2 + 1)$, from which the solution can be read off. This does not work for general values of k, and there were no techniques available for solving general cubics in the times of Diophantus.

Moreover, even if we know how to solve cubics we do not know how to choose the sides of the square and the cube in such a way that the resulting cubic has a rational solution.

1.3   Bachet

We do not know whether or how much Diophantus was studied in antiquity. Hypatia (355–415), the daughter of Theon of Alexandria (335–405), is often said to have written a comment on Diophantus *Arithmetica*; this story seems to be based on a misguided interpretation by Tannery. Diophantus was studied by many Muslim (and a few Byzantine) scientists; in Western Europe, Diophantus remained unknown until Johannes Regiomontanus from Königsberg (Lower Franconia, Bavaria) discovered a copy of six of the 13 books in a library in Venice in 1463.

The first edition of the *Arithmetica* was prepared a century later by Wilhelm Holtzmann (1532–1576) under the name of Guilielmus Xylander; based on this work Claude Gaspard Bachet de Mériziac (1581–1638) published an improved version in 1621—not only the text had to be translated from Greek into Latin, Bachet (like Xylander before him) also had to correct many corrupted passages that had crept into the manuscript over the centuries, and he tried to make the text accessible to his readers by detailed comments.

In his edition of Diophantus' *Arithmetica* Bachet asked whether the equation $y^2 + 2 = x^3$ that showed up in Problem VI.19 possesses other rational solutions except the one given by Diophantus, and he answered this question in the positive by presenting a method that allowed him to find a new solution of such an equation from a known one.

Bachet achieved his result using the Diophantine technique of clever substitutions, which we may interpret geometrically (see Fig. 1.3): If we intersect the curve $y^2 = x^3 - 2$ (such a curve is called an elliptic curve) and its tangent

$$y = \tfrac{27}{10}(x - 3) + 5$$

in $P(3, 5)$, then we obtain a second point of intersection $\left(\tfrac{129}{100}, \tfrac{383}{1000}\right)$.

Of course Bachet did not think of tangents at all, and he did not determine the tangent using analytic means: Differential calculus had not yet been discovered, and neither did coordinate systems exist, which came into being under the hands of Pierre Fermat and René Descartes. Bachet rather chose his Diophantine substitution

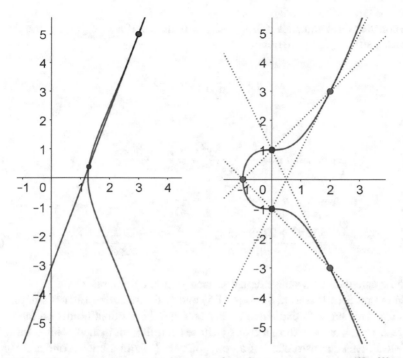

Fig. 1.3 Left: Tangent method on the elliptic curve $y^2 = x^3 - 2$; construction of $\left(\frac{129}{100}, \frac{383}{1000}\right)$ from $(3, 5)$. Right: The elliptic curve $y^2 = x^3 + 1$ with the five integral points $(-1, 0)$, $(0, \pm 1)$ and $(2, \pm 3)$, two lines through three points and the tangents in $(2, \pm 3)$

in such a way that a linear equation resulted, which then necessarily has a rational solution.

Bachet knew, as did his readers, that this calculation is a "proof by example," that is, this solution is general in the sense that it may be applied without any problems to any equation of the form $y^2 + k = x^3$. In fact, if we set $y_1 = y - \eta$ and $x_1 = x - r\eta$, then the equation $y_1^2 + k = x_1^3$ yields

$$y^2 - 2y\eta + \eta^2 + k = x^3 - 3rx^2\eta + 3r^2x\eta^2 - r^3\eta^3.$$

Since $y^2 + k = x^3$ this means that

$$-2y\eta + \eta^2 = -3rx^2\eta + 3r^2x\eta^2 - r^3\eta^3,$$

and after dividing through by $\eta \neq 0$ we obtain

$$r^3\eta^2 + (1 - 3r^2x)\eta + 3rx^2 - 2y = 0. \tag{1.4}$$

The constant term vanishes if $3rx^2 - 2y = 0$, that is, if $r = \frac{2y}{3x^2}$. Plugging this into (1.4) and solving for η we obtain

$$\eta = -\frac{27}{8y^3}x^6 + \frac{9}{2y}x^3,$$

hence

$$x_1 = x - r\eta = \frac{9x^4 - 8y^2x}{4y^2} = \frac{x^4 + 8kx}{4y^2}, \tag{1.5}$$

$$y_1 = y - \eta = y + \frac{27}{8y^3}x^6 - \frac{9}{2y}x^3$$

$$= \frac{8y^4 + 27x^6 - 36y^2x^3}{8y^3} = \frac{-x^6 + 20kx^3 + 8k^2}{8y^3}. \tag{1.6}$$

In these calculations we have replaced each y^2 in the numerator by $x^3 - k$.

In the modern literature the Eqs. (1.5) and (1.6) are called Bachet's duplication formula. This name is explained by the fact that the rational points on the elliptic curve $E : y^2 = x^2 - k$ form a group (whose neutral point is the "point at infinity"). In fact, Bachet's construction of a new point on E from a known one corresponds algebraically to the addition of the original point to itself, or, more precisely, to the multiplication of this point by -2; indeed, in our notation we have $-2(x, y) = (x_1, y_1)$ and $2(x, y) = (x_1, -y_1)$.

We also point out that Bachet could not solve the problem of finding a solution of the equation $y^2 + 2 = x^3$ except by trial and error; today we know a lot more about solving equations $y^2 = x^3 - k$ in integers or rational numbers, but we do not know an algorithm that produces a solution in finitely many steps or shows that there is no solution (there are such algorithms for quadratic equations such as $ax^2 + by^2 + cz^2 = 0$ and the Pell equation $y^2 = Nx^2 + 1$). Bachet's accomplishment is having shown how to find (in general infinitely many) rational points from a known one.

1.4　Fermat

In Fermat's time, the problems studied in number theory were about perfect numbers (numbers that are equal to the sum of their proper divisors such as $2^{p-1}(2^p - 1)$ for prime numbers $2^p - 1$), amicable numbers (pairs of numbers such as 220 and 284, for which the sum of the proper divisors of one number is equal to the other) and figurate numbers (patterns in the sequences of triangular numbers and square numbers). It was the study of Bachet's edition of Diophantus' *Arithmetica* that made Pierre Fermat (1607–1665) start his own investigations in number theory. On the margin of a page in his copy he wrote his remark that the equation $x^n + y^n = z^n$

is not solvable in natural numbers for any exponent $n \geq 3$, and even claimed to have a wonderful proof which the margin of his book was too small to contain. Since he never made this claim public (it was published posthumously by his son Samuel) and since Fermat was not exactly suffering from modesty we may assume that he eventually discovered that his idea for proving the case $n = 4$ could not be transferred to other exponents n.

In his copy of Diophantus' *Arithmetica*, Fermat also made the following remark concerning Bachet's equation $y^2 + 2 = x^3$:

> Is there another square in integers apart from 25 that, increased by 2, gives a cube? This seems difficult to investigate; but I can show by a rigorous demonstration that 25 is the only square that is smaller by 2 than a cube. In rational numbers, Bachet's method yields many such squares, but the theory of integers, which is very beautiful and very subtle, was so far known to nobody, neither to Bachet, or to any other author whose works I have seen.

In his letter to Carcavi written in 1657, Fermat tried to make Carcavi believe that he was able to prove the following assertions using infinite descent:

- There is no cube that can be decomposed into two cubes.
- There is only one square which is 2 less than a cube, namely 25.
- There exist only two squares that, when you add 4, give a cube, namely 4 and 121.
- All squared powers[6] of 2, increased by 1, are prime numbers.

The last claim is Fermat's conjecture that all numbers of the form $2^{2^n} + 1$ are prime, which Euler disproved by observing that $F_5 = 2^{32} + 1 = 641 \cdot 6700417$. It seems to me that Fermat did not know how to prove any of these claims but was convinced that the key to their proofs was infinite descent.

1.4.1 *Integral Solutions of* $y^2 + 2 = x^3$

We now will have a closer look at Bachet's method of constructing rational points on elliptic curves $y^2 = x^3 - k$. To this end, let $P(\frac{m}{M}, \frac{n}{N})$ be such a rational point, and assume that the fractions are written in reduced form and with $M, N > 0$. It follows from $(\frac{n}{N})^2 = (\frac{m}{M})^3 - k$ that $n^2 M^3 = m^3 N^2 - k M^3 N^2$. Since N^2 divides the right side, N^2 must also divide $n^2 M^3$. But n and N are coprime, so we can conclude that[7] $N^2 \mid M^3$. In a similar way we obtain $M^3 \mid N^2$. Thus the natural numbers M^3 and N^2 divide themselves, hence we must have $M^3 = N^2$. This is only possible if M is a square and N a cube, and thus there exists a natural number e with $M = e^2$ and $N = e^3$.

---

[6]Fermat means the sequence $2, 2^2 = 4, 4^2 = 16, 16^2 = 256$, etc.

[7]The notation $a \mid b$ stands for "a divides b".

Proposition 1.6 *Each rational point on the elliptic curve $y^2 = x^3 - k$ has the form* $(\frac{m}{e^3}, \frac{n}{e^2})$, *where* $\gcd(m, e) = \gcd(n, e) = 1$.

If we now plug $x = \frac{m}{e^3}$ and $y = \frac{n}{e^2}$ into Bachet's duplication formula, we find

$$x_1 = \frac{\frac{m^4}{e^{12}} + 8k\frac{m}{e^3}}{4\frac{n^2}{e^4}} = \frac{m^4 + 8kme^9}{4n^2e^8},$$

$$y_1 = \frac{-\frac{m^6}{e^{18}} + 20k\frac{m^3}{e^9} + 8k^2}{8\frac{n^3}{e^6}} = \frac{-m^6 + 20km^3e^9 + 8k^2e^{18}}{8n^3e^{12}}.$$

Thus if $(\frac{m}{e^3}, \frac{n}{e^2})$ is a rational point on E, for which m and n are both odd and e is even, then $m_1 = m^4 + 8kme^9$ and $n_1 = -m^6 + 20km^3e^9 + 8k^2e^{18}$ are again odd, and $e_1 = 2ne^4$ is not only even, but divisible by a much higher power of 2 than e. This shows

Proposition 1.7 *Bachet's method applied to the point* $(3, 5)$ *on the elliptic curve* $E : y^2 = x^3 - 2$ *yields only points whose coordinates have even denominator (when written in lowest terms) and thus does not produce any point with integral coordinates.*

The proof of this proposition may have been within Fermat's reach despite the very modest tools he had at his disposal. But it does not follow from this proposition that there are no integral solutions of $y^2 + 2 = x^3$ except $(3, \pm 5)$ since Bachet's method does not yield all rational solutions. Similarly, Bachet's method applied to the equation $y^2 + 4 = x^3$ and the integral point $(2, 2)$ does not yield any integral points beyond $(5, 11)$ (see Exer. 1.8.).

1.4.2   The Fermat Equation $x^4 + y^4 = z^2$

Already Diophantus showed that there exist Pythagorean triples in which one leg or the hypotenuse is a square number, i.e., that the Diophantine equations $a^4 + b^2 = c^2$ and $a^2 + b^2 = c^4$ have nontrivial solutions (see Exercise 1.2). Fermat asked why Diophantus did not discuss the question of finding Pythagorean triples in which two sides are square numbers, and he answered this question by observing that this is due to the unsolvability of the problem:

Theorem 1.8 *The integral solutions of the Diophantine equation $x^4 + y^4 = z^2$ are the obvious solutions with $x = 0$ or $y = 0$. In particular, the equation $x^4 + y^4 = z^4$ does not possess any nontrivial solutions in integers.*

This theorem is one of the few for which Fermat left at least a sketch of a proof. His correspondence partner Frénicle de Bessy published a detailed presentation of the proof, but it is not known just how big Fermat's contribution to Frenicle's publication actually was.

The proof is based on an application of infinite descent: Starting with a hypothetical solution (x, y, z) in natural numbers one constructs a new solution (x_1, y_1, z_1) that is "smaller" (in a suitable way) than the original solution. Since natural numbers cannot decrease indefinitely, this will lead to a contradiction.

For the proof of Fermat's claim we assume that there is a solution (x, y, z) of $x^4 + y^4 = z^2$ in natural numbers with $xy \neq 0$. If p is a common divisor of x and z, then $p \mid y$, hence $p^4 \mid z^2$ and $p^2 \mid z$; but then we may cancel p^4, and applying this reasoning repeatedly we arrive at a solution (x, y, z) in which x, y, and z are pairwise coprime.

Clearly x and y have different parity, and we may assume that x is even and y is odd. According to Proposition 1.5 there exist natural numbers a, b with $x^2 = 2ab$, $y^2 = a^2 - b^2$ and $z = a^2 + b^2$. Since y is odd, a and b have different parity. If a is even and b is odd, then we obtain $1 \equiv y^2 = a^2 - b^2 \equiv 0 - 1 \equiv -1 \bmod 4$: Contradiction. Thus a is odd and b is even, and applying Proposition 1.5 to the equation $b^2 + y^2 = a^2$ we obtain the existence of integers $c, d \in \mathbb{N}$ with $b = 2cd$, $y = c^2 - d^2$ and $a = c^2 + d^2$. This gives us $x^2 = 4cd(c^2 + d^2)$, hence $(x/2)^2 = cd(c^2 + d^2)$. Now c, d and $c^2 + d^2$ are pairwise coprime (a common factor would divide a and b, hence x and y) and their product is a square. Applying the Square Lemma 1.4 twice (first to the pair cd and $c^2 + d^2$, then to c and d) we find that these factors must be squares, up to possible factors ± 1. By choosing c and d positive we obtain $c = e^2$, $d = f^2$ and $c^2 + d^2 = g^2$ for $e, f, g \in \mathbb{N}$.

But now we have $e^4 + f^4 = g^2$, hence we have found a new solution of the equation $x^4 + y^4 = z^2$. Since

$$z = a^2 + b^2 = (c^2 + d^2)^2 + 4c^2d^2 > g^4 \geq g,$$

this solution is smaller than our original solution. In other words: To every solution $(x, y, z) \in \mathbb{N}^3$ with $xy \neq 0$ there exists another solution $(e, f, g) \in \mathbb{N}^3$ with $0 < g < z$ (if we had $g = 0$, it would follow that $e = f = 0$ and thus $b = 0$, hence $x = 0$: Contradiction). Thus there cannot exist a solution $(x, y, z) \in \mathbb{N}^3$ with $xy \neq 0$ since after finitely many steps we would obtain an integral solution (e, f, g) with $0 < g < 1$. This proves Fermat's claim.

As impressive as this proof is, it only uses descent and a repeated application of the Square Lemma.

1.5 Euler

Leonhard Euler (1707–1783) was perhaps the most productive mathematician of all times; almost half of his books and articles were published after he had become blind. It was Christian Goldbach who infected Euler with the virus of number theory in 1729 (see [89]) by asking him in his first letter whether Euler knew about Fermat's conjecture that all numbers of the form $2^{2^n} + 1$ are prime. Euler was not impressed, but Goldbach did not let go. He showed Euler how to prove that no Fermat number $> 2^{2^4} + 1$ is divisible by any prime number below 100, and when Euler eventually discovered that each prime factor of $2^{2^n} + 1$ has the form $p = 2^n k + 1$ and that $2^{2^5} + 1$ is divisible by $541 = 40 \cdot 16 + 1$, he was hooked. Euler was the only prominent mathematician[8] of his time who studied number theory until Lagrange appeared on the mathematical stage.

For explaining the idea behind Euler's attempt at proving Fermat's conjecture that $x = 3$ and $y = 5$ are the only solutions of $y^2 = x^3 - 2$ in natural numbers, we will look into his *Algebra* [38]. Euler's proof contained a gap, but it displays his originality.

In order to understand Euler's reasoning it is necessary to study Euler's proof of one of the most beautiful results of elementary number theory, the Two-Squares Theorem, according to which each prime number of the form $4n + 1$ can be written as a sum of two squares.

1.5.1   The Two-Squares Theorem

The first statement that every prime number of the form $p = 4n + 1$ can be written as a sum of two squares shows up in Albert Girard's (1595–1632) edition of Simon Stevin's (1548–1620) *Arithmetique* published in 1625. This edition contains the first four books of the *Arithmetica* Diophantus translated by Stevin, and the fifth and sixth book translated by Girard. In connection with Problem V.12, Girard writes:

Determination of a number that can be divided into two squares of integers.

   I. Each square number.
  II. Each prime number that exceeds a multiple of 4 by a unit.
 III. The products of such numbers.
  IV. And the double of each of these.

The first proof that every prime number of the form $p = 4n + 1$ can be written uniquely as a sum of two squares is due to Fermat, and the first published proof to Euler [37]. Euler approaches his proof slowly and thoughtfully and he explains why, apart from $p = 2 = 1^2 + 1^2$, only primes of the form $4n + 1$ can be sums of two squares.

His first considerations are concerned with the representation of $2p$ as sums of two squares: If $p = a^2 + b^2$, then

$$2p = 2a^2 + 2b^2 = (a - b)^2 + (a + b)^2 \tag{1.7}$$

---

[8]There were numerous less known mathematicians interested in Diophantine problems or the investigation of perfect and amicable numbers.

is also a sum of two squares. If, conversely, $2p = c^2 + d^2$ is a sum of two squares, then c and d must be both odd, hence

$$p = \left(\frac{c+d}{2}\right)^2 + \left(\frac{c-d}{2}\right)^2 \tag{1.8}$$

is a sum of two squares of integers.

Euler's goal in the last part of his *Algebra* was convincing his readers that quadratic irrationalities were the actual reason for the existence of such identities. In the present case (1.7) and (1.8) may be explained via the multiplication and the division by $1 + i$ in integers of the form $a + bi$: Taking the norm[9] of

$$(1 + i)(a + bi) = (a - b) + (a + b)i$$

yields the identity (1.7). Conversely, (1.8) is a consequence of

$$\frac{c + di}{1 + i} = \frac{(c + di)(1 - i)}{(1 + i)(1 - i)} = \frac{c + d + (d - c)i}{2}.$$

Euler now plays the same game with two odd prime numbers: If $p = a^2 + b^2$ and $q = c^2 + d^2$, then

$$pq = (a^2 + b^2)(c^2 + d^2) = (ac + bd)^2 + (ad - bc)^2.$$

This identity also follows easily by using complex numbers:

$$(a - bi)(c + di) = (ac + bd) + (ad - bc)i,$$

whereas the product $(a + bi)(c + di)$ leads to the representation

$$pq = (ac - bd)^2 + (ad + bc)^2. \tag{1.9}$$

The existence of two expressions for pq as sums of two squares suggests that products of two primes of the form $4n + 1$ can always be written as sums of two squares in two different ways. In the second part of his article Euler uses this idea to develop an algorithm for finding the prime factors of sums of two squares.

The idea behind Euler's proof of the Two-Squares Theorem is reversing the process above. We have seen (see Eqs. (1.7) and (1.8)) that a prime number p is a sum of two squares if and only if $2p$ is. It is therefore a natural idea to show that some multiple mp of a prime $p \equiv 1 \bmod 4$ is a sum of two squares, and then to reduce m via the product formula (1.9) until we end up with $m = 1$.

---

[9]The norm of a Gaussian integer $x + iy$ is $(x + iy)(x - iy) = x^2 + y^2$.

Euler follows this idea using induction (Fermat had chosen the equivalent method of infinite descent):

(1) The prime numbers $p = 4n + 1$ are those for which -1 is a quadratic residue; it follows from $x^2 \equiv -1 \bmod p$ that $x^2 + 1 = mp$ for some natural number m.
(2) For each prime number $p = 4n + 1$ there is a multiple mp that can be written as a sum of two squares, where we may choose $m < p$.
(3) If m is even, then $\frac{m}{2} p$ is also a sum of two squares.
(4) Each odd prime factor q of m is sum of two squares by induction hypothesis, and then $\frac{m}{q} p$ is a sum of two squares.

Claim (2) follows from (1): If $p = 4n + 1$, then $x^2 \equiv -1 \bmod p$ has a solution, hence p divides $x^2 + 1$ and so $x^2 + 1 = mp$. By picking the residue class of x modulo p between $-\frac{p}{2}$ and $\frac{p}{2}$ we obtain $mp = x^2 + 1 < \frac{p^2}{4} + 1 < p^2$ and thus $m < p$. Since we have already proved (3) it only remains to show (1) and (4).

The basis of Euler's proof of the first claim (1) is Fermat's Theorem that $a^{p-1} \equiv 1 \bmod p$ for all integers a coprime to p. For primes $p = 4n + 1$ this shows that for all integers a and b coprime to p, the expression

$$a^{p-1} - b^{p-1} = a^{4n} - b^{4n} = (a^{2n} - b^{2n})(a^{2n} + b^{2n})$$

is divisible by p. Since the expression in the last bracket is a sum of two squares $(a^n)^2 + (b^n)^2$, everything boils down to showing that a and b can be chosen in such a way that p does not divide the expression $a^{2n} - b^{2n}$ in the first bracket.

As Euler was to find out later, this can be seen quite easily: Just pick $b = 1$ and show that there is at least one integer a not divisible by p for which $a^{2n} - 1$ is not divisible by p. In fact, if all integers $a = 1, 2, \ldots, p - 1$ had the property that p is a divisor of $a^{2n} - 1$, then the polynomial $x^{2n} - 1$ would have more than $2n = \frac{p-1}{2}$ roots modulo p, which is impossible.

It remains to prove the last claim, the induction step. We assume that each prime $q = 4n + 1$ less than p is a sum of two squares and then show that p is also a sum of two squares.

Thus let $mp = x^2 + y^2$ for an integer $m < p$. If d is a common divisor of x and y, then m is divisible by d^2, and division by d^2 yields $m_1 p = x_1^2 + y_1^2$, where x_1 and y_1 are coprime. Moreover we may assume that m_1 is odd.

Now write $mp = a^2 + b^2$, where m is odd and where a and b are coprime. From $a^2 + b^2 \equiv 0 \bmod m$ we obtain the congruence $(a/b)^2 \equiv -1 \bmod m$. In particular, $(a/b)^2 \equiv -1 \bmod q$ for each prime divisor q of m. Since $q < p$, we know by induction assumption that $q = c^2 + d^2$. Thus $\frac{a^2+b^2}{c^2+d^2}$ is an integer. Therefore, the numbers

$$c^2(a^2 + b^2) = a^2 c^2 + b^2 c^2 \quad \text{and} \quad a^2(c^2 + d^2) = a^2 c^2 + a^2 d^2$$

are divisible by $q = c^2 + d^2$, hence so is their difference

$$b^2c^2 - a^2d^2 = (bc - ad)(bc + ad).$$

Since $q = c^2 + d^2$ is prime, q divides one of the two factors by Euclid's Lemma 1.2. Changing the sign of d if necessary we may assume that q divides $bc - ad$. Then

$$mpq = (a^2 + b^2)(c^2 + d^2) = (bc - ad)^2 + (ac + bd)^2,$$

and here the left side as well as the square $(bc - ad)^2$ are divisible by q. Thus q also divides $ac + bd$, and canceling q^2 yields

$$\frac{m}{q} \cdot p = \left(\frac{bc - ad}{q}\right)^2 + \left(\frac{ac + bd}{q}\right)^2.$$

If $\frac{m}{q} = 1$, then we are done; if not, consider a prime factor q_1 of $\frac{m}{q}$ and repeat the last step. After finitely many steps we have found a presentation of p as a sum of two squares.

As a corollary of Euler's investigation we observe

Theorem 1.9 (Euler's Decomposition Theorem) *If $m = x^2 + y^2$ is a sum of two coprime squares, and if $m = p_1 p_2 \cdots p_t$ is the prime factorization of m, then we can choose integers $x_j, y_j \in \mathbb{Z}$ for which $p_j = x_j^2 + y_j^2$, and the decomposition of m into two squares can be obtained by a repeated application of the identity (1.9) to the decompositions of the primes p_j as sums of two squares.*

$$x^2 + y^2 = (x_1^2 + y_1^2)(x_2^2 + y_2^2) \cdots (x_t^2 + y_t^2).$$

The condition that x and y be coprime is necessary: Although $3^2 + 3^2 = 2 \cdot 3 \cdot 3$, the number 3 cannot be written as a sum of two squares.

For the induction proof to go through it was important that there is a a multiple mp of p for which mp is a sum of two squares and $m < p$. Such a multiple existed because it follows from $|x|, |y| < \frac{p}{2}$ that $x^2 + y^2 < \frac{1}{2}p^2$. This step also works for numbers of the form $x^2 + 2y^2$, and even for numbers of the form $x^2 + 3y^2$ if we avoid that x and y are both odd. Thus for numbers of the form $x^2 + 2y^2$ and $x^2 + 3y^2$ there are similar decomposition theorems. For numbers of the form $x^2 + 5y^2$, however, our proof does not work any more, and as a matter of fact, the decomposition theorem does not hold: Although $1^2 + 5 \cdot 1^2 = 6$, neither 2 nor 3 can be written in the form $x^2 + 5y^2$.

It is hard to believe that Euler did not see this; even though he did not formulate the decomposition theorem anywhere, it follows from his work that he must have believed the following result to be true:

Theorem 1.10 (Euler's Decomposition Theorem) *If $N = x^2 + my^2$ for coprime integers x and y and positive squarefree numbers m, if $N = p_1 \cdots p_t$ is a product* *of prime numbers, and if each prime factor p_j has the form $p_j = x_j^2 + my_j^2$, then the signs of the y_j can be chosen in such a way that the decomposition $N = x^2 + my^2$ is obtained by a repeated application of the identity*

$$(a^2 + mb^2)(c^2 + md^2) = (ac - mbd)^2 + m(ad + bc)^2$$

from

$$(x_1^2 + my_1^2)(x_2^2 + my_2^2) \cdots (x_t^2 + my_t^2).$$

The usual counterexamples, which we will discuss repeatedly in the next few chapters, do not apply to this result, which can be proved using tools available to Euler. For solving Diophantine equations one needs a stronger version of the decomposition theorem, namely the analogue of Theorem 1.9 for numbers of the form $x^2 + 2y^2$. The strong version would follow from the weak one if we could show that prime divisors of numbers of the form $x^2 + my^2$ with $\gcd(x, y) = 1$ again have this form. But as we have already seen, this is false already for $m = 5$.

Euler eventually must have realized that there is a serious problem with his approach. In one of the many posthumous papers of Euler [39, Art. 44] we find the following question:

The formula $181^2 + 7 = 32^3$ is worthy of our whole attention; although $32 = 5^2 + 7$ it is not true that

$$181 + \sqrt{-7} = (5 + \sqrt{-7})^3,$$

although we have

$$181 + \sqrt{-7} = \frac{1 - 3\sqrt{-7}}{8}(5 + \sqrt{-7})^3.$$

We also remark that

$$\frac{1 + 3\sqrt{-7}}{8} \cdot \frac{1 - 3\sqrt{-7}}{8} = 1.$$

which shows that the development into imaginary factors requires further investigations.

This problem "worthy of our whole attention" disappears by introducing algebraic integers of the form $\frac{a + b\sqrt{-7}}{2}$ with $a \equiv b \bmod 2$. In particular, Euler's question remained open until Dedekind succeeded in clearing up the notion of algebraic integers (see Exercise 2.25).

1.5.2 Euler's Algebra

The part of Euler's exposition in his *Algebra* that is of interest to us in connection with the arithmetic of surds begins in § 162, where he investigates the factorization of the expression $ax^2 + bxy + cy^2$. Euler recalls that such an expression is a square if the discriminant $b^2 - 4ac = 0$, a product of rational linear factors if $b^2 - 4ac$ is a square, and a product of irrational (and even complex) factors otherwise.

The first two cases are quickly dealt with; in the third case, Euler gets rid of the term bxy by completing the square, and he ends up with the problem of factoring expressions of the form $ax^2 + cy^2$.

In § 168 he begins with the simplest case $x^2 + y^2$, where the method used for solving the first two cases (where the discriminant was a square) does not work:

> In this case, the method above does not apply since this expression cannot be written as a product of two rational factors; yet the irrational factors into which this expression can be factored and which represent this product $(x + y\sqrt{-1})(x - y\sqrt{-1})$ may be just as useful for us. If the expression $xx + yy$ has actual factors, then the irrational factors must again have factors, since if they did not have any divisors, then its product could not have any. But since these factors are irrational and even complex, and since the numbers x and y do not have a common divisor, they cannot have any rational factors; rather they have to be irrational and even imaginary of the same kind.

Behind these considerations is the observation that composite values of $x^2 + y^2$ may be explained by factorizations of $x + yi$. For example, $1^2 + 8^2 = 65 = 5 \cdot 13$ is composite, and Euler's claim is that this is a consequence of the fact that $1 + 8i$ can be factored. In fact, we have $1 + 8i = (-1 + 2i)(3 - 2i)$.

As we have seen above, Euler has proved this implicitly in his proof that each prime of the form $p = 4n + 1$ is a sum of two squares.

1.5.3 Bachet's Equation $y^2 + 2 = x^3$

Euler discusses the equation $y^2 + 2 = x^3$ in § 193 of his *Algebra*:

> It is required to find square numbers of integers that, if 2 is added, become cubes, as happens in the case of the square 25. We ask whether there are any other such squares.

Euler's solution is the following:

> Since $x^2 + 2$ must be a cube and 2 is the double of a square, we first look for the cases in which the expression $x^2 + 2y^2$ becomes a cube, which happens according to what we have seen above in § 188 when $x = p^3 - 6pq^2$ and $y = 3p^2q - 2q^3$. Since here we have $y = \pm 1$ we must have $3p^2q - 2q^3 = q(3p^2 - 2q^2) = \pm 1$, and thus q must divide 1. Therefore $q = 1$, and so $3p^2 - 2 = \pm 1$; if the positive sign holds, then $3p^2 = 3$ and $p = 1$, hence $x = 5$; the negative sign yields an irrational value for p, which is impossible here. This implies that 25 is the only square in integers with the desired property.

Here Euler uses his decomposition theorem: If $x^2 + 2 = m^3$, then there is a representation $m = p^2 + 2q^2$ such that $x^2 + 2 = (p^2 + 2q^2)^3$.

One can often read that Euler used unique factorization for integers of the form $a + b\sqrt{-2}$ in his solution of the Bachet-Fermat equation $y^2 + 2 = z^3$; as a matter of fact, primes of the form $x + y\sqrt{-2}$ do not occur anywhere in Euler's work, and it would be more precise to say that Euler's proof had a gap that we can fill with the unique factorization theorem for such numbers. What is true is that Euler tried to transfer the Square Lemma (Theorem 1.4) and its cubic analog to numbers of the form $a + b\sqrt{c}$.

1.5.4 The Cubic Fermat Equation

In § 243 of his *Algebra*, Euler discusses the following

Theorem. It is not possible to find two cubes whose sum or difference is a cube.

Of course it is sufficient to prove the impossibility of $x^3 + y^3 = z^3$ in integers, since $x^3 - y^3 = z^3$ is equivalent to $x^3 = y^3 + z^3$.

Euler first proves that x and y may be assumed to be coprime and odd. Then their sum and difference are even, and setting $x = p + q$ and $y = p - q$ he finds

$$x^3 + y^3 = 2p(p^2 + 3q^2),$$

and Fermat's claim boils down to the question whether the product $2p(p^2 + 3q^2)$ can be a cube or not. Elementary congruences show that p must be divisible by 4, so that $\frac{p}{4}(p^2 + 3q^2)$ is a cube.

We can easily show that $\frac{p}{4}$ and $p^2 + 3q^2$ are either coprime or have greatest common divisor 3. In the first case, $p^2 + 3q^2$ must be a cube:

Now let us make $pp + 3qq$ a cube, which may be achieved, as we have shown above, by setting

$$p + q\sqrt{-3} = (t + u\sqrt{-3})^3 \text{ and } p - q\sqrt{-3} = (t - u\sqrt{-3})^3.$$

This makes $pp + 3qq = (tt + 3uu)^3$ into a cube, and now we find $p = t^3 - 9tuu = t(tt - 9uu)$ and $q = 3ttu - 3u^3 = 3u(tt - uu)$.

Next Euler uses the condition that

$$2p = 2t(t^2 - 9u^2) = 2t(t - 3u)(t + 3u)$$

must be a cube in order to find a contradiction using infinite descent. In fact, the coprimality of the factors implies that $2t = e^3$, $t - 3u = f^3$ and $t + 3u = g^3$ must be cubes. This implies

$$e^3 = 2t = (t - 3u) + (t + 3u) = f^3 + g^3,$$

and we have found a new solution (e, f, g) of the cubic Fermat equation, which easily can be shown to be smaller than the solution we started with unless $xyz = 0$.

The problematic point in Euler's proof is the following, as was pointed out to Euler by the Berlin mathematician and calculator Abraham Wolff in a letter to Euler written on August 9, 1770:

> The difficulty lies in the fact that I lack the trick by which I can convince myself that if $pp + 3qq = (tt + 3uu)^3$, that is
>
> $$(p + q\sqrt{-3}) \cdot (p - q\sqrt{-3}) = (t + u\sqrt{-3})^3(t - u\sqrt{-3})^3,$$
>
> the value of $p + q\sqrt{-3}$ must necessarily be $(t + u\sqrt{-3})^3$.

It is not known whether Euler answered this letter; but he certainly knew that others felt there was a gap.

1.5.5 Euler and the Problem of Units

In § 188, Euler discusses the question of how to make expressions such as $ax^2 + cy^2$ into a cube. To this end he uses numbers of the form $x\sqrt{a} + y\sqrt{-c}$. Euler obtains his solutions by setting

$$x\sqrt{a} + y\sqrt{-c} = (p\sqrt{a} + q\sqrt{-c})^3.$$

This yields

$$x = ap^3 - 3cpq^2 \quad \text{and} \quad y = 3ap^2q - cq^3.$$

It is correct that these values of x and y satisfy the equation

$$ax^2 + cy^2 = z^3 \tag{1.10}$$

for $z = ap^2 + cq^2$. But what Euler uses (and needs) in his applications is the converse, namely that *each* solution is given by these equations. Euler's occasional remarks concerning the coprimality of the coefficients show that he has seen that this converse must be proved. Numerous examples that perhaps were known to Euler show, however, that such a proof cannot be as simple as Euler may have thought.

Euler's digression into the theory of numbers of the form $x\sqrt{a} + y\sqrt{c}$ cannot be avoided: For example, we have $7^2 - 10 \cdot 2^2 = 9 = 3^2$, yet $7 + 2\sqrt{10}$ is not a square of a number of the form $a + b\sqrt{10}$. In the present case, the obstacle may be overcome by considering numbers of the form $a\sqrt{2} + b\sqrt{5}$, because now

$$7 + 2\sqrt{10} = (\sqrt{2} + \sqrt{5})^2.$$

In Euler's *Algebra*, this example does not occur. He came across these numbers when trying to make expressions such as $2x^2 - 5y^2$ into cubes. In this case, Euler observes, it is not sufficient to set

$$x\sqrt{2} + y\sqrt{5} = (p\sqrt{2} + q\sqrt{5})^3;$$

rather one has to consider equations of the form

$$x\sqrt{2} + y\sqrt{5} = (p\sqrt{2} + q\sqrt{5})^3 (3 + \sqrt{10})^n$$

for arbitrary integers n. Since $-(3 + \sqrt{10})(3 - \sqrt{10}) = 1$, the powers of $3 + \sqrt{10}$ are divisors of 1, or, in words that Euler could not have used, units in the ring of numbers of the form $x + y\sqrt{10}$. In this connection, see Exercise 9.19.

1.6   Gauss

The final step in the direction of algebraic number theory was taken by Carl Friedrich Gauss (1777–1855). Gauss is one of the greatest mathematicians of all time. He was only 18 years old when he solved a 2000-year-old problem by showing that a regular polygon with 17 sides can be constructed using ruler and compass— he obtained the proof by developing the (algebraic!) theory of cyclotomy, which he included in his *Disquisitiones Arithmeticae*, one of the most famous textbooks on number theory. The *Disquisitiones* also contained the first complete proofs of the quadratic reciprocity law. Another fundamental discovery by Gauss was elliptic functions (doubly periodic functions $\mathbb{C} \longrightarrow \mathbb{C}$, obtained by inverting elliptic integrals that Euler and Legendre had studied extensively and which appear in the computation of the circumference of ellipses—whence their name), about which he published almost nothing at all.

In his *Disquisitiones Arithmeticae* [43] published in 1801, Gauss gave a quite modern presentation of elementary number theory (in the sense that he covered congruences, unique factorization, residue class groups and primitive roots), erected on a safe foundation: After the proof of Euclid's Lemma 1.2 in [43, Art. 14] he states and proves the theorem of unique factorization. Before Gauss, the uniqueness of prime factorization had been known, but it was not regarded as a fundamental property: As in Euclid's *Elements*, it was rather regarded as an auxiliary result, e.g., for finding all the divisors of numbers of the form $2^{p-1}(2^p - 1)$ for primes $2^p - 1$ in the construction of perfect numbers. The observation that all divisors of a number N are obtained exactly once by multiplying the divisors of the prime powers dividing N is essentially equivalent to unique factorization as far as the content is concerned. For Gauss (and the number theorists after him), unique factorization was a principle on which elementary number theory is founded. In particular, Gauss realized that unique factorization may be used to *prove* results about integers. Only after this conceptual progress did it become possible to ask whether unique factorization holds for numbers of the form $a + b\sqrt{-2}$.

Exactly this question, for numbers of the form $a + bi$, was discussed by Gauss a quarter of a century later in his second memoir [45] on biquadratic residues:

> After having begun to think about this topic already in 1805 we soon became convinced that the natural source of a general theory has to be found in an extension of the field of arithmetic [...].
>
> In fact, whereas the higher arithmetic in the questions discussed so far deals only with real numbers, the theorems concerning biquadratic residues only appear in their whole simplicity and natural beauty if the field of arithmetic is extended to the imaginary numbers, so that its objects are, without any restriction, the numbers of the form $a+bi$, where as usual i denotes the imaginary unit $\sqrt{-1}$, and where a and b run through all real integers between $-\infty$ and $+\infty$.

Gauss uses the numbers $a + bi$ not only as servants for finding identities, as Euler and Lagrange have done, but develops the arithmetic in this domain ab ovo: He defines divisibility, units, prime numbers and shows, with the help of binary quadratic forms, that the integers $a + bi$ can be factored uniquely, up to unit factors and ordering, into prime elements. Both Dirichlet (1805–1859) and Jacobi (1804–1851) were surprised and highly impressed by the idea of allowing these numbers $a + bi$ as *modules*, which allowed Gauss to transfer his theory of congruences to these numbers.

Dirichlet later even extended Gauss's theory of binary quadratic forms to the ring of "Gaussian integers" (parts of his lectures on the elementary arithmetic of this ring have survived as lecture notes by Gustav Arendt [3]), and he realized that unique prime factorization in such domains is a consequence of the existence of a Euclidean algorithm.

For a few number fields, unique factorization could be proved in this way, but it was not clear what to do with number rings in which unique factorization does not hold. It was more or less taken for granted that a general theory would have to be based on a generalization of the theory of binary quadratic forms to forms of higher degree. Eisenstein (1823–1852) developed a theory of cyclic cubic fields in the language of cubic forms, and Dirichlet worked out the analytic class number formula for cyclotomic fields using the language of forms before Kummer (1810–1893) succeeded in creating an arithmetic of cyclotomic number fields based on his notion of ideal numbers. Dedekind (1831–1916) extended Kummer's ideas to general number fields using his theory of ideals.

1.7   Kummer and Dedekind

In Euclid's books on number theory (volumes 7, 8, and 9 of his *Elements*), a prime number was a natural number greater than 1 without any proper divisors. It was Kummer who first realized that this definition of primality was not the right one in algebraic number fields. In his article [71] he wrote:

I have noticed that, even if α cannot be decomposed into complex factors, it might not have the true nature of a complex prime number, since in general it lacks the first and most important property of primes, namely that the product of two primes is not divisible by any prime number different from them.

Instead of looking at Kummer's examples taken from cyclotomic number fields, we will verify Kummer's claims in suitable quadratic rings. We begin by presenting a non-example, namely the factorization

$$6 = 2 \cdot 3 = (2 + \sqrt{-2})(2 - \sqrt{-2}) \tag{1.11}$$

in $\mathbb{Z}[\sqrt{-2}]$ (by $\mathbb{Z}[\sqrt{m}]$ we denote the set of numbers of the form $a + b\sqrt{m}$ with $a, b \in \mathbb{Z}$). These factorizations look different but may be explained by the fact that the factors can be decomposed further, just as those in $12 = 2 \cdot 6 = 3 \cdot 4$. In fact, we have

$$2 = -\sqrt{-2}^2, \qquad\qquad 2 + \sqrt{-2} = \sqrt{-2} \cdot (1 - \sqrt{-2}),$$
$$3 = (1 + \sqrt{-2})(1 - \sqrt{-2}), \qquad 2 - \sqrt{-2} = -\sqrt{-2} \cdot (1 + \sqrt{-2}).$$

Thus the two factorizations result from combining the factors in the "prime decomposition"

$$6 = -\sqrt{-2}^2 (1 + \sqrt{-2})(1 - \sqrt{-2})$$

in two different ways.

The example

$$6 = 2 \cdot 3 = (1 + \sqrt{-5})(1 - \sqrt{-5})$$

in the ring $R = \mathbb{Z}[\sqrt{-5}]$ looks similar to the one above, but is different. It is true that 2 is not prime in R because the product is divisible by 2, yet 2 divides none of the two factors $1 \pm \sqrt{-5}$ since the elements $(1 \pm \sqrt{-5})/2$ do not belong to $R = \mathbb{Z}[\sqrt{-5}]$).

This factorization cannot be refined. In fact, in $\mathbb{Z}[\sqrt{-5}]$, the element 2 is irreducible, that is, it cannot be written as a product of nonunits. For proving this claim we need the norm map defined by $N(x + y\sqrt{-5}) = x^2 + 5y^2$. Since $N(x + y\sqrt{-5}) = (x + y\sqrt{-5})(x - y\sqrt{-5})$, the norm is multiplicative, i.e., we have $N(\alpha)N(\beta) = N(\alpha\beta)$ for elements $\alpha, \beta \in R$.

Now assume that 2 does have a factorization into nonunits, say $2 = \alpha\beta$ with $\alpha, \beta \in R$. Taking the norm of both sides we obtain $4 = N(2) = N(\alpha)N(\beta)$. The smallest norms in R are $N(\pm 1) = 1$, $N(\pm 2) = 4$, and $N(\pm\sqrt{-5}) = 5$. Thus the only solutions of the equation $4 = N(\alpha)N(\beta)$ with $\alpha, \beta \in R$ are $\alpha = \pm 2$ and $\beta = \pm 1$, or $\beta = \pm 2$ and $\alpha = \pm 1$. But these are trivial factorizations since the elements

±1 are units since they obviously divide 1. The result of these considerations is that

$$6 = 2 \cdot 3 = (1 + \sqrt{-5})(1 - \sqrt{-5}) \tag{1.12}$$

are two essentially different factorizations of 6 into irreducible elements of R.

Richard Dedekind was one of the pioneers of abstract algebra. Notions such as ring, field, and ideal are due to him. The invasion of algebra by these originally number theoretic concepts took place in the 1920s under the massive influence of Emmy Noether (1882–1935).

When Dedekind transferred the arithmetic of the ordinary integers to algebraic number rings in [28, III, p. 239], he gave a lot of thought to the question of how best to define the notion of a prime number in a number ring.

In order to explain why the classical definition of prime as irreducible is not suitable for doing arithmetic in number rings he considers the ring $\mathbb{A}$ of all algebraic integers and observes that the usual definition is of no use here:

> If we would define prime numbers as elements that do not possess any divisors essentially different from itself and units, then it is easily seen that such a number does not exist at all [in $\mathbb{A}$]; for if α is an algebraic integer, then it always has infinitely many essentially different divisors, for example, the numbers $\sqrt{\alpha}$, $\sqrt[3]{\alpha}$, $\sqrt[4]{\alpha}$, etc., which [...] are algebraic integers.

Dedekind goes on to say that, on the other hand, it is easy to define the notion of coprime elements without using the decomposition into irreducibles:

> Two non-zero algebraic integers α and β are called coprime if each element divisible by α and β is also divisible by $\alpha\beta$.

Today we are accustomed to definitions such as this one; for constructivists such as Leopold Kronecker this was a bad definition: it does not allow you to decide a priori whether two given integers are coprime since the definition requires checking infinitely many conditions. In fact we have to verify that the infinitely many integers divisible by α and β are also divisible by the product $\alpha\beta$. It is therefore clear that for computing with such numbers, one has to find an algorithm that allows us to decide in finitely many steps whether two given elements are coprime or not.

A little later (p. 250) Dedekind observes that in algebraic number rings there often exist essentially different decompositions of elements into irreducible elements and then continues as follows:

> This contradicts the notion of the character of primality that holds in the number theory of the rational integers to such an extent that we shall not accept an irreducible element as a prime; thus we need to look out for a stronger criterion than the inadequate irreducibility, similar to what we did earlier for the notion of coprimality [...], by not decomposing the integer we are investigating but by studying how it behaves as a module:
>
> An integer μ shall be called prime if it is not a unit, and if every product $\eta\rho$ divisible by μ has at least one factor η or ρ divisible by μ.

Here Dedekind has turned a characteristic theorem on primes, namely Euclid's Lemma 1.2, into a definition, because this property is well suited for building a theory.

1.7.1 From Ideal Numbers to Ideals

The fact that in some algebraic number fields there exist irreducible elements that
lack the defining properties of primes implies that the theorem of unique prime
factorization does not hold in such rings. In the case of quadratic number rings it was
possible to justify calculations needed for solving certain Diophantine equations by
invoking the language of binary quadratic forms; for example, Dirichlet proved the
unsolvability of the quintic Fermat equation $x^5 + y^5 = z^5$ in positive integers by
using the theory of the quadratic forms $x^2 - 5y^2$.

In order to be able to say something about certain Diophantine equations even
if the corresponding number field does not have unique factorization, Kummer
invented the notion of an "ideal" prime number. His basic idea is, from today's
point of view, a very modern one: Investigate an algebraic structure by studying
homomorphisms into simpler structures (see [81]).

Let us once more consider the ring $R = \mathbb{Z}[\sqrt{-5}]$. We have seen above that
the elements 2 and 3 are irreducible in R, but not prime. If there was an element
π of norm 2, then we could consider the residue class ring of R modulo π; this
quotient ring would have two elements, because it can be shown that the number
of residue classes modulo an element of R is equal to its norm. Reduction modulo
π thus would give us a ring homomorphism $f : R \longrightarrow \mathbb{Z}/2\mathbb{Z}$. Kummer realized
that such a ring homomorphism exists even when there is no element of norm 2. In
fact, all we have to do is set $f(a + b\sqrt{-5}) = a + b + 2\mathbb{Z}$. Thus although there
is no prime element π of norm 2, we can work modulo π by simply applying f.
Such ring homomorphisms (or, less anachronistically, such procedures for attaching
a residue class to each element) were called "ideal primes" by Kummer.

Heinrich Jung has shown in [69] how to develop the whole theory of quadratic
number fields based on this notion of ideal primes as ring homomorphisms. The
only obstacle in this approach is the fact that it is not at all obvious how to multiply
ideal numbers. Dedekind later replaced these ideal numbers by the kernels of the
associated ring homomorphisms and called them ideals. In his theory, the product
of two ideals is simply the ideal generated by the products of the elements from each
ideal.

1.8   Exercises

1.1. Already the Babylonians, about 4000 years ago, knew how to calculate the
space diagonal of a door, and composed problems from integral solutions of
equations such as $x^2 + y^2 + z^2 = w^2$. To this end they looked for Pythagorean
triples such as (3, 4, 5) and (5, 12, 13), in which the hypotenuse of one
triangle is equal to the leg of the other one, and then obtained the solution
$3^2 + 4^2 + 12^2 = 13^2$.

Show how to find infinitely many such solutions.

1.2. Show that there are infinitely many Pythagorean triples (a, b, c) in which a or c is a square number.

1.3. The vectors $\begin{pmatrix} 2 \\ 2 \\ 1 \end{pmatrix}$ and $\begin{pmatrix} 4 \\ 4 \\ 7 \end{pmatrix}$ with lengths 3 and 9, respectively, suggest that there exist infinitely many vectors $\begin{pmatrix} a \\ a \\ b \end{pmatrix}$ whose length $\sqrt{a^2 + a^2 + b^2}$ is an integer. Clearly this holds if and only if $2a^2 + b^2 = c^2$ is the square of an integer c.

   Write this equation in the form $2a^2 = c^2 - b^2 = (c - b)(c + b)$ and conclude that setting $c - b = 4s^2$ and $c + b = 2r^2$ yields solutions. Deduce that $a = 2rs, b = r^2 - 2s^2$ and $c = r^2 + 2s^2$.

   Parametrize the ellipse $x^2 + 2y^2 = 1$ also using the lines through the point $(-1, 0)$.

1.4. Parametrize the unit sphere $x^2 + y^2 + z^2 = 1$ using lines through the point $(-1, 0, 0)$.

1.5. If a and b are represented by the form $x^2 - my^2$, then so is their product. The content of this identity was already known to Brahmagupta, who used it for solving the equation $x^2 = my^2 + 1$ in integers.

1.6. Derive Bachet's duplication formula using analytic geometry. The slope of the tangent in a point may be obtained by implicit differentiation: $2yy' = 3x^2$ implies $y' = \frac{3x^2}{2y}$.

1.7. Show that Bachet's duplication formula applied twice to the point $(3, 5)$ on $y^2 = x^3 - 2$ yields the rational point

$$\left(\frac{2340922881}{7660^2}, \frac{113259286337292}{7660^3} \right).$$

1.8. Apply Bachet's duplication formula to the point $(2, 2)$ on the elliptic curve $y^2 + 4 = x^3$ and show that the only integral point resulting by repeated duplication is $(5, 11)$.

1.9. Show that there is no Pythagorean triple whose legs are prime numbers.

   Show moreover that if (p, b, q) is a Pythagorean triple in which one leg p and the hypotenuse q are primes, then $b = q - 1$.

1.10. Show that if (a, b, c) is a Pythagorean triple, then so is $(t - a, t - b, t + c)$. Similarly, consider the triple $(t + a, t + b, 2t - c)$ and find more ways of constructing a new Pythagorean triple from a known one.

   Show moreover that this method also works for sums of three squares.

1.11. Find a counterexample to the following statement: If p is prime and $kp = a^2 + mb^2$, and if k can be written in the form $k = c^2 + md^2$, then $p = e^2 + mf^2$.

1.12. Show that there are infinitely many primes of the form $p = 4n + 1$ as well as of the form $q = 4n + 3$.

   Hint: As in Euclid's proof, consider the integers $N_1 = (p_1 \cdots p_t)^2 + 1$ and $N_3 = 4q_1 \cdots q_t - 1$.

1.13. The following trick due to Ernst Trost [123, 124] is simple but often remarkably useful. Given a Diophantine equation $at^2 + bt + c = 0$ with a

rational solution t, the solution formula $t_{1,2} = \frac{-b \pm \sqrt{\Delta}}{2a}$ for quadratic equations
tells us that the discriminant $\Delta = b^2 - 4ac$ must be a square.

This almost trivial observation is turned into a useful method by the
following trick: If $x^4 - 4y^4 = z^2$ is solvable in integers (or rationals), then
the quadratic equation $x^4 - tz^2 - 4y^4t^2 = 0$ has a rational solution for $t = 1$.
Thus the discriminant Δ of the quadratic equation in t must be a square, i.e.,
$\Delta = z^4 + 16x^4y^4 = w^2$ must be solvable (in rational numbers and thus in
integers). Since the only solutions of this equation are the trivial solutions
with $xy = 0$ or $z = 0$, the only solutions of the original equation are $y = 0$
(and $z = x^2$).

Show using Trost's discriminant trick that the only integral solutions of the
equation $x^4 - 2y^2 = 1$ are the trivial solutions $(x, y) = (\pm 1, 0)$.

1.14. Show using Trost's discriminant trick that if $y^2 = x^3 - dx$ has nontrivial
rational solutions, then so does $y^4 + 4dx^4 = w^2$. Hint: $dxt^2 + y^2t - x^3 = 0$.

1.15. Show that -2 is a quadratic residue modulo prime numbers $p \equiv 1, 3 \bmod 8$.
Hint: For primes $p \equiv 1 \bmod 8$, Euler's proof that -2 is a quadratic residue
modulo p works fine. For primes $p \equiv 3 \bmod 8$ you have to show that -1 and
2 are quadratic nonresidues modulo p.

1.16. Transfer Euler's proof of the Two-Squares Theorem to primes of the form
$x^2 + 2y^2$.

1.17. Show that the representations $21^2 = 11^2 + 5 \cdot 8^2 = 19^2 + 5 \cdot 4^2$ of $21^2 = 441$
may be explained by the two representations of 21 by the form $x^2 + 5y^2$.
What about $21^2 = 6^2 + 5 \cdot 9^2 = 14^2 + 5 \cdot 7^2$?

1.18. Is it possible to solve the equation $y^2 + 2 = x^3$ in integers using methods
from elementary number theory?

1.19. Euler shows that setting $y\sqrt{2} + \sqrt{5} = (a\sqrt{2} + b\sqrt{5})^3$ does not lead to the
solution $(x, y) = (3, 4)$ of the equation $2y^2 - 5 = x^3$. Verify this and deduce
the solution by invoking factors of the form $(3 + \sqrt{10})^n$.

1.20. Let $q \equiv 3, 7, 11 \bmod 16$ be a prime number. Show that the equation $a^4 -
qb^4 = 1$ has $(a, b) = (\pm 1, 0)$ as its only integral solutions.

1.21. Show that the Diophantine equation $y^2 = x^3 + 7$ does not have a solution in
integers.
Hint: First show that x must be odd and then consider the equation $y^2 + 1 = (x + 2)(x^2 - 2x + 4)$.

1.22. Show that the Diophantine equation $y^2 = x^3 - 17$ does not have a solution in
integers.

1.23. Let $k = A^3 + B^2$ for natural numbers $A \equiv 3 \bmod 4$ and $B \equiv 0 \bmod 2$, and
assume that B is not divisible by a prime number $q \equiv 3 \bmod 4$. Show that the
Diophantine equation $y^2 = x^3 - k$ does not have an integral solution with y
even.

1.24. Let $k = A^3 + B^2$ for coprime integers A and B, and assume that $A \equiv 2 \bmod 4$
and that B is not divisible by any prime $q \equiv 3 \bmod 4$. Show that $y^2 = x^3 - k$
does not have any integral solutions.

1.25. Prove that the Diophantine equation $y^2 = x^3 - k$ always has an integral solution if $k = A^3 - B^2$. Use sage to find all integral points on $y^2 = x^3 + 17$ (observe that $17 = 2^3 + 3^2$).

1.26. (Hermite) Set

$$a = p + qi, \quad b = r + si, \quad c = -r + si, \quad d = p - qi,$$
$$a' = p' + q'i, \ b' = r' + s'i, \ c' = -r' + s'i, \ d' = p' - q'i$$

in the product

$$\begin{vmatrix} a & b \\ c & d \end{vmatrix} \cdot \begin{vmatrix} a' & b' \\ c' & d' \end{vmatrix} = \begin{vmatrix} aa' + bc' & ab' + bd' \\ ca' + dc' & cb' + dd' \end{vmatrix}$$

of the determinants $(ad - bc) \cdot (a'd' - b'c')$ and derive Euler's product formula for sums of four squares.

1.27. The Pythagorean equation $x^2 + y^2 = z^2$ can be written in factored form $y^2 = z^2 - x^2 = (z - x)(z + x)$. The same is true for $x^2 + y^2 = 2z^2$ after multiplying it through by 2: $4z^2 = 2x^2 + 2y^2 = (x + y)^2 + (x - y)^2$ yields $(x + y)^2 = (2z)^2 - (x - y)^2 = (2z + x - y)(2z - x + y)$.

Euler realized that such a decomposition is possible for the equation $ax^2 + by^2 = cz^2$ whenever this equation has a rational solution. Prove this claim.

Chapter 2
Quadratic Number Fields

In this chapter we provide the foundations for doing arithmetic in quadratic number rings. We will explain what a quadratic number field is, and which elements we will regard as "integers." In addition, we will visualize certain aspects of the arithmetic of quadratic number fields geometrically by introducing Pell conics.

2.1 Quadratic Number Fields

This book deals with the arithmetic of quadratic number fields, and in this and the next section we will present the main actors in our play. In Chap. 4 we will give a precise definition of what we mean by notions such as divisibility, units, and prime elements, and only then will we return to the question how to put Euler's solution of the Diophantine equation $y^2 + 2 = x^3$ in integers onto a solid foundation and apply his reasoning to other examples.

Let $m \in \mathbb{Z} \setminus \{0, 1\}$ be a squarefree integer; then the set

$$k = \mathbb{Q}(\sqrt{m}) = \{a + b\sqrt{m} : a, b \in \mathbb{Q}\}$$

of numbers of the form $a + b\sqrt{m}$, where a and b are rational numbers, is called a *quadratic number field* (the fact that k is actually a field is proved in Exercise 2.1). We call k real or complex quadratic according as $m > 0$ or $m < 0$.

The element $\alpha = a + b\sqrt{m} \in k$ is a root of the quadratic polynomial $P_\alpha(x) = x^2 - 2ax + a^2 - mb^2 \in \mathbb{Q}[x]$; its other root $\alpha' = a - b\sqrt{m}$ is called the *conjugate* of α. Moreover we call

$$
\begin{aligned}
N\alpha &= \alpha\alpha' &&= a^2 - mb^2 && \text{the } \textit{norm} \text{ of } \alpha, \\
\operatorname{Tr}\alpha &= \alpha + \alpha' &&= 2a && \text{the } \textit{trace} \text{ of } \alpha, \text{ and} \\
\operatorname{disc}(\alpha) &= (\alpha - \alpha')^2 &&= 4mb^2 && \text{the } \textit{discriminant} \text{ of } \alpha.
\end{aligned}
$$

© The Author(s), under exclusive license to Springer Nature Switzerland AG 2021
F. Lemmermeyer, *Quadratic Number Fields*, Springer Undergraduate
Mathematics Series, https://doi.org/10.1007/978-3-030-78652-6_2

The conjugate, the norm, the trace, and the discriminant of $\alpha = \frac{3+\sqrt{5}}{2} \in \mathbb{Q}(\sqrt{5})$, for example, are

$$\alpha' = \frac{3-\sqrt{5}}{2}, \quad N\alpha = \frac{3^2-5}{4} = 1, \quad \mathrm{Tr}\,\alpha = 3 \quad \text{and} \quad \mathrm{disc}\,(\alpha) = 5.$$

As we have seen, Euler first used numbers of the form $a + b\sqrt{c}$ for solving Diophantine equations in ordinary integers. In order to get equations in integers from relations in quadratic number rings we need maps $R \longrightarrow \mathbb{Z}$. Since we will mainly exploit multiplicative relations (decomposition into factors, divisibility, units), maps respecting the multiplicative structure such as the norm are particularly important (see, e.g., Exercise 2.7, and for the proof of the proposition below, Exercise 2.6).

Proposition 2.1 *For all $\alpha, \beta \in k$ we have*

$$N(\alpha\beta) = N\alpha\, N\beta \quad \text{and} \quad \mathrm{Tr}(\alpha + \beta) = \mathrm{Tr}\,\alpha + \mathrm{Tr}\,\beta.$$

Moreover $N\alpha = 0$ if and only if $\alpha = 0$, and disc $(\alpha) = 0$ if and only if $\alpha \in \mathbb{Q}$.

The map $\sigma : k \longrightarrow k : \alpha \longmapsto \sigma(\alpha) := \alpha'$ is called the *nontrivial automorphism* of $k/\mathbb{Q}$. Since $\sigma \circ \sigma = \mathrm{id}$ (the identity map; observe that $\sigma \circ \sigma(\alpha) = \sigma(\sigma(\alpha)) = \sigma(\alpha') = \alpha'' = \alpha$), $\{\mathrm{id}, \sigma\}$ is a group of order 2 with respect to composition, called the *Galois group* of $k/\mathbb{Q}$ and denoted by $\mathrm{Gal}\,(k/\mathbb{Q})$. Instead of $\sigma(\alpha)$ we will often write[1] α^σ.

The Galois group of a field extension is named after Évariste Galois (1811–1832), a French mathematician who died in a duel at age of 20. Galois revolutionized algebra by introducing group theoretic methods into the theory of the solution of polynomial equations by radicals. Over time, "Galois theory" evolved from a theory of polynomials to a theory of field extensions (apart from Richard Dedekind, Ernst Steinitz must be mentioned in this connection). The immense importance for the arithmetic of number fields even in such a simple case as that of quadratic extensions, where the Galois group has only two elements, will become clear in Chap. 9.

2.1.1 Quadratic Extensions as Vector Spaces

If $k \subseteq K$ are fields, then we may interpret K as a k-vector space. The vectors are the elements of K (these elements form an additive group), the scalars are the elements of k, and scalar multiplication is the usual multiplication of elements of k

---

[1]In extensions with non-abelian Galois groups one has to distinguish carefully between these notations since $\sigma\tau(\alpha)$ is often meant to be $\sigma(\tau(\alpha))$, whereas $\alpha^{\sigma\tau} = (\alpha^\sigma)^\tau$.

with elements of K inside K. The dimension of K as a k-vector space is called the *degree* of the field extension and is denoted by $(K : k) := \dim_k K$.

Quadratic extensions $K/\mathbb{Q}$, where $K = \mathbb{Q}(\sqrt{m}\,)$, have degree 2 since $\{1, \sqrt{m}\,\}$ is a $\mathbb{Q}$-basis of K because each element of K can be written uniquely as a $\mathbb{Q}$-linear combination of 1 and $\sqrt{m}$. In Exercise 2 we will investigate connections between different $\mathbb{Q}$-bases.

The interpretation of quadratic field extensions as vector spaces allows us to interpret the norm and the trace as maps that show up naturally in linear algebra. If we identify $1 = \binom{1}{0}$ and $\sqrt{m} = \binom{0}{1}$ and write $\alpha = a + b\sqrt{m} = \binom{a}{b}$, then multiplication by α is a linear map that can be described by a 2×2-matrix M_α whose columns are formed by the images of the basis vectors. Therefore this matrix is given by $M_\alpha = \left(\begin{smallmatrix} a & mb \\ b & a \end{smallmatrix}\right)$ since $\sqrt{m} \cdot (a + b\sqrt{m}) = mb + a\sqrt{m}$. Now observe that the determinant of M_α is the norm of α and that the trace of M_α is the trace of α (see Exercise 2.12). This observation explains once more that the norm is multiplicative and the trace additive.

2.2 Rings of Integers

In order to ask (and answer) questions concerning the arithmetic of quadratic number rings we first have to identify the "integers" in our fields. The obvious solution would be working in the rings $\mathbb{Z}[\sqrt{m}\,]$ of elements of the form $a + b\sqrt{m}$ with $a, b \in \mathbb{Z}$. This choice is not always the right one, as will become clear later (see, e.g., Exercise 2.15).

When Kummer studied higher reciprocity laws and Fermat's Last Theorem in cyclotomic fields he worked in the more or less obvious rings

$$\mathbb{Z}[\zeta] = a_0 + a_1\zeta + a_2\zeta^2 + \ldots + a_{p-1}\zeta^{p-1},$$

where ζ is a primitive p-th root of unity (see Exercise 2.45), and used Gaussian periods in their subfields. If $p = 3$, then $\mathbb{Q}(\zeta) = \mathbb{Q}(\sqrt{-3}\,)$ since we can choose $\zeta = \frac{-1+\sqrt{-3}}{2}$; in this case we see that the ring $\mathbb{Z}[\zeta]$ is strictly larger than $\mathbb{Z}[\sqrt{-3}\,]$.

Dirichlet solved the quintic Fermat equation $x^5 + y^5 = z^5$ using elements in the ring $\mathbb{Z}[\sqrt{5}\,]$; he did not consider elements of the form $\frac{p+q\sqrt{5}}{2}$ although this was more or less suggested by equations such as

$$P + Q\sqrt{5} = \frac{(\phi + \psi\sqrt{5}\,)^5}{2^4},$$

which would look a lot more symmetric if they were written in the form

$$\frac{P + Q\sqrt{5}}{2} = \left(\frac{\phi + \psi\sqrt{5}}{2}\right)^5.$$

Dirichlet also proved his unit theorem (a generalization of the solvability of the Pell equation) in rings of the form $\mathbb{Z}[\alpha]$, where α is a root of a monic polynomial

$$x^n + a_{n-1}x^{n-1} + \ldots + a_1 x + a_0$$

with coefficients $a_j \in \mathbb{Z}$.

It is not clear whether the question how to define algebraic integers was perceived as a problem before Dedekind gave the definition. The quote that

> Talent hits a target no one else can hit;
> Genius hits a target no one else can see

is credited to Schopenhauer. This definition of genius certainly applies to Dedekind.

The correct idea is to look for a ring that is as large as possible, but which does not contain any rational numbers except the ordinary integers. More precisely, we will denote by $\mathcal{O}$ the maximal ring[2] in $K = \mathbb{Q}(\sqrt{m})$ with the following properties:

- $\mathcal{O} \cap \mathbb{Q} = \mathbb{Z}$: The integral elements of the subfield of the rational numbers in K are exactly the ordinary integers.
- $\mathcal{O}^\sigma = \mathcal{O}$: If $\alpha = r + s\sqrt{m}$ is integral, then so is its conjugate $\alpha^\sigma = r - s\sqrt{m}$.

If $\alpha = r + s\sqrt{m}$ is in $\mathcal{O}$, then so is $\alpha^\sigma = r - s\sqrt{m}$ by the second condition, hence $\operatorname{Tr}\alpha = \alpha + \alpha^\sigma = 2r$ and $N\alpha = \alpha\alpha^\sigma = r^2 - ms^2$ are elements of $\mathcal{O} \cap \mathbb{Q}$ and therefore must be ordinary integers. Thus if α is an *algebraic integer*, then the monic polynomial

$$P_\alpha(x) = (x - \alpha)(x - \alpha^\sigma) = x^2 - \operatorname{Tr}(\alpha)x + N\alpha = x^2 - 2rx + r^2 - ms^2 \in \mathbb{Q}[x]$$

must have integral coefficients. We will call $\alpha \in K$ an algebraic integer if $P_\alpha(x)$ has coefficients in $\mathbb{Z}$. More generally, algebraic integers are roots of monic polynomials

$$x^n + a_{n-1}x^{n-1} + \ldots + a_1 x + a_0$$

with coefficients in $\mathbb{Z}$. The numbers $\sqrt{2}$, $\sqrt{-3}$ and $\frac{1+\sqrt{5}}{2}$, for example, are algebraic integers because they are roots of the monic polynomials $x^2 - 2$, $x^2 + 3$ and $x^2 - x - 1$, respectively, all of which have integral coefficients. On the other hand, $\frac{1}{\sqrt{2}}$ and $\frac{1+\sqrt{3}}{2}$ are algebraic numbers, but not algebraic integers because they are roots of the monic polynomials $x^2 - \frac{1}{2}$ and $x^2 - x - \frac{1}{2}$, respectively. It can be shown that algebraic numbers form a field, and that the algebraic integers form an integral domain (or simply a domain from now on).

---

[2]It is not clear a priori that such a maximal ring always exists.

The set of all integral elements in a number field k is called the ring $\mathcal{O}_k$ of (algebraic) integers in k. For quadratic number fields we will show that this set is actually a ring after having characterized these integers.

Theorem 2.2 *The integral elements in the quadratic number field $k = \mathbb{Q}(\sqrt{m})$ are given by*

$$\mathcal{O}_k = \begin{cases} \{a + b\sqrt{m} : a, b \in \mathbb{Z}\} & \text{if } m \equiv 2, 3 \bmod 4, \\ \{\frac{a+b\sqrt{m}}{2} : a, b \in \mathbb{Z}, a \equiv b \bmod 2\} & \text{if } m \equiv 1 \bmod 4. \end{cases}$$

Proof Assume that $\alpha = r + s\sqrt{m}$ is an algebraic integer with $r, s \in \mathbb{Q}$; then $\operatorname{Tr} \alpha = 2r$ and $N\alpha = r^2 - ms^2$ are ordinary integers. If we plug $2r \in \mathbb{Z}$ into the second equation, then we find that $4ms^2$ must be an integer. Since m is squarefree, $4s^2$ and thus finally $2s$ must be an integer. In fact, write $4s^2 = x^2/y^2$ for coprime integers $x, y \in \mathbb{Z}$; since $4ms^2$ is an integer, we find $y^2 \mid mx^2$; since $\gcd(x, y) = 1$ we find $y^2 \mid m$, and since m is squarefree this implies $y = \pm 1$.

Thus we may write $2r = a$ and $2s = b$ for integers $a, b \in \mathbb{Z}$. Now we exploit once more the fact that $N\alpha = r^2 - ms^2$ is an integer and find that $a^2 - mb^2 \equiv 0 \bmod 4$.

- If $m \equiv 2 \bmod 4$, then $2 \mid a$, $4 \mid a^2$ and $2 \mid b$, hence $r, s \in \mathbb{Z}$: Each algebraic integer has the form $r + s\sqrt{m}$ with $r, s \in \mathbb{Z}$.
- If $m \equiv 3 \bmod 4$, then $0 \equiv a^2 - mb^2 \equiv a^2 + b^2 \bmod 4$; this is only possible if a and b are even, and as above this implies that r and s must be integers.
- If $m \equiv 1 \bmod 4$, then we obtain the congruence $0 \equiv a^2 - mb^2 \equiv a^2 - b^2 \bmod 4$, which holds if and only if $a \equiv b \bmod 2$. Thus the algebraic integers in this case have the form $\frac{1}{2}(a + b\sqrt{m})$, where a and b are either both even or both odd. It is easily verified that these numbers are indeed algebraic integers.

This completes the proof. □

The field $k = \mathbb{Q}(\sqrt{m})$ consists of all $\mathbb{Q}$-linear combinations of 1 and $\sqrt{m}$. Does something similar hold for the ring $\mathcal{O}_k$ of integers, that is, does there exist an $\omega \in \mathcal{O}_k$ such that every $\alpha \in \mathcal{O}_k$ is a $\mathbb{Z}$-linear combination of 1 and ω? In this case we write $\mathcal{O}_k = \mathbb{Z} \oplus \omega\mathbb{Z}$ and call $\{1, \omega\}$ an *integral basis*. The answer to our question is in fact positive:

Corollary 2.3 *We have $\mathcal{O}_k = \mathbb{Z} \oplus \omega\mathbb{Z}$ for*

$$\omega = \begin{cases} \sqrt{m}, & \text{if } m \equiv 2, 3 \bmod 4; \\ \frac{1+\sqrt{m}}{2}, & \text{if } m \equiv 1 \bmod 4. \end{cases}$$

In particular, $\mathcal{O}_k$ is a ring.

Proof Only in the second case there is something to show. Assume therefore that $m \equiv 1 \bmod 4$ and $\alpha = \frac{1}{2}(a + b\sqrt{m})$ with $a \equiv b \bmod 2$; setting $c = \frac{a-b}{2}$ and $d = b$ we find $\alpha = c + d\omega$ with $c, d \in \mathbb{Z}$; the proof of the converse is just as simple.

The fact that $\mathcal{O}_k$ is a ring is now easily seen to be true by showing that the sum, difference, and the product of two elements of the form $a + b\omega$ with $a, b \in \mathbb{Z}$ again have this form. To this end we have to show that the product of two elements has this form, and this boils down to showing that $\omega^2 = r + s\omega$ for integers r and s. But clearly $\omega^2 = m = m + 0\omega$ for $m \equiv 2, 3 \bmod 4$, and $\omega^2 = \frac{1+m+2\sqrt{m}}{4} = \frac{m-1}{4} + \omega$ for $m \equiv 1 \bmod 4$. $\qquad\square$

The number $\Delta = \operatorname{disc} k := \left| \left(\begin{smallmatrix} 1 & \omega \\ 1 & \omega' \end{smallmatrix} \right) \right|^2 = (\omega - \omega')^2$ is called the *discriminant*[3] of the quadratic number field k. We find

$$\operatorname{disc} k = \begin{cases} 4m & \text{if } m \equiv 2, 3 \bmod 4, \\ m & \text{if } m \equiv 1 \bmod 4. \end{cases}$$

It is easily seen that $\{1, \frac{\Delta + \sqrt{\Delta}}{2}\}$ is an integral basis for any quadratic number field.

Our next result justifies our choice of the ring of integers in quadratic number fields:

Proposition 2.4 *The rational numbers contained in $\mathcal{O}_k$ are the ordinary integers:* $\mathcal{O}_k \cap \mathbb{Q} = \mathbb{Z}$.

Proof Clearly $\mathbb{Z} \subseteq \mathcal{O}_k \cap \mathbb{Q}$, so we have to prove the reverse inclusion. Assume therefore that $\alpha \in \mathcal{O}_k$; then $\alpha = \frac{1}{2}(a + b\sqrt{m})$ with $a \equiv b \bmod 2$. If $\alpha \in \mathbb{Q}$, then we must have $b = 0$; since b is even, so is a hence $\alpha = \frac{a}{2} \in \mathbb{Z}$. $\qquad\square$

It can be shown that $\mathcal{O}_k$ is the maximal subring of k with the property that $\mathcal{O}_k \cap \mathbb{Q} = \mathbb{Z}$; for this reason, $\mathcal{O}_k$ is often called the *maximal order* of k. A domain $\mathcal{O} \subset \mathcal{O}_k$ is called an *order* if $\mathcal{O}$ properly contains the ring $\mathbb{Z}$, i.e., if $\mathbb{Z} \subsetneq \mathcal{O} \subseteq \mathcal{O}_k$. By Proposition 2.4 we deduce immediately that each order $\mathcal{O}$ has the property $\mathcal{O} \cap \mathbb{Q} = \mathbb{Z}$.

Examples of number fields that are not quadratic are pure cubic number fields

$$\mathbb{Q}(\sqrt[3]{2}) = \{a + b\sqrt[3]{2} + c\sqrt[3]{4} : a, b, c \in \mathbb{Q}\},$$

which have degree 3, and cyclotomic fields

$$\mathbb{Q}(\zeta) = \{a_0 + a_1\zeta + a_2\zeta^2 + \ldots + a_{p-2}\zeta^{p-2} : a_j \in \mathbb{Q}\},$$

where ζ is a root of $\frac{x^p - 1}{x - 1} = 1 + x + \ldots + x^{p-1}$ and $p \geq 5$ is prime, and which have degree $p - 1$. We will occasionally use these fields as examples that lie outside of the scope of this book, and in the last chapter we will show for a deeper understanding of quadratic number fields we cannot avoid studying cyclotomic fields.

---

[3]The discriminant of a quadratic number field does not depend on the choice of the integral basis; see Exercise 2.3.

2.3 The Unit Circle

The elements of a quadratic number field with norm 1 form a group with respect to multiplication, since if $N\alpha = 1$ and $N\beta = 1$, then clearly $N(\alpha\beta) = 1$ and $N(\alpha/\beta) = 1$. The elements $x + yi$ with norm 1 in the field $\mathbb{Q}(i)$ are characterized by $N(x + yi) = x^2 + y^2 = 1$, i.e., the corresponding points (x, y) lie on the unit circle. Elements with norm 1 may be easily constructed by forming the quotient of two elements with the same norm: Thus $\frac{m+ni}{m-ni}$ has norm $\frac{m^2+n^2}{m^2+n^2} = 1$, and from

$$\frac{m + ni}{m - ni} = \frac{(m + ni)^2}{(m - ni)(m + ni)} = \frac{m^2 - n^2 + 2mni}{m^2 + n^2}$$

we can read off the parametrization

$$x = \frac{m^2 - n^2}{m^2 + n^2}, \quad y = \frac{2mn}{m^2 + n^2}$$

of the rational points on the unit circle. The fact that we get all rational points on the unit circle in this way, i.e., that all elements of norm 1 can be written as quotients $\frac{m+ni}{m-ni}$, is the content of Hilbert's Theorem 90, which will be important in Chap. 9.

It is a natural question whether the group structure of rational points on the unit circle given by the multiplication of the corresponding elements in $\mathbb{Q}(i)$ can be interpreted geometrically. This is indeed the case (see Fig. 2.1):

Theorem 2.5 *The elements $a + bi \in \mathbb{Q}(i)$ with norm 1 correspond to the rational points (x, y) on the unit circle $x^2 + y^2 = 1$. If $P(a, b)$ and $Q(c, d)$ are two rational points, then we obtain the point R corresponding to the product $(a + bi)(c + di)$ as follows:*

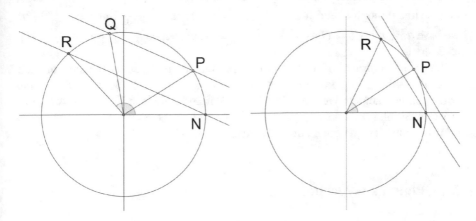

Fig. 2.1 Addition on the unit circle $P \oplus Q = R$ and $2P = R$

- *If P and Q are distinct, R is the second point of intersection of the unit circle and the parallel to PQ through the point N(1, 0).*
- *If P = Q, then R is the second point of intersection of the unit circle and the line through N that is parallel to the tangent in P.*

The point R corresponding to the product $(a + bi)(c + di) = ac - bd + (ad + bc)i$ has coordinates $(ac - bd, ad + bc)$. We have to show that the lines NR and PQ are parallel; to this end we first assume that the x-coordinates of P and Q are distinct. We then have to show that the slopes are equal:

$$\frac{d - b}{c - a} = \frac{ad + bc}{ac - bd - 1}.$$

Clearing the denominators we find

$$(d - b)(ac - bd - 1) = (ad + bc)(c - a),$$

which is equivalent to

$$(a^2 + b^2 - 1)d = (c^2 + d^2 - 1)b.$$

The last equation holds since $a^2 + b^2 = c^2 + d^2 = 1$.

If $P \neq Q$, but both points have the same x-coordinate, then the line PQ is parallel to the y-axis, and the parallel to PQ through N is a tangent to the circle in N; thus in this case $R = N$. Algebraically this corresponds to the product

$$(a + bi)(a - bi) = a^2 + b^2 = 1.$$

If finally $P = Q$, then the tangent is orthogonal to the line connecting the origin with P, and thus has slope $m = -\frac{a}{b}$. On the other hand, $(a + bi)^2 = a^2 - b^2 + 2abi$, i.e., the line through N and $R(a^2 - b^2, 2ab)$ has slope $\frac{2ab}{a^2 - b^2 - 1}$. Since $a^2 = 1 - b^2$ we have $a^2 - b^2 - 1 = (1 - b^2) - b^2 - 1 = -2b^2$, hence $\frac{2ab}{a^2 - b^2 - 1} = \frac{2ab}{-2b^2} = -\frac{a}{b}$ as desired.

Since the argument of a product of two complex numbers is the sum of their arguments, the group law on the unit circle is based on the addition of the corresponding angles: We have $P \oplus Q = R$ if and only if $\sphericalangle NOP + \sphericalangle NOQ = \sphericalangle NOR$. Similar remarks apply for the group law on the elements with norm 1 in arbitrary complex quadratic number fields.

2.4 Platon's Hyperbola

The points (x, y) corresponding to elements $\alpha = x + y\sqrt{m}$ with norm $N\alpha = x^2 - my^2 = 1$ in real quadratic number fields lie on a hyperbola. Whereas in complex quadratic number fields there can only be finitely many integral points on the norm-1 ellipses for simple geometric reasons (and in fact only the points $(\pm 1, 0)$ except when $\Delta = -3$ or $\Delta = -4$), the situation is fundamentally different in real quadratic number fields.

As a simple example consider the elements of norm 1 in $\mathbb{Z}[\sqrt{2}]$, that is, numbers $x + y\sqrt{2}$ with $x^2 - 2y^2 = 1$. It is easy to see that $3 + 2\sqrt{2}$ is such an element, and that $(3, 2)$ is an integral point on the hyperbola $\mathcal{H} : x^2 - 2y^2 = 1$. Since $N(1, 0)$ is another integral point, we can define a geometric group law on the set of integral (or rational) points on $\mathcal{H}$ by calling a point $R = P \oplus Q$ the sum of the points P and Q if R is the second point of intersection of the parallel to PQ through N with the hyperbola $\mathcal{H}$ (see Fig. 2.2).

Just as in the case of the unit circle we find

Theorem 2.6 *The numbers $a + b\sqrt{2} \in \mathbb{Q}(\sqrt{2})$ with norm 1 correspond bijectively to the rational points $P(a, b)$ on the hyperbola $\mathcal{H} : x^2 - 2y^2 = 1$. If $P(a, b)$ and $Q(c, d)$ are two such points, then we obtain the point R corresponding to the product $(a + b\sqrt{2})(c + d\sqrt{2})$ as the second point of intersection of the parallel to PQ through $N(1, 0)$ with the hyperbola $\mathcal{H}$ if P and Q are distinct, and as the second point of intersection of the tangent in P if $P = Q$.*

The proof is similar to the one for the unit circle. But as we shall see in a moment, the hyperbola $\mathcal{H}$ contains infinitely many integral points, whereas the unit circle only contains four such points. These integral points on $\mathcal{H}$ arise from $P(3, 2)$ by repeated addition. The point $n \cdot P$ corresponds to the element $(3 + 2\sqrt{2})^n$. We claim

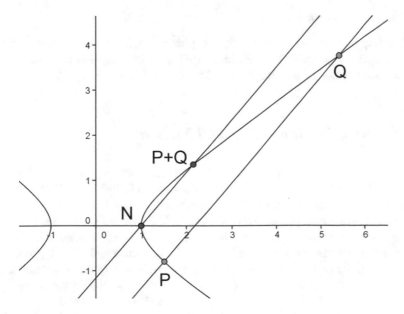

Fig. 2.2 Addition of points on Platon's hyperbola

that the only integral points on the right branch of the hyperbola are given by the integral multiples of P, which correspond to the powers $(3 + 2\sqrt{2})^n$ with $n \in \mathbb{Z}$.

To this end let Q be an arbitrary integral point on the upper right branch of $\mathcal{H}$, and assume that Q does not have the form nP. Since the x-coordinates of nP are not bounded, there must exist a natural number n such that Q lies properly between nP and $(n + 1)P$. Subtracting nP shows that $Q \ominus nP$ is an integral point lying properly between $N(1, 0)$ and $P(3, 2)$; but such a point does not exist.

The integral points on the lower right branch are obtained by reflection at the x-axis, which corresponds geometrically to conjugation, i.e., to multiplication of the exponent by -1. Thus every integral point on the right branch of the hyperbola is an integral multiple of P.

Since the integral points on the left branch of the hyperbola $\mathcal{H}$ are obtained by a reflection at the y-axis, which corresponds algebraically to multiplication by -1, we have shown:

Theorem 2.7 *The units of norm* 1 *in the ring* $\mathbb{Z}[\sqrt{2}]$ *are given by*

$$\varepsilon = (-1)^m (3 + 2\sqrt{2})^n$$

with $0 \leq m \leq 1$ *and* $n \in \mathbb{Z}$.

From these elements we obtain all units with norm -1 *via multiplication by* $1 + \sqrt{2}$. *Since* $3 + 2\sqrt{2} = (1 + \sqrt{2})^2$, *each unit in* $\mathbb{Z}[\sqrt{2}]$ *has the form*

$$\varepsilon = (-1)^m (1 + \sqrt{2})^n$$

with $0 \leq m \leq 1$ *and* $n \in \mathbb{Z}$.

The map $\varepsilon \mapsto (n, m)$ induces an isomorphism between the unit group in $\mathbb{Z}[\sqrt{2}]$ and the abstract group $\mathbb{Z}/2\mathbb{Z} \oplus \mathbb{Z}$. In Chap. 7 we will show that the unit group of $\mathbb{Z}[\sqrt{m}]$ for any nonsquare integer $m \geq 2$ is isomorphic to $\mathbb{Z}/2\mathbb{Z} \oplus \mathbb{Z}$.

2.4.1 Platon's Side and Diagonal Numbers

We have already mentioned that Euler was initiated to number theory by his friend Christian Goldbach (1690–1764). In one of his letters to Euler (see [89]) Goldbach claimed not only to have proved Fermat's theorem that 1 is the only triangular number that is a fourth power, but that actually 1 was the only square among them. Triangular numbers are numbers of the form $T_n = \frac{n(n+1)}{2}$; the reason behind their name is the fact that T_n pebbles may always be arranged in the form of a triangle (see [85]). Euler replied immediately that there are infinitely many squares among the triangular numbers. In fact, setting $T_n = m^2$ and completing the square gives $(2n + 1)^2 - 2(2m)^2 = 1$, hence $x = 2n + 1$ and $y = 2m$ satisfy the equation $x^2 - 2y^2 = 1$. The smallest solution in positive integers clearly is $(x, y) = (3, 2)$, which leads to $(m, n) = (1, 1)$. The next solution is $(x, y) = (17, 12)$, which yields the triangular number $T_8 = 36$, which clearly is a square.

These pairs of numbers (x, y) are called Platon's side and diagonal numbers. Platon (427–347) remarked that the square with side $s = 5$ has a diagonal that differs not much from $d = 7$. In fact, this diagonal has length $\sqrt{2 \cdot 5^2} = \sqrt{50}$ by the Theorem of Pythagoras, whereas $7^2 = 49$. The approximation $\sqrt{2} \approx \frac{7}{5}$ thus comes from the equation $7^2 - 2 \cdot 5^2 = -1$. Theon of Smyrna (ca. 70–135 A.D.; Smyrna is today called Izmir) explained that if x_n and y_n are numbers with $x_n^2 - 2y_n^2 = \pm 1$, then $x_{n+1}^2 - 2y_{n+1}^2 = \mp 1$, where we have set

$$x_{n+1} = x_n + 2y_n \quad \text{and} \quad y_{n+1} = x_n + y_n.$$

As we have seen above we obtain the integral solutions of the equation $x^2 - 2y^2 = \pm 1$ by setting

$$x_n + y_n \sqrt{2} = \pm (1 + \sqrt{2})^n.$$

If we choose the positive sign, then

$$x_n + y_n \sqrt{2} = (1 + \sqrt{2})^n, \qquad x_n - y_n \sqrt{2} = (1 - \sqrt{2})^n,$$

and this implies

$$x_n = \frac{(1 + \sqrt{2})^n + (1 - \sqrt{2})^n}{2}, \quad y_n = \frac{(1 + \sqrt{2})^n - (1 - \sqrt{2})^n}{2\sqrt{2}}.$$

2.5 Fibonacci's Hyperbola

In this section we will discuss a few connections between Fibonacci numbers and certain quadratic irrationalities, and will derive Binet's[4] Formula. Fibonacci (1170–1250), also named Leonardo of Pisa, was the son of a merchant from Pisa. During his education in North African countries he became familiar with the Hindu-Arabic numbers. In his famous book *Liber Abaci* he presented these numbers and methods for computing with them.

The Fibonacci numbers U_n named after him show up in this book and are defined recursively by

$$U_1 = U_2 = 1, \quad U_{n+1} = U_n + U_{n-1} \quad \text{for } n \geq 2.$$

---

[4]Binet published his formula in 1843; it was already known to Daniel Bernoulli in 1728—see [11, p. 90].

Thus the first few Fibonacci numbers are

$$1, 1, 2, 3, 5, 8, 13, 21, 34, 55, 89, 144, 233, \ldots,$$

and it is a natural question to ask whether there is an explicit formula for U_n.

2.5.1 Generating Functions

Generating functions are a powerful tool for investigating sequences of numbers. Here we will only use generating functions of the form $f(q) = \sum a_n q^n$ associated with a sequence (a_n). In the case of Fibonacci numbers, the generating function is given by

$$f(q) = \sum_{n=1}^{\infty} U_n q^n.$$

The recursion formula $U_{n+1} = U_n + U_{n-1}$ then provides us with the relation

$$f(q) - qf(q) - q^2 f(q) = q. \tag{2.1}$$

In fact we have

$$
\begin{aligned}
f(q) &= q + q^2 + 2q^3 + 3q^4 + \ldots + \quad U_n q^n + \ldots \\
qf(q) &= \quad\ \ q^2 + \ q^3 + 2q^4 + \ldots + U_{n-1} q^n + \ldots \\
q^2 f(q) &= \quad\qquad\ \ q^3 + \ q^4 + \ldots + U_{n-2} q^n + \ldots,
\end{aligned}
$$

and this clearly implies (2.1). Solving for $f(q)$ we obtain

$$f(q) = \frac{q}{1 - q - q^2}. \tag{2.2}$$

At this point we recall the dictum of Erich Hecke, who wrote in [60, p. 201] that the

> precise knowledge of the behaviour of an analytic function in the neighbourhood of its singular points is a source of number-theoretic theorems.

In the present case, the poles of f are given by $q = \frac{1}{\omega}$ and $q = \frac{1}{\omega'}$, where $\omega = \frac{1+\sqrt{5}}{2}$ and $\omega' = \frac{1-\sqrt{5}}{2}$. The computation of the partial fraction decomposition of a rational function $f(q) = \frac{A(q)}{B(q)}$ is simplified by employing Euler's formulas. If B is a monic polynomial with simple roots, and if $\deg A < \deg B$, then we can set

$$\frac{A(q)}{B(q)} = \sum_j \frac{a_j}{q - b_j} \tag{2.3}$$

with $a_j, b_j \in \mathbb{C}$. For determining the coefficients a_k we multiply (2.3) by $q - b_k$ and let $q \to b_k$. On the right side we obtain

$$\lim_{q \to b_k} (q - b_k) \sum_j \frac{a_j}{q - b_j} = a_k$$

since clearly

$$\lim_{q \to b_k} \frac{q - b_k}{q - b_j} = \begin{cases} 1 & \text{if } k = j, \\ 0 & \text{if } k \neq j. \end{cases}$$

In order to evaluate the left side, we use L'Hospital's rule and find

$$\lim_{q \to b_k} (q - b_k) \frac{A(q)}{B(q)} = \lim_{q \to b_k} \frac{A(q) + (q - b_k) A'(q)}{B'(q)} = \frac{A(b_k)}{B'(b_k)}.$$

This shows

Proposition 2.8 (Euler's Formulas) *Let $A(q)$ and $B(q)$ be polynomials in $\mathbb{C}[q]$, where B is assumed to have only simple roots. Then the coefficients a_k in the partial fraction decomposition (2.3) are determined by*

$$a_k = \frac{A(b_k)}{B'(b_k)}. \tag{2.4}$$

Thus the partial fraction decomposition of f is given by

$$f(q) = \frac{q}{1 - q - q^2} = \frac{1}{\sqrt{5}} \left(\frac{1}{1 - \omega q} - \frac{1}{1 - \omega' q} \right).$$

Developing this into a geometric series we obtain

$$f(q) = \frac{1}{\sqrt{5}} \left(1 + \omega q + \omega^2 q^2 + \omega^3 q^3 + \ldots - 1 - \omega' q - \omega'^2 q^2 - \omega'^3 q^3 - \ldots \right)$$

$$= \frac{1}{\sqrt{5}} \left((\omega - \omega') q + (\omega^2 - \omega'^2) q^2 + (\omega^3 - \omega'^3) q^3 + \ldots \right).$$

Comparing the coefficients of q^n here and in the definition of the generating function yields

Theorem 2.9 (Binet's Formula) *The Fibonacci numbers U_n admit the explicit representation*

$$U_n = \frac{\omega^n - \omega'^n}{\omega - \omega'}, \tag{2.5}$$

where ω and ω' are the roots of the quadratic equation $x^2 - x - 1 = 0$.

2.5.2 Group Law

It is hardly surprising that the Fibonacci numbers show up in connection with the hyperbola $\mathcal{F} : x^2 - xy - y^2 = 1$ since the denominator of the function $f(q)$ is $Q(1, q)$, where $Q(x, y) = x^2 - xy - y^2$ is a quadratic form with discriminant 5.

Theorem 2.10 *The group law on the hyperbola $\mathcal{F} : x^2 - xy - y^2 = 1$ with neutral element $N(1, 0)$, in which the sum of two points P and Q is the second point of intersection of the parallel to PQ through N with $\mathcal{F}$, is given by the equation $(x_1, y_1) \oplus (x_2, y_2) = (x_3, y_3)$ with*

$$x_3 = x_1 x_2 + y_1 y_2, \quad y_3 = x_1 y_2 + x_2 y_1 - y_1 y_2.$$

The simple proof is left to the readers as Exercise 2.33.

By computing multiples of the integral point $P = (2, 1)$ on the Fibonacci hyperbola (see Fig. 2.3) we obtain the points

$$2P = (5, 3), \quad 3P = (13, 8), \quad 4P = (34, 21), \ldots.$$

Using induction it is easily proved that $kP = (U_{2k+1}, U_{2k})$ for all $k \in \mathbb{N}$.

As in the case of Platon's hyperbola we can show that all integral points on the right branch of the Fibonacci hyperbola are integral multiples of $(2, 1)$.

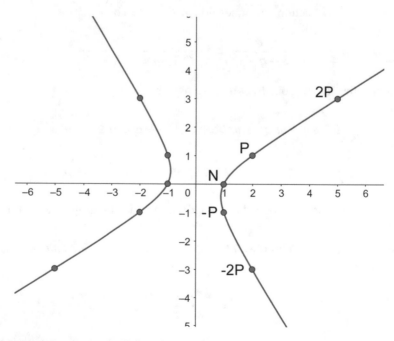

Fig. 2.3 Integral points on the Fibonacci hyperbola

2.6 Vieta Jumping

We can also generate infinitely many integral points on the Fibonacci hyperbola using a technique that has become known as "Vieta jumping" in recent years, and we can then show that there are no others.

The fundamental observation is the following: If $P = (x, y)$ is any integral point on the Fibonacci hyperbola, then there is a second integral point $P^* = (x, y')$ with the same x-coordinate. This is because for a fixed value of x, the quadratic equation $x^2 - xy - y^2 = 1$ in y has two solutions y_1, y_2, and that if y_1 is an integer, so is $y_2 = -x - y_1$. For the same reason there must be an integral point $P_* = (x', y)$ with the same y-coordinate as P.

Vieta jumping on conics is connected with the group law; in our case, $P \oplus P^* = (1, -1)$ and $P \oplus P_* = (-1, 0)$, as is easily seen from the geometric interpretation of the group law (see Fig. 2.4).

In order to show that all integral points on the Fibonacci hyperbola have the form kP or $kP \oplus (-1, 0)$ we consider an arbitrary integral point $Q(x, y)$. If $x > y \geq 1$, then $Q_* = (x', y)$ is an integral point with $y' < x$; if $y > x > 1$, on the other hand, then $Q^* = (x, y')$ is an integral point with $y < x'$. Repeating this descent eventually leads to an integral point with $x = \pm 1$, thus one of the four points $(\pm 1, 0)$ or $(\pm 1, \mp 1)$. Conversely we have to show that all points arising by the two operations P^* and P_* from $P(1, 0)$ have the form kP or $kP \oplus (-1, 0)$. We will leave the details once more to the reader (see Exercise 2.34).

2.6.1 The IMO Problem

The following problem due to Stephan Beck was posed at the International Mathematical Olympiad in 1988.

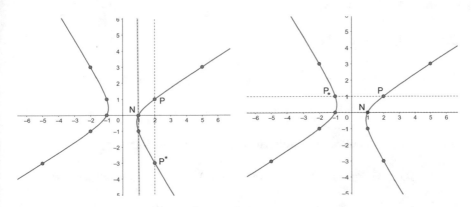

Fig. 2.4 $P \oplus P^* = (1, -1)$ and $P \oplus P_* = (-1, 0)$

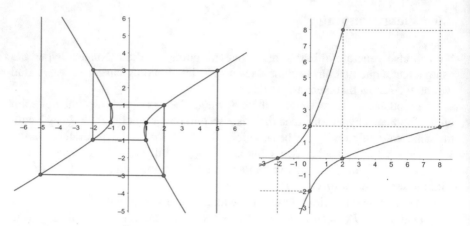

Fig. 2.5 Vieta jumping on the Fibonacci hyperbola (left) and on $\mathcal{C}_4 : x^2 - 4xy + y^2 = 4$ (right)

Let a and b be positive integers such that $ab + 1$ divides $a^2 + b^2$. Prove that $\frac{a^2+b^2}{ab+1}$ is a perfect square.

For the proof, assume that $P(a, b)$ is an integral point on the conic $\mathcal{C}_k : x^2 - kxy + y^2 = k$, and that k is not a square. Since k is not a square, we must have $k \geq 2$ (and as a matter of fact $k \geq 3$, since $k = 2$ implies $2 = (x - y)^2$, which is impossible in integers) (Fig. 2.5).

Next we claim that as long as $a \neq b$ we can find an integral point (a', b') on $\mathcal{C}_k$ lying in the first quadrant with $a' + b' < a + b$. Applying this step sufficiently often we obtain an integral point of the form (A, A); but then $A^2 = \frac{k}{2-k}$ implies $k = 1$ contradicting our assumptions.

The construction of (a', b') is easy: Assume that $b > a$; then $P^*(a, b')$ with $b' = ka - b$ is an integral point on $\mathcal{C}_k$, and $ab' = a^2 - k$ shows that $b' < a$. If $a > b$, then $P_* = (kb - a, b)$ has the desired properties. This proves our claims.

2.6.2 Markov's Equation

In [64], Adolf Hurwitz investigated Diophantine equations such as this one:

$$x_1^2 + x_2^2 + x_3^2 = kx_1x_2x_3.$$

If (x_1, x_2, x_3) is an integral solution, then so are, by Vieta jumping,

$$(kx_2x_3 - x_1, x_2, x_3), \quad (x_1, kx_1x_3 - x_2, x_3) \quad \text{and} \quad (x_1, x_2, kx_1x_2 - x_3).$$

For $k = 3$, this equation has the obvious integral solution $(1, 1, 1)$, and Vieta jumping gives rise to a whole tree of integral solutions.

For more on Markov's equation, its history and unsolved problems connected with it, see Aigner [1].

2.6.3 Summary

We have introduced the following notions, which will be fundamental for the following chapters:

- quadratic number fields
- norms, traces, and discriminants
- Galois groups of quadratic extensions of $\mathbb{Q}$
- rings of integers (maximal order)
- integral bases

For an introduction to the theory of group laws on conics see [86].

2.7 Exercises

2.1. Show that a quadratic number field $k = \mathbb{Q}(\sqrt{m})$, where m is a squarefree integer $\neq 1$, is a field.

2.2. Show that elements $\alpha, \beta \in K = \mathbb{Q}(\sqrt{m})$ form a $\mathbb{Q}$-basis of K if and only if the 2×2-matrix M defined by $\left(\begin{smallmatrix}\alpha\\\beta\end{smallmatrix}\right) = M\left(\begin{smallmatrix}1\\\sqrt{m}\end{smallmatrix}\right)$ is a matrix in the group $\mathrm{GL}_2(\mathbb{Q})$, i.e., if and only if $\det M \neq 0$.

2.3. Show that elements $\alpha, \beta \in \mathcal{O}_K$, where $\mathcal{O}_K$ is the ring of integers of $K = \mathbb{Q}(\sqrt{m})$, form an integral basis of $\mathcal{O}_K$ if and only if the 2×2-matrix M defined by $\left(\begin{smallmatrix}\alpha\\\beta\end{smallmatrix}\right) = M\left(\begin{smallmatrix}1\\\omega\end{smallmatrix}\right)$ is a matrix in the group $\mathrm{SL}_2(\mathbb{Z})$, i.e., if and only if the matrix has integral entries and determinant $\det M = \pm 1$.

2.4. Verify the equation

$$\begin{pmatrix} U_n & U_{n+1} \\ U_{n+1} & U_{n+2} \end{pmatrix} = \begin{pmatrix} 0 & 1 \\ 1 & 1 \end{pmatrix}^{n+1}$$

for Fibonacci numbers U_n. Diagonalize $T = \left(\begin{smallmatrix}0&1\\1&1\end{smallmatrix}\right)$ (i.e., find an invertible matrix $S \in M_2(\mathbb{C})$ with $D = S^{-1}TS = \left(\begin{smallmatrix}\alpha&0\\0&\beta\end{smallmatrix}\right)$) and observe that $T^n = (S^{-1}DS)^n = S^{-1}D^nS$. Since it is very easy to take powers of diagonal matrices, one now obtains a formula for the numbers U_n.

2.5. Prove that $p \mid U_{p\pm1}$ by expanding ω^p using the binomial theorem. Also show that for primes $p \equiv \pm1 \bmod 5$ we have $p \mid U_{p-1}$, for primes $p \equiv \pm2 \bmod 5$, on the other hand, $p \mid U_{p+1}$. The last result is due to Lagrange.

Joseph Louis Lagrange (1736–1813) was a French mathematician with Italian origins. In number theory, he is known for his proofs of the Four-

Squares Theorem (each natural number is the sum of at most four square numbers) and the solvability of the Pell equation, as well as for his theory of reduction of binary quadratic forms.

Hint: Show that the congruence $(a + b)^p \equiv a^p + b^p \bmod p$ holds in arbitrary rings.

2.6. Prove Proposition 2.1. In particular if $\alpha = a + b\sqrt{m} \in \mathbb{Q}(\sqrt{m})$, where m is not a square, show the following:

1. $\mathrm{Tr}(\alpha) = 0$ if and only if $a = 0$.
2. $\mathrm{disc}\,\alpha = 0$ if and only if $b = 0$.
3. $N\alpha = 0$ if and only if $a = b = 0$.

2.7. Show that if $\alpha \mid \beta$ in $\mathcal{O}_k$, then $N\alpha \mid N\beta$ in $\mathbb{Z}$.

2.8. Let $x^2 + px + q = 0$ be a quadratic equation with the solutions ω and ω'. Show that $\mathrm{disc}\,\omega = (\omega - \omega')^2 = p^2 - 4q$ coincides with the discriminant of the quadratic equation. What happens in case of the equation $ax^2 + bx + c = 0$?

2.9. Let m be a nonzero integer. Show that the following assertions are equivalent:

1. $\mathbb{Q}(\sqrt{m}) = \{a + b\sqrt{m}, a, b \in \mathbb{Q}\}$ is a field.
2. $x^2 - m$ is irreducible in $\mathbb{Q}[x]$.
3. The integer m is not a square in $\mathbb{Q}$.
4. $N(a + b\sqrt{m}) = a^2 - mb^2 = 0$ implies $a = b = 0$.

2.10. Let m be a squarefree integer and $K = \mathbb{Q}(\sqrt{m})$. Show that the square root $\sqrt{b}$ of an integer b is an element of K if and only if either $b = r^2$ is a square or $b = s^2m$ for some integer s.

2.11. Show that $\sigma : k \longrightarrow k$ is a ring homomorphism, i.e., show that $\sigma(\alpha + \beta) = \sigma(\alpha) + \sigma(\beta)$ and $\sigma(\alpha\beta) = \sigma(\alpha)\sigma(\beta)$ for all $\alpha, \beta \in k$. Show moreover that $\alpha \in k$ is in $\mathbb{Q}$ if and only if $\alpha = \sigma(\alpha)$.

2.12. Let $K/\mathbb{Q}$ be a quadratic extension. Verify that K is a $\mathbb{Q}$-vector space.

Show that multiplication by $\alpha = a + b\sqrt{m} \in K$ is a $\mathbb{Q}$-linear map $K \to K$; show that, with respect to the $\mathbb{Q}$-basis $\{1, \sqrt{m}\}$ of K, the map is described by $x \mapsto Ax$, where $x = \binom{r}{s}$ describes the element $r + s\sqrt{m}$ and where A is given by $A = \left(\begin{smallmatrix} a & mb \\ b & a \end{smallmatrix}\right)$.

Show that $N\alpha = \det A$ and $\mathrm{Tr}\,\alpha = \mathrm{Tr}\,A$, and that norm and trace do not depend on the choice of the basis.

2.13. Show that an element α of a quadratic number field is integral if and only if $\alpha' = \sigma(\alpha)$ is integral.

2.14. Show that if $\{1, \omega\}$ is an integral basis of $\mathcal{O}_k$, then so is $\{1, \omega - a\}$ for any integer $a \in \mathbb{Z}$.

Show more generally: If $\{\omega_1, \omega_2\}$ is an integral basis and if a, b, c, d are integers such that $ad - bc = 1$, then $\{a\omega_1 + b\omega_2, c\omega_1 + d\omega_2\}$ is also an integral basis.

2.15. Determine all $m < 0$ for which the ring $\mathcal{O}_k$ of integers in $k = \mathbb{Q}(\sqrt{m})$ contains an element of norm 2 or 3.

2.16. An abelian group M is called a G-module, if the group G acts on M, that is, if there is a map $G \times M \longrightarrow M : (g, m) \longmapsto gm$ with the following properties:

1. $g(m + m') = gm + gm'$,
2. $(gg')m = g(g'm)$,
3. $1m = m$

for all $g, g' \in G$ and all $m, m' \in M$. Show that the Galois group $G = \mathrm{Gal}\,(k/\mathbb{Q})$ of a quadratic number field k acts on the abelian groups $k, k^\times$ and $\mathcal{O}_k$ via $(\sigma, \alpha) \longmapsto \sigma(\alpha)$.

2.17. Solve the equation $x^2 + y^2 = 2z^2$ using "Euler's trick": Write the equation in the form $(x + y)^2 + (x - y)^2 = (2z)^2$.

2.18. An integral basis of the form $\{\omega, \sigma(\omega)\}$ is called a *normal integral basis*. Show that $\mathcal{O}_k$ has a normal integral basis of and only if $m \equiv 1 \bmod 4$, i.e., if and only if disc k is odd.

2.19. Show that $\mathbb{Q}(\sqrt[3]{2}) = \{a + b\sqrt[3]{2} + c\sqrt[3]{4} : a, b, c \in \mathbb{Q}\}$ is a field, but that the subset of all elements of the form $a + b\sqrt[3]{2}$ is not a field.

2.20. Although $2^2 + 5 \cdot 1^2 = 9$ is a square, it does not result from a decomposition $2 + \sqrt{-5} = (a + b\sqrt{-5})^2$. Show, however, that

$$2 + \sqrt{-5} = i\left(\frac{i + \sqrt{5}}{1 + i}\right)^2 = -\left(\frac{1 - \sqrt{-5}}{\sqrt{2}}\right)^2.$$

Explain the relation $31^2 - 26 \cdot 6^2 = 5^2$ by a similar decomposition of $31 + 6\sqrt{26}$.

2.21. The norm of $17 + 4\sqrt{15}$ is a square. Show that the square root of $17 + 4\sqrt{15}$ has the form $a\sqrt{3} + b\sqrt{5}$, and find more examples.

2.22. Let m be a positive integer. Show that if $a^2 - mb^2 = c^2$ and $1 \leq a \leq m$, then $a + b\sqrt{m}$ cannot be the square of a number of the form $r + s\sqrt{m}$ with $r, s \in \mathbb{Z}$ and $s \neq 0$. Show moreover that such examples exist for every composite positive integer m.

2.23. Consider the quadratic number fields $K = \mathbb{Q}(\sqrt{-m})$ with squarefree $m = u^2 - 4$ for an odd integer $u \geq 3$. Show that $2^2 + m = u^2$ is a counterexample to the Square Lemma in $\mathbb{Z}[\sqrt{-m}]$.

2.24. An entry in Joseph Liouville's notebook, probably written while the French mathematicians struggled with Gabriel Lamé's purported proof of Fermat's Last Theorem, contains the following equation:

$$169 = 13 \cdot 13 = (4 + 3\sqrt{-17})(4 - 3\sqrt{-17}).$$

Show that this is a counterexample to the Square Product Theorem in $\mathbb{Z}[\sqrt{-17}]$.

2.25. Show that Euler's problems with the equation $32 = 5^2 + 7$ are due to the fact that he worked in the ring $\mathbb{Z}[\sqrt{-7}]$ instead of in the ring of integers

$\mathbb{Z}[\frac{1+\sqrt{-7}}{2}]$. Verify that

$$\frac{5+\sqrt{-7}}{2} = \left(\frac{-1-\sqrt{-7}}{2}\right)^3,$$

and factorize $\frac{181+\sqrt{-7}}{2}$ similarly.

2.26. Use the fact that addition of points on the unit circle corresponds to the addition of angles to derive the addition formulas for trigonometric functions.

2.27. Project the points on the unit circle from the point $Z(-1, 0)$ to the tangent t in N, and associate the point Z with the "point at infinity" on t. Which group law on t is induced by the group law on the unit circle under this projection?

2.28. The inverse of the duplication formula $2(x, y) = (x^2 - y^2, 2xy)$ for rational points on the unit circle corresponds to taking the square root of the complex number $x + yi$ corresponding to the point (x, y). Show that the two solutions of $\frac{1}{2}(x, y)$, where $x, y > 0$, are given by $\left(\varepsilon\sqrt{\frac{1+x}{2}}, \varepsilon\sqrt{\frac{1-x}{2}}\right)$ for $\varepsilon = \pm 1$.

Convince yourself that a repeated application of halving points to $\cos\frac{\pi}{4} = \sin\frac{\pi}{4} = \frac{1}{2}\sqrt{2}$ yields the formulas

$$\cos\frac{\pi}{8} = \frac{1}{2}\sqrt{2+\sqrt{2}}, \qquad\qquad \sin\frac{\pi}{8} = \frac{1}{2}\sqrt{2-\sqrt{2}},$$

$$\cos\frac{\pi}{16} = \frac{1}{2}\sqrt{2+\sqrt{2+\sqrt{2}}}, \qquad \sin\frac{\pi}{16} = \frac{1}{2}\sqrt{2-\sqrt{2+\sqrt{2}}},$$

etc.

2.29. Show that the group law on the hyperbola $xy = 1$ with neutral element $N(1, 1)$ is given by $(x_1, y_1) \oplus (x_2, y_2) = (x_1 x_2, y_1 y_2)$.

2.30. Show that the group law on the parabola $y = x^2$ with neutral element $N(0, 0)$ is given by $(x_1, y_1) \oplus (x_2, y_2) = (x_1 + x_2, y_1 + y_2 + 2x_1 x_2)$.

2.31. Show that the generating function $f(q)$ of the Fibonacci numbers satisfies the functional equation

$$f\left(\frac{1}{q}\right) = f(-q).$$

2.32. Show that for Fibonacci numbers U_n we have

$$\lim_{n\to\infty} \frac{U_{n+1}}{U_n} = \frac{\sqrt{5}+1}{2}.$$

2.33. Prove Theorem 2.10: Show that

1. $P_3(x_3, y_3)$ lies on the Fibonacci hyperbola $x^2 - xy - y^2 = 1$, and that
2. the slope of the line through P_1 and P_2 is equal to that through P_3 and N.

2.34. Determine all integral points on the Fibonacci hyperbola using Vieta jumping.

2.35. Consider the Lucas–Lehmer hyperbola $x^2 - 3y^2 = 1$. Show that the group law with neutral element $N(1, 0)$ is given by

$$(x_1, y_1) + (x_2, y_2) = (x_1 x_2 + 3y_1 y_2, x_1 y_2 + x_2 y_1).$$

Show that the integral points on this hyperbola are the multiples of $P(2, 1)$ and their negatives. Show in addition that $2^k P = (x_k, y_k)$ with $x_{k+1} = 2x_k^2 - 1$.

2.36. Let n be an odd natural number. Show that n is prime if and only if there is an integer a with $a^{n-1} \equiv 1 \bmod n$ and $a^k \not\equiv 1 \bmod n$ for each proper divisor k of $n - 1$.

Deduce that $n = 2^m + 1$ is prime if and only if $3^{(n-1)/2} \equiv -1 \bmod n$ (this is called Pépin's test).

We can formulate this primality test in the language of conics. An odd integer n is prime if and only if there is a point P on the hyperbola $xy = 1$ defined over $\mathbb{Z}/n\mathbb{Z}$ for which $(n - 1)P = (1, 1)$ and $kP \neq (1, 1)$ for each proper divisor k of $n - 1$.

For $n = 17$ and $P = (3, 6)$ (the coordinates have to be read modulo 17), for example, we have $2P = (9, 2)$, $4P = (13, 4)$, $8P = (-1, -1)$ and $16P = (1, 1)$, and this proves that 17 is prime.

For more on primality tests using conics see Hambleton [50]. Factorization algorithms based on the arithmetic of Pell conics are studied in Eelkema [33]. We also mention a proof of the quadratic reciprocity law based on Pell conics due to Hambleton and Scharaschkin [52].

2.37. Let p be a prime number with $\left(\frac{3}{p}\right) = -1$. Show that the points modulo p on the conic $x^2 - 3y^2 = 1$ form a cyclic group of order $p + 1$.

Show moreover that $p = 2^q - 1$ is prime if and only if $\frac{p+1}{2}P = (-1, 0)$ for $P = (2, 1)$. Show also that this is equivalent to $\frac{p+1}{4}P = (0, b)$ for a suitable b modulo p.

2.38. Find all integral points on the Beck conic $x^2 - 4xy + y^2 = 4$.

For $Q = (a, b)$ let $Q^* = (a, b')$ and $Q_* = (a', b)$ denote the points derived from Q by Vieta jumping. With $P = (2, 0)$ and $T = (0, 2)$ show that $P \oplus P = T_*$ and $P \oplus T = P_*$.

2.39. Find all integral points on the conic $x^2 + y^2 - 3xy + 1 = 0$.

2.40. (Romanian Team Selection Test 1991) Let a and b be positive integers. Prove that if the number $\frac{a+1}{b} + \frac{b}{a}$ is an integer, then it is equal to 3.

2.41. Vieta jumping works for Platon's hyperbola $\mathcal{H} : x^2 - 2y^2 = 1$ after a coordinate transformation. Use the substitution $x = Y + Y$, $y = Y$ for finding all integral points on $\mathcal{H}$ using Vieta jumping.

2.42. Transform the hyperbola $x^2 - 3y^2 = 1$ using $x = X + 2Y$ and $y = Y$, and determine all integral points on these conics using Vieta jumping.

2.43. Determine all integral points on the hyperbolas $x^2 - (n^2 - 1)y^2 = 1$ using Vieta jumping.

2.44. Show that algebraic integers form a ring using the example $\alpha = \sqrt{3}$ and $\beta = \sqrt[3]{2}$, i.e., find monic polynomials with integral coefficients whose roots are $\alpha + \beta$ and $\alpha\beta$, respectively.

2.45. Let ζ be a primitive n-th root of unity, i.e., an algebraic number with the property that n is the smallest positive exponent satisfying $\zeta^n = 1$.

Show that the set $\mathbb{Z}[\zeta]$ consisting of all elements $\alpha = \sum_{j=0}^{n-1} a_j \zeta^j$ with $a_j \in \mathbb{Z}$ forms a ring.

2.46. Let α be a root of an irreducible monic polynomial f of degree n and with integral coefficients, and let $K = \mathbb{Q}(\alpha)$ be the smallest field extension of $\mathbb{Q}$ containing α. Show that K consists of all expressions $\omega = a_0 + a_1\alpha + a_2\alpha^2 + \ldots + a_{n-1}\alpha^{n-1}$ with $a_j \in \mathbb{Q}$.

The conjugates of α are the roots $\alpha_1 = \alpha$, α_2, $\ldots$, α_n of f, and the conjugates of ω are $\omega_j = a_0 + a_1\alpha_j + a_2\alpha_j^2 + \ldots + a_{n-1}\alpha_j^{n-1}$. Define the norm of ω to be the product of its conjugates: $N(\omega) = \omega_1\omega_2\cdots\omega_n$. Show that $N(\omega)$ is an integer, and that ω is a unit if and only if $N(\omega) = \pm 1$.

2.47. A natural number n is called *powerful* if $p \mid n$ for some prime p implies that $p^2 \mid n$; in other words: if the exponent of each prime in the prime factorization of n is at least 2.

Show that there are infinitely many consecutive powerful numbers; the smallest example is $(8, 9)$.

Chapter 3
The Modularity Theorem

In the last chapter we have investigated a few Pell conics such as $x^2 - 2y^2 = 1$ and $x^2 - xy - y^2 = 1$. For finding all integral points on Pell conics $Q(x, y) = 1$, where $Q(x, y) = ax^2 + bxy + cy^2$ is a binary quadratic form, it is natural to ask whether this equation has solutions in rational numbers or in residue class rings.

The general philosophy behind this way of investigating a mathematical problem in the integers is to study the object in question in simpler rings such as the field of rational numbers or finite fields.[1]

3.1 Pell Conics Over Fields

The solvability of the Pell equation $x^2 - my^2 = 1$ in integers is a nontrivial problem. Describing the solutions in fields, in particular in the field $\mathbb{Q}$ of rational numbers or in finite fields $\mathbb{F}_p = \mathbb{Z}/p\mathbb{Z}$, is a rather simple problem. In this section, we will explain how to find all rational solutions of a Pell equation; in the next section we will discuss the solutions in finite fields, which will lead us to the definition of Kronecker symbols. Quadratic number fields will not play a big role in this chapter, which belongs to elementary number theory; but Kronecker symbols will turn out to play a central role in the arithmetic of quadratic number fields later on.

---

[1] Beginners in mathematics may find it hard to believe that mathematicians think of finite fields (and even p-adic numbers) as being simpler objects than integers. One possible way of measuring the simplicity of structures A and B is counting homomorphisms from A and B into structures C. For example, there are many homomorphisms from $\mathbb{Z}$ to finite fields $\mathbb{F}_p$, whereas the only homomorphisms from $\mathbb{F}_p$ to $\mathbb{Z}$ or to finite fields are either the trivial homomorphism mapping everything to 0 or (in the case of $\mathbb{F}_p \longrightarrow \mathbb{F}_p$) an isomorphism.

© The Author(s), under exclusive license to Springer Nature Switzerland AG 2021
F. Lemmermeyer, *Quadratic Number Fields*, Springer Undergraduate
Mathematics Series, https://doi.org/10.1007/978-3-030-78652-6_3

3.1.1 Parametrization of Conics

Let m be a nonzero integer, and consider the Pell equation $x^2 - my^2 = 1$. This equation has two trivial solutions $(\pm 1, 0)$, and like each quadratic equation describing a smooth curve (parabolas, ellipses and hyperbolas), it has infinitely many rational solutions as soon as it has a single one. In fact, there is a simple geometric method of finding these rational solutions. The basic idea is that a line with rational slope t through one known rational point, say $Q(-1, 0)$, will intersect the conic in a second point P, and this point must have rational coordinates because of Vieta's formulas.

Let us go through this procedure step by step. The line through $Q(-1, 0)$ with slope t is given (in a standard Cartesian coordinate system) by the equation $y = t(x + 1)$. Intersecting this line with the Pell conic $x^2 - my^2 = 1$ yields the equation $x^2 - 1 - mt^2(x + 1)^2 = 0$, which we can write in the form

$$(x + 1)(x - 1 - mt^2(x + 1)) = 0.$$

This product is 0 if either $x_1 = 1$ or if the second factor equal to 0, which leads to

$$x = \frac{1 + mt^2}{1 - mt^2}, \quad y = t(x + 1) = \frac{2t}{1 - mt^2}.$$

If $m = n^2$ is a square, we have to exclude the values $t = \pm\frac{1}{n}$. Since we have, by the choice of our coordinate system, not considered the line $x = -1$ with slope ∞, this parametrization gives every rational point on the Pell conic $\mathcal{P} : x^2 - my^2 = 1$ with the exception of Q itself.

If we set $t = \frac{s}{r}$ for integers $r \neq 0$ and s, the parametrization derived above yields

Theorem 3.1 *The rational points $(x, y) \neq (-1, 0)$ on the Pell conic $\mathcal{P} : x^2 - my^2 = 1$, where m is a nonsquare integer, are given by*

$$x = \frac{r^2 + ms^2}{r^2 - ms^2}, \quad y = \frac{2rs}{r^2 - ms^2}. \tag{3.1}$$

Finding the rational points on a Pell conic $\mathcal{C}$ is thus a rather easy task. It is much more difficult to find the integral points on $\mathcal{C}$.

3.1.2 Pell Conics Over Finite Fields

The parametrization of rational points on a Pell conic $\mathcal{P} : x^2 - my^2 = 1$ carries over word for word to arbitrary fields:

Theorem 3.2 *Let m be a nonzero integer and p an odd prime number not dividing m. Every point $P(x, y) \neq (-1, 0)$ on the Pell conic $\mathcal{P} : x^2 - my^2 = 1$ with $x, y \in \mathbb{F}_p$ is given by*

$$x_t = \frac{1 + mt^2}{1 - mt^2}, \quad y_t = \frac{2t}{1 - mt^2},$$

where t is an arbitrary element of $\mathbb{F}_p$ with $mt^2 \neq 1$.

We say that a conic $Q(x, y) = ax^2 + bxy + cy^2 + dx + ey + f = 0$ is defined over a field F if $Q \in F[x, y]$ has coefficients in F. A point (ξ, η) on the conic is called an F-rational point if $\xi, \eta \in F$. We can find all $\mathbb{F}_5$-rational points on the Pell conic $x^2 - 2y^2 = 1$, for example, by plugging $t = 0, 1, 2, 3, 4 \in \mathbb{F}_5$ into the parametrization in the theorem above; we find (keeping in mind that we are working modulo 5):

$$x_0 = 1, \qquad\qquad y_0 = 0,$$
$$x_1 = 2, \qquad\qquad y_1 = -2,$$
$$x_2 = -\frac{9}{7} \equiv -2, \qquad\qquad y_2 = -2,$$
$$x_3 = -\frac{19}{17} \equiv -2, \qquad\qquad y_3 = 2,$$
$$x_4 = 2, \qquad\qquad y_4 = 2.$$

This shows that there are, together with $(-1, 0)$, exactly six $\mathbb{F}_5$-rational points on the Pell conic $x^2 - 2y^2 = 1$.

Counting the number of $\mathbb{F}_p$-rational points on an arbitrary Pell conic $x^2 - my^2 = 1$ is not hard (we are still assuming that $p \nmid 2m$). The number of $\mathbb{F}_p$-rational points on $\mathcal{P}$ depends on whether the equation $mt^2 = 1$ has a solution in $\mathbb{F}_p$. Such a solution exists if and only if m is a square in $\mathbb{F}_p^\times$; in fact, if $m = n^2$ in $\mathbb{F}_p^\times$, then $mt^2 = 1$ for $t = \pm\frac{1}{n}$. In this case, the parametrization yields $\mathbb{F}_p$-rational points for $p - 2$ values of p, so including $(-1, 0)$ there are $p - 1$ points in $\mathcal{P}(\mathbb{F}_p)$. If m is not a square in $\mathbb{F}_p$, then all p values of t yield points in $\mathcal{P}(\mathbb{F}_p)$, so in this case there are $p + 1$ points in $\mathcal{P}(\mathbb{F}_p)$.

Theorem 3.3 *Let m be an integer and p an odd prime not dividing 2m. Then the number of $\mathbb{F}_p$-rational point on the Pell conic $\mathcal{P} : x^2 - my^2 = 1$ is given by*

$$\#\mathcal{P}(\mathbb{F}_p) = \begin{cases} p - 1 & \text{if } m \text{ is a square in } \mathbb{F}_p^\times, \\ p + 1 & \text{otherwise.} \end{cases}$$

Whether an integer m is a square modulo an odd prime p or not is described by Legendre symbols, which we introduce next.

3.2   The Symbols of Legendre, Kronecker, and Jacobi

Let p be an odd prime number and a an integer. If a is not divisible by p, then we say that a is a quadratic residue modulo p if the congruence $x^2 \equiv a \bmod p$ is solvable in integers, and a quadratic nonresidue otherwise. Then we define the Legendre symbol $\left(\frac{a}{p}\right)$ by

$$\left(\frac{a}{p}\right) = \begin{cases} 0 & \text{if } p \mid a, \\ +1 & \text{if } p \nmid a, \ a \text{ is a quadratic residue modulo} p, \\ -1 & \text{if } p \nmid a, \ a \text{ is a quadratic nonresidue modulo} p. \end{cases} \tag{3.2}$$

Algebraically, the Legendre symbol provides us with a group homomorphism from the group $(\mathbb{Z}/p\mathbb{Z})^\times$ of coprime residue classes modulo p to the group $\{-1, +1\}$. This follows easily from the fact that $\left(\frac{a}{p}\right)$ only depends on the residue class of a mod p and the multiplicativity of the Legendre symbol:

Proposition 3.4 *The Legendre symbol is multiplicative:*

$$\left(\frac{ab}{p}\right) = \left(\frac{a}{p}\right)\left(\frac{b}{p}\right).$$

This property is a consequence of the fact that $(\mathbb{Z}/p\mathbb{Z})^\times$ is cyclic. Recall that an integer g is called a primitive root modulo N if each coprime residue class a modulo N has the form $a \equiv g^k \bmod N$. It is known that there exist primitive roots modulo every prime. A primitive root modulo an odd prime p is always a quadratic nonresidue: If $g \equiv h^2 \bmod p$, then $g^{\frac{p-1}{2}} \equiv h^{p-1} \equiv 1 \bmod p$, which would imply that the powers of g represent at most half of the coprime residue classes modulo p.

For the same reason, all odd powers g^{2k+1} are quadratic nonresidues since if $g^{2k+1} \equiv h^2 \bmod p$, then $g \equiv (g^{-k}h)^2 \bmod p$ would be a quadratic residue. Thus g^k is a quadratic residue modulo p if and only if k is even. But now multiplicativity follows: If, e.g., a and b are quadratic nonresidues modulo p, then $a \equiv g^k$ and $b \equiv g^h \bmod p$ for odd exponents k and h, hence $ab \equiv g^{k+h} \bmod p$ is a quadratic residue.

The existence of primitive roots also implies Euler's criterion:

Proposition 3.5 (Euler's Criterion) *For all integers a not divisible by the prime p we have*

$$\left(\frac{a}{p}\right) \equiv a^{\frac{p-1}{2}} \bmod p.$$

If a is a quadratic residue, then $a \equiv g^{2k} \bmod p$, hence $a^{\frac{p-1}{2}} \equiv g^{(p-1)k]} \equiv 1 \bmod p$; if a is a quadratic nonresidue, then $a \equiv g^{2k+1} \bmod p$, hence $a^{\frac{p-1}{2}} \equiv g^{(p-1)k} g^{\frac{p-1}{2}} \equiv g^{\frac{p-1}{2}} \equiv -1 \bmod p$. This follows from the fact that $x \equiv g^{\frac{p-1}{2}} \bmod$

p is a solution of the congruence $x^2 \equiv 1 \bmod p$, hence p divides $(x - 1)(x + 1)$, and since $\mathbb{F}_p$ is a field, we must have $x \equiv \pm 1 \bmod p$. Since g is a primitive root, we cannot have $x \equiv 1 \bmod p$.

3.2.1 Kronecker Symbol

The Kronecker symbol is a slight modification of the Legendre symbol and will turn out to be useful for describing the behavior of prime numbers in quadratic number fields. The numerator of a Kronecker symbol is restricted to discriminants Δ of quadratic number fields. For odd prime numbers p, the Kronecker symbol $(\frac{\Delta}{p})$ coincides with the ordinary Legendre symbol. If Δ is odd, we set, in addition, $(\frac{\Delta}{2}) = +1$ or -1 according as $\Delta \equiv 1 \bmod 8$ or $\Delta \equiv 5 \bmod 8$. In other words, we define $(\frac{\Delta}{2}) = (\frac{2}{\Delta})$, where $(\frac{\Delta}{2})$ is a Kronecker symbol and $(\frac{2}{\Delta})$ a Legendre symbol.

3.2.2 Gauss's Lemma

Gauss's Lemma in the theory of quadratic residues is an elementary technique for studying properties of Legendre and Kronecker symbols. Let $p = 2m + 1$ denote an odd prime number; then any set of integers $\{a_1, \ldots, a_m\}$ with the property that every coprime residue class is represented by an element of the form $\pm a_j$ is called a *half system* modulo p. The standard half system modulo $p = 2m + 1$ is the set $A_p = \{1, 2, 3, \ldots, m\}$.

Now let a be an integer coprime to p, and write, for each $1 \leq j \leq m$,

$$a \cdot a_j \equiv \varepsilon_j a_j'$$

for $\varepsilon_j = \pm 1$ and some $a_j' \in A_p$. Taking the product over all m such congruences yields

$$a^m \prod a_j = \prod \varepsilon_j \prod a_j'.$$

Since no a_j' occurs twice, we must have $\prod a_j = \prod a_j'$, and since this product is coprime to p, it follows that

$$a^m \equiv \prod \varepsilon_j \bmod p.$$

By Euler's criterion we have

$$a^{\frac{p-1}{2}} \equiv \left(\frac{a}{p}\right) \bmod p,$$

and this implies

Lemma 3.6 (Gauss's Lemma) *Let $p = 2m + 1$ be an odd prime number and $A = \{a_1, \ldots, a_m\}$ a half system modulo p. If we write $a \cdot a_j \equiv \varepsilon_j a'_j$ for each $1 \le j \le m$ with $\varepsilon_j = \pm 1$ and $a'_j \in A$, then the Legendre symbol $(\frac{a}{p})$ is given by*

$$\left(\frac{a}{p}\right) = \prod \varepsilon_j.$$

For determining $(\frac{2}{7})$, for example, we take the half system $\{1, 2, 3\}$ modulo 7 and write

$$2 \cdot 1 \equiv +2 \bmod 7,$$
$$2 \cdot 2 \equiv -3 \bmod 7,$$
$$2 \cdot 3 \equiv -1 \bmod 7,$$

hence $(\frac{2}{7}) = (-1)^2 = +1$.

Next we determine a few Legendre and Kronecker symbols. We begin with $(\frac{-1}{p}) = (\frac{-4}{p})$. The value of this symbol follows immediately from Euler's criterion:

$$\left(\frac{-1}{p}\right) \equiv (-1)^{\frac{p-1}{2}} \bmod p \quad \text{implies the equation} \quad \left(\frac{-1}{p}\right) = (-1)^{\frac{p-1}{2}}.$$

Since the power of -1 on the right side only depends on the residue class of p mod 4, we find

Proposition 3.7 *We have $(\frac{-1}{p}) = (-1)^{\frac{p-1}{2}}$. In particular, the Legendre symbol $(\frac{-1}{p})$ for primes $p \ge 3$ only depends on the residue class of p mod 4; in fact, for positive prime numbers p we have*

$$\left(\frac{-1}{p}\right) = \begin{cases} +1 & \text{if } p \equiv 1 \bmod 4, \\ -1 & \text{if } p \equiv 3 \bmod 4. \end{cases}$$

In order to become familiar with Gauss's Lemma we now use it for giving a second proof of this proposition. To this end we write $p = 2n + 1$ and multiply the representatives of the half system $\{1, 2, \ldots, n\}$ by -1:

$$-1 \cdot 1 \equiv -1 \bmod p,$$
$$-1 \cdot 2 \equiv -2 \bmod p,$$
$$\cdots \qquad \cdots$$
$$-1 \cdot n \equiv -n \bmod p.$$

Gauss's Lemma then tells us that $(\frac{-1}{p}) = (-1)^n = (-1)^{\frac{p-1}{2}}$.

In a similar way we can now determine the Legendre symbol $(\frac{2}{p})$. We first assume that $p = 4m + 1$ and write

$$2 \cdot 1 \equiv 2 \bmod p,$$
$$2 \cdot 2 \equiv 4 \bmod p,$$
$$\cdots \qquad \cdots$$
$$2 \cdot m \equiv 2m \bmod p,$$
$$2 \cdot (m + 1) \equiv 2m + 2 \equiv -(2m - 1) \bmod p,$$
$$2 \cdot (m + 2) \equiv 2m + 4 \equiv -(2m - 3) \bmod p,$$
$$\cdots \qquad \cdots$$
$$2 \cdot 2m \equiv 4m \equiv -1 \bmod p.$$

This shows that

$$\left(\frac{2}{p}\right) = (-1)^m = \begin{cases} +1 & \text{if } p \equiv 1 \bmod 8, \\ -1 & \text{if } p \equiv 5 \bmod 8. \end{cases}$$

Repeating this calculation for primes $p = 4m + 3$ will show that

$$\left(\frac{2}{p}\right) = \begin{cases} -1 & \text{if } p \equiv 3 \bmod 8, \\ +1 & \text{if } p \equiv 7 \bmod 8. \end{cases}$$

Thus we have proved

Proposition 3.8 *The Kronecker symbol* $(\frac{8}{p}) = (\frac{2}{p})$ *only depends on the residue class of p modulo* 8 *and is given by*

$$\left(\frac{2}{p}\right) = (-1)^{\frac{p^2-1}{8}} = \begin{cases} +1 & \text{if } p \equiv \pm 1 \bmod 8, \\ -1 & \text{if } p \equiv \pm 3 \bmod 8. \end{cases}$$

3.2.3   Composite Moduli

The Jacobi symbol $(\frac{a}{m})$ is defined for odd integers $m > 1$ and coincides with the Legendre symbol if m is prime. For composite m it is defined via multiplicativity of the denominator: if $m = \prod p$ is a product of odd primes, then we set

$$\left(\frac{a}{m}\right) = \prod_p \left(\frac{a}{p}\right).$$

We also generalize the Kronecker symbol $(\frac{a}{m})$ to all positive integers m by multiplicativity.

Observe that if $(\frac{a}{m}) = -1$, then the congruence $a \equiv x^2 \bmod m$ is not solvable. In fact, $(\frac{a}{m}) = -1$ implies that there is a prime p dividing m with $(\frac{a}{p}) = -1$, hence a is a quadratic nonresidue modulo p and therefore modulo m. On the other hand, $(\frac{a}{m}) = +1$ does not imply that a is a quadratic residue modulo m for composite values of m, as the example $(\frac{2}{15}) = (\frac{2}{3})(\frac{2}{5}) = (-1)^2 = +1$ shows.

We now ask whether Gauss's Lemma also holds for composite values of m. The answer is positive, but there is a catch: If we only consider residue classes coprime to m, then the sign is trivial in too many cases.

3.2.4 Zolotarev and Frobenius

Gauss's Lemma requires the choice of a half system, but the resulting quadratic character of a modulo p does not depend on this choice. Zolotarev and Frobenius[2] have found a modification of Gauss's Lemma that does not require choosing a half system. Let n be an odd integer and a an integer coprime to n. Then multiplication by a induces a permutation π_a of the residue classes modulo n. Each permutation of finitely many objects can be written (in many different ways) as a product of transpositions (permutations that switch two elements). The sign of a permutation is -1 or $+1$ according as this number of transpositions is odd or even.

For describing permutations we can use the matrix and the cycle notation. The permutation $\pi = \left(\begin{smallmatrix} 1 & 2 & 3 \\ 2 & 1 & 3 \end{smallmatrix}\right)$ of the set $\{1, 2, 3\}$ maps 1 to 2 and 2 to 1, thus switches 1 and 2, and leaves 3 fixed. We can write π also as the product of the cycles $(1\ 2)$ and (3), where the cycle $(1\ 2)$ maps 1 to 2 and 2 to the beginning 1 of the cycle, whereas (3) leaves 3 (and all the other elements) fixed. We can even omit (3) and simply write $\pi = (1\ 2)$ when we demand that elements that do not occur in a cycle are fixed.

Multiplication by 2 on $\mathbb{Z}/7\mathbb{Z}$ induces the permutation $\pi_2 = \left(\begin{smallmatrix} 0 & 1 & 2 & 3 & 4 & 5 & 6 \\ 0 & 2 & 4 & 6 & 1 & 3 & 5 \end{smallmatrix}\right)$. We can also write π_2 as a product of cycles: $\pi_2 = (124)(365)$. Decomposing these cycles into transpositions (here we read from right to left; see Exercise 3.7) we find $\pi_2 = (12)(24)(36)(65)$. Thus π_2 has sign $+1$.

We now define the Zolotarev symbol $[\frac{a}{n}]$ for odd integers $n > 1$ by

$$\left[\frac{a}{n}\right] = \operatorname{sign} \pi_a.$$

---

[2]See [136] and [41]; our presentation is a simplification of the one given in [62].

Clearly $[\frac{a}{n}]$ only depends on the residue class of a modulo n. Since multiplication by ab is multiplication by a followed by multiplication by b, we have

$$\left[\frac{ab}{n}\right] = \left[\frac{a}{n}\right]\left[\frac{b}{n}\right].$$

Thus the Zolotarev symbol is multiplicative in the numerator.

Proposition 3.9 *The Zolotarev symbol is multiplicative in the denominator: If m and n are odd integers and if a is coprime to m and n, then*

$$\left[\frac{a}{m}\right]\left[\frac{a}{n}\right] = \left[\frac{a}{mn}\right].$$

This follows from the isomorphism $\mathbb{Z}/mn\mathbb{Z} \simeq \mathbb{Z}/m\mathbb{Z} \times \mathbb{Z}/n\mathbb{Z}$, i.e., the Chinese Remainder Theorem. If $\alpha : A \longrightarrow A$ and $\beta : B \longrightarrow B$ are permutations, then $\alpha \times \beta$ denotes the induced permutation on $A \times B$. Clearly (see Exercise 3.8)

$$\operatorname{sign}(\alpha \times \beta) = (\operatorname{sign}\alpha)^{\#B} \cdot (\operatorname{sign}\beta)^{\#A}. \tag{3.3}$$

Now multiplication by a induces permutations α on $\mathbb{Z}/m\mathbb{Z}$ and β on $\mathbb{Z}/n\mathbb{Z}$, and this gives us a permutation $\alpha \times \beta$ on $\mathbb{Z}/m\mathbb{Z} \times \mathbb{Z}/n\mathbb{Z}$.

Since $\gcd(m, n) = 1$, any integer k can be written in the form $k = xn + ym$ for integers x, y that are uniquely determined modulo m and modulo n, respectively. Associating $k \bmod mn$ with the pair $(x \bmod m, y \bmod n)$ induces the isomorphism $\mathbb{Z}/mn\mathbb{Z} \simeq \mathbb{Z}/m\mathbb{Z} \times \mathbb{Z}/n\mathbb{Z}$.

Multiplication by a on $\mathbb{Z}/mn\mathbb{Z}$ induces multiplication by a on $\mathbb{Z}/m\mathbb{Z} \times \mathbb{Z}/n\mathbb{Z}$ since $ak = a(xn + ym) = (ax)n + (ay)m$. Thus

$$\left[\frac{a}{mn}\right] = \left[\frac{a}{m}\right]^n \left[\frac{a}{n}\right]^m = \left[\frac{a}{m}\right]\left[\frac{a}{n}\right]$$

for odd coprime integers m and n.

It remains to evaluate $[\frac{a}{n}]$ for prime powers $n = p^k$. We will show that the Zolotarev symbol $[\frac{a}{n}]$ coincides with the Jacobi symbol $(\frac{a}{n})$.

Proposition 3.10 (Zolotarev's Lemma) *For all odd integers n, the symbols of Zolotarev and Jacobi coincide:*

$$\left[\frac{a}{n}\right] = \left(\frac{a}{n}\right).$$

We first prove the claim for an odd prime modulus p. We choose a primitive root g modulo p; then multiplication by g induces the permutation

$$\pi_g = \begin{pmatrix} g & g^2 & g^3 & \cdots & g^{p-2} & g^{p-1} \\ g^2 & g^3 & g^4 & \cdots & g^{p-1} & g \end{pmatrix}$$

on $(\mathbb{Z}/p\mathbb{Z})^\times$ whose sign is given by $\operatorname{sign} \pi_g = (-1)^{p-2} = -1$. This can be seen by writing the permutation as a product of cycles:

$$\pi_g = (g g^2 \dots g^{p-1}) = (g g^2)(g^2 g^3) \cdots (g^{p-2} g^{p-1}).$$

By multiplicativity, the sign of the permutation induced by multiplication by g^r is $(-1)^r$; thus the permutation induced by multiplication by a modulus a has sign $+1$ if and only if a is a square modulo p; since the residue class $0 \bmod p$ is fixed, this is the content of the equation $[\frac{a}{p}] = (\frac{a}{p})$.

Now consider the case where $n = p^k$ is an odd prime power. Assume that we already know that the sign of the permutation induced by multiplication by an integer a coprime to p on $\mathbb{Z}/p^{k-1}\mathbb{Z}$ is $[\frac{a}{p^{k-1}}]$. Observe that this multiplication preserves the set $A = (\mathbb{Z}/n\mathbb{Z})^\times$ of residue classes coprime to p as well as the set $B = \mathbb{Z}/n\mathbb{Z} \setminus (\mathbb{Z}/n\mathbb{Z})^\times$ of multiples of p. The action of multiplication by a on B is the same as on the set $B' = \{\frac{b}{p} : b \in B\}$, and this implies that the sign of this permutation is simply $[\frac{a}{p^{k-1}}]$.

The sign of the permutation on A is $[\frac{a}{p}]$, which can be proved in exactly the same way as for $k = 1$. Equation (3.3) now tells us that the sign of the permutation π_a is given by

$$\left[\frac{a}{p}\right]\left[\frac{a}{p^{k-1}}\right] = \left[\frac{a}{p^k}\right].$$

We now prove that Zolotarev's Lemma is equivalent[3] to Gauss's Lemma. Consider a half system $a_1, \dots, a_m$ modulo $n = 2m + 1$. The remaining nonzero residue classes are $n - a_m, \dots, n - a_1$. If we write $a \cdot a_i \equiv \varepsilon_i a_j$ for $\varepsilon_j \in \{-1, +1\}$, then Gauss's Lemma says that $(\frac{a}{n}) = \prod \varepsilon_i$.

We will give a proof "by example": Consider the nonzero residue classes modulo $n = 15$ and the permutation induced by multiplication with 2. We will write the $n-1$ residue classes in pairs $(a_i, n - a_i)$ as follows:

$$1 \quad 2 \quad 3 \quad 4 \quad 5\,6\,7$$
$$14 \ 13 \ 12 \ 11 \ 10\,9\,8$$

We now multiply all residue classes by 2 and reduce modulo 15:

$$2 \quad 4\,6\,8\,10\,12\,14$$
$$13 \ 11\,9\,7 \quad 5 \quad 3 \quad 1$$

---

[3]This proof is essentially due to Frobenius [41].

The vertical pairs coincide with the original pairs except that some pairs are flipped. This is because if $(a, p - a)$ is such a pair and if $2a \equiv b \bmod p$, then $2(p - a) \equiv p - b \bmod p$.

Now we perform the permutation γ that interchanges the entries at the top and at the bottom if the number on top is larger than the one at the bottom; observe that the number of swaps is the number of sign changes in Gauss's Lemma:

$$
\begin{array}{ccccccc}
2 & 4\,6\,7 & 5 & 3 & 1 \\
13 & 11\,9\,8 & 10 & 12 & 14
\end{array}
$$

Finally we apply a permutation σ that puts the vertical pairs in the original order. Since we are always changing the place of two residue classes at the same time, $\operatorname{sign}\sigma = +1$:

$$
\begin{array}{ccccccc}
1 & 2 & 3 & 4 & 5\,6\,7 \\
14 & 13 & 12 & 11 & 10\,9\,8
\end{array}
$$

Now $\sigma\gamma\pi_a$ is the trivial permutation, hence $\operatorname{sign}\sigma\gamma\pi_a = 1$. Since $\operatorname{sign}\sigma = 1$ we deduce that $\operatorname{sign}\gamma = \operatorname{sign}\pi_a$, and this proves the equivalence of Gauss's and Zolotarev's Lemma. In particular we have

Proposition 3.11 *Gauss's Lemma holds for composite odd values of $m = 2n+1$: If $\{a_1, \ldots, a_n\}$ is a half system modulo m and $a \cdot a_j \equiv \varepsilon_j a'_j \bmod m$ for suitable signs $\varepsilon_j = \pm 1$, then $\left(\frac{a}{m}\right) = \prod \varepsilon_j$.*

3.2.5 A Few Applications

Using Gauss's Lemma for composite values we now can determine the value of $\left(\frac{-1}{m}\right)$ for all positive odd integers m:

Proposition 3.12 *The value of the Jacobi symbol $\left(\frac{-1}{m}\right)$, where m is an odd natural number, is given by*

$$
\left(\frac{-1}{m}\right) = \begin{cases} -1 & \text{if } m \equiv 3 \bmod 4, \\ +1 & \text{if } m \equiv 1 \bmod 4. \end{cases}
$$

The proof via Gauss's Lemma for prime moduli (see Prop. 3.7) carries over word for word. The following result follows painlessly from this proposition:

Proposition 3.13 *Let Δ be a quadratic discriminant, and set $N = |\Delta|$. Then*

$$
\left(\frac{\Delta}{N-1}\right) = \operatorname{sgn}(\Delta) = \begin{cases} +1 & \text{if } \Delta > 0, \\ -1 & \text{if } \Delta < 0. \end{cases} \tag{3.4}
$$

Assume first that N is even (and thus divisible by 4 since Δ is a discriminant). Then

$$\left(\frac{\Delta}{N-1}\right) = \begin{cases} \left(\frac{N}{N-1}\right) = \left(\frac{1}{N-1}\right) = +1 & \text{if } \Delta = N > 0, \\ \left(\frac{-N}{N-1}\right) = \left(\frac{-1}{N-1}\right) = -1 & \text{if } \Delta = -N < 0, \end{cases}$$

where we have used Prop. 3.12.

If N is odd, then either $N = \Delta \equiv 1 \bmod 4$, or $N = -\Delta \equiv 3 \bmod 4$. If $N \equiv 1 \bmod 4$, then $N - 1 = 2^j n$ for some odd integer n and $\left(\frac{N}{N-1}\right) = \left(\frac{N}{2}\right)^j \left(\frac{N}{n}\right) = \left(\frac{2}{N}\right)^j \left(\frac{n}{N}\right) = \left(\frac{N-1}{N}\right) = \left(\frac{-1}{N}\right) = +1$; if N is negative, then $j = 1$, and the same calculation yields $\left(\frac{-N}{N-1}\right) = \left(\frac{-N}{2}\right)\left(\frac{-N}{n}\right) = \left(\frac{2}{N}\right)\left(\frac{n}{N}\right) = \left(\frac{N-1}{N}\right) = \left(\frac{-1}{N}\right) = -1$.

As a corollary we observe that if $\Delta < 0$, then there always exists a prime number $p < |\Delta|$ such that $\left(\frac{\Delta}{p}\right) = -1$. This is also true for positive discriminants (see Theorem 3.21), but in this case we seem to need more than just the modularity of the Kronecker symbol $\left(\frac{-4}{\cdot}\right)$.

3.3   Euler's Modularity Conjecture

We have seen above that the Kronecker symbol $\left(\frac{-4}{m}\right) = \left(\frac{-1}{m}\right)$ only depends on the residue class of m mod 4. Similarly, the Kronecker symbol $\left(\frac{8}{m}\right) = \left(\frac{2}{m}\right)$ only depends on the residue class of m mod 8; we have proved this only for prime values of m, but the proof via Gauss's Lemma also works for composite values of m.

If we look at the Legendre symbol $\left(\frac{12}{m}\right)$, then a few quick calculations provide us with the following table:

p	5	7	11	13	17	19	23	25	29	31
$\left(\frac{3}{p}\right)$	-1	-1	$+1$	$+1$	-1	-1	$+1$	$+1$	-1	-1

The pattern is obvious: The values have period 12. Numerical experiments with other small integers a suggest the following conjecture due to Euler:[4]

Theorem 3.14 (Euler's Modularity Conjecture) *For each nonzero integer a there exists a modulus N such that the Jacobi symbol $\left(\frac{a}{m}\right)$ for natural numbers m only depends on the residue class of m modulo N. In other words: For all natural numbers m and n we have*

$$\left(\frac{a}{m}\right) = \left(\frac{a}{n}\right) \quad if \quad m \equiv n \bmod N. \tag{3.5}$$

In fact, we can always choose $N = 4|a|$. If $a > 0$, Eq. (3.5) also holds if $m \equiv -n \bmod N$.

---

[4]See, e.g., [36].

Euler formulated this conjecture for prime numbers m and n, and of course without using Legendre or Jacobi symbols.

The following result holds in many similar situations in which some notion of modularity shows up:

Proposition 3.15 *If Euler's Modularity Conjecture for $(\frac{a}{m})$ for the moduli N_1 and N_2, then it also holds modulo $N = \gcd(N_1, N_2)$.*

Proof Assume that the Jacobi symbol $(\frac{a}{\cdot})$ is modular for the moduli N_1 and N_2, and let $N = \gcd(N_1, N_2)$. If m is a natural number coprime to $2a$, then we have to show that $(\frac{a}{m}) = (\frac{a}{m+N})$.

To this end we write $N = rN_1 - sN_2$, where we assume without loss of generality that $r, s > 0$ (if not, we simply switch N_1 and N_2). Then

$$\left(\frac{a}{m}\right) = \left(\frac{a}{m + rN_1}\right) \qquad \text{modularity modulo} N_1$$

$$= \left(\frac{a}{m + rN_1 - sN_2}\right) \qquad \text{modularity modulo} N_2$$

$$= \left(\frac{a}{m + N}\right) \qquad N = rN_1 + sN_2.$$

This completes the proof. $\qquad\qquad\qquad\qquad\qquad\qquad\qquad\qquad\qquad\qquad$ □

This property allows us to define the *conductor* of the Kronecker symbol as the smallest positive integer N for which $(\frac{\Delta}{\cdot})$ is modular.

We have already seen that the Kronecker symbol $(\frac{-4}{\cdot})$ is defined modulo 4; since $-1 = (\frac{-4}{3}) \neq (\frac{-4}{5}) = +1$, the conductor cannot be a proper divisor of 4 and thus is equal to 4. In a similar way we can see that the Kronecker symbols $(\frac{8}{\cdot})$ and $(\frac{-8}{\cdot})$ have conductor 8. We will prove below that the Kronecker symbol $(\frac{\Delta}{m})$ has conductor $N = |\Delta|$.

Next we show that it is sufficient to prove Euler's Modularity Conjecture for $a = -1$ and prime values of a:

Proposition 3.16 *Assume that the Jacobi symbols $(\frac{a}{\cdot})$ and $(\frac{b}{\cdot})$ are defined modulo $4|a|$ and $4|b|$, respectively. Then the Jacobi symbol $(\frac{ab}{\cdot})$ is defined modulo $4|ab|$.*

Proof We have to show that $(\frac{ab}{m}) = (\frac{ab}{m+4|ab|})$ for all natural numbers m coprime to ab. In fact we have

$$\left(\frac{ab}{m + 4|ab|}\right) = \left(\frac{a}{m + 4|ab|}\right)\left(\frac{b}{m + 4|ab|}\right) \qquad \text{multiplicativity of Jacobi symbols}$$

$$= \left(\frac{a}{m}\right)\left(\frac{b}{m}\right) \qquad \text{modularity of} (\frac{a}{\cdot}) \text{ and} (\frac{b}{\cdot})$$

$$= \left(\frac{ab}{m}\right) \qquad \text{multiplicativity of Jacobi symbols}$$

This completes the proof. $\qquad\qquad\qquad\qquad\qquad\qquad\qquad\qquad\qquad\qquad$ □

As a corollary we obtain

Corollary 3.17 *If Euler's Modularity Conjecture holds for $a = -1$ and for prime values of a, then it holds in general.*

3.3.1 The Quadratic Reciprocity Law

This Modularity Theorem is equivalent to the quadratic reciprocity law and should be seen as its essential content. Legendre's formulation of the reciprocity law, which determines the value of the product $(\frac{p}{q})(\frac{q}{p})$, is an historical accident.

Kronecker was the first to emphasize that the heart of the quadratic reciprocity law is not Legendre's formula

$$\left(\frac{p}{q}\right)\left(\frac{q}{p}\right) = (-1)^{\frac{p-1}{2} \cdot \frac{q-1}{2}}. \tag{3.6}$$

In connection with higher reciprocity laws and the class fields of complex multiplication [70] he pointed out that Euler's formulation catches the essence of quadratic reciprocity better than that of Legendre:

> Very early on Euler had made the observation that the prime divisors of quadratic forms with discriminant D are contained in certain linear forms $mD + \alpha$, but only in 1783 he formulated this observation, which was highly important for the development of number theory, in the remarkable way which gave rise to the name reciprocity law.[†] The elegance of the correlation, which was—rightly—emphasized, pushed the meaning and the aim of Euler's original observation to the background. When I recently found a specific new law by applying the arithmetic theory of singular modules to the power residues of complex numbers I was reminded of this first formulation with which Euler had published the essential content of the quadratic reciprocity law; and since this law in the theory of power residues is particularly important not only because of its analogy with the historical point of departure but also because it suggests a new phase of the development of reciprocity laws, I would like to present this observation briefly to the Academy today.
>
> [†] Compare my remarks in the Monatsbericht from April 1875, p. 268. [Werke II, p. 3–4].

Euler's Modularity Theorem is equivalent to the quadratic reciprocity law in the form given by Legendre (we formulate it more generally for the Jacobi symbol):

Theorem 3.18 (Quadratic Reciprocity Law) *Let m and n be odd coprime natural numbers. Then*

$$\left(\frac{m}{n}\right)\left(\frac{n}{m}\right) = (-1)^{\frac{m-1}{2} \cdot \frac{n-1}{2}}. \tag{3.7}$$

In addition, there are the supplementary laws

$$\left(\frac{-1}{m}\right) = (-1)^{\frac{m-1}{4}} \quad and \quad \left(\frac{2}{m}\right) = (-1)^{\frac{m^2-1}{8}}.$$

It is not difficult to prove the equivalence of Euler's Modularity Theorem (3.5) and Legendre's quadratic reciprocity law.

Reciprocity Implies Modularity We have to show that, for positive integers m and a, we have $(\frac{a}{m}) = (\frac{a}{a+4m})$. Assume first that a is odd; then we have

$$\left(\frac{a}{m+4a}\right) = (-1)^{\frac{a-1}{2} \cdot \frac{m-1}{2}} \left(\frac{m+4a}{a}\right) = (-1)^{\frac{a-1}{2} \cdot \frac{m-1}{2}} \left(\frac{m}{a}\right) = \left(\frac{a}{m}\right)$$

because $m + 4a \equiv m \bmod 4$.

If $a = 2b$ is even we may assume that a is squarefree, so b is odd. Then

$$\left(\frac{a}{m+4a}\right) = \left(\frac{2}{m+8b}\right)\left(\frac{b}{m+8b}\right) = (-1)^{\frac{m-1}{2} \cdot \frac{b-1}{2}} \left(\frac{2}{m}\right)\left(\frac{m+8b}{b}\right)$$

$$= (-1)^{\frac{m-1}{2} \cdot \frac{b-1}{2}} \left(\frac{2}{m}\right)\left(\frac{m}{b}\right) = \left(\frac{2}{m}\right)\left(\frac{b}{m}\right) = \left(\frac{a}{m}\right).$$

Modularity Implies Reciprocity The modularity of the Kronecker symbol with conductor 4 implies that $(\frac{-1}{m})$ only depends on the residue class of m mod 4. Since $(\frac{-1}{3}) = -1$ and $(\frac{-1}{5}) = +1$, we have $(\frac{-1}{m}) = +1$ or -1 according as $m \equiv 1 \bmod 4$ or $m \equiv -1 \bmod 4$. But this is the exact content of the first supplementary law.

Similarly, the second supplementary law follows from the fact that $(\frac{2}{m})$ only depends on the residue class of m modulo 8; this implies that $(\frac{2}{m}) = +1$ when $m \equiv \pm 1 \bmod 8$ and $(\frac{2}{m}) = -1$ otherwise, which is the second supplementary law.

For deriving (3.7) in the case $m \equiv n \bmod 4$ and $m > n$ from the modularity theorem, for example, we set $a = \frac{m-n}{4}$ and verify that

$$\left(\frac{a}{m}\right) = \left(\frac{4a}{m}\right) = \left(\frac{m-n}{m}\right) = \left(\frac{-n}{m}\right), \quad \left(\frac{a}{n}\right) = \left(\frac{4a}{n}\right) = \left(\frac{m-n}{n}\right) = \left(\frac{m}{n}\right).$$

This implies $(\frac{-n}{m}) = (\frac{m}{n})$. If $m \equiv 1 \bmod 4$, we have $(\frac{-1}{m}) = +1$, hence $(\frac{m}{n}) = (\frac{m}{n})$, which is the quadratic reciprocity law in this case. If $m \equiv n \equiv 3 \bmod 4$, on the other hand, then $(\frac{m}{n}) = (\frac{-n}{m}) = (-1)^{\frac{m-1}{2}}(\frac{-n}{m}) = (-1)^{\frac{m-1}{2} \cdot \frac{n-1}{2}}(\frac{-n}{m})$, since here $\frac{m-1}{2} \equiv \frac{n-1}{2} \equiv 1 \bmod 2$.

If $m \equiv -n \bmod 4$ we set $a = \frac{m+n}{4}$ and find, using $m = 4a - n$ and $n = 4a - m$:

$$\left(\frac{m}{n}\right)\left(\frac{n}{m}\right) = \left(\frac{4a-n}{n}\right)\left(\frac{4a-m}{m}\right) = \left(\frac{a}{n}\right)\left(\frac{a}{m}\right) = 1,$$

where the last equality follows from the modularity conjecture since $m \equiv -n \bmod 4a$ and $a > 0$.

3.3.2   Proof of Euler's Modularity Conjecture

Before we turn to the proof of Euler's Modularity Conjecture we look at yet another special case.

Proposition 3.19 *The Kronecker symbol* $(\frac{12}{\cdot})$ *is modular with conductor* 12.

We proceed as in the proof of Proposition 3.8. Since $(\frac{12}{\cdot}) = (\frac{3}{\cdot})$, we apply Gauss's Lemma to $a = 3$ and the modulus m. We choose $A = \{1, 2, \ldots, \frac{m-1}{2}\}$ as our half system and then have to count the number of integers k for which either $\frac{m}{2} < 3k < m$, $\frac{3m}{2} < 3k < 2m$ or $\frac{5m}{2} < 3k < 3m$. Dividing through by 3 we find that we have to count the number of integers k in the intervals $[\frac{m}{6}, \frac{m}{3}]$, $[\frac{m}{2}, \frac{2m}{3}]$ and $[\frac{5m}{6}, m]$.

What happens to the number of integers k in the interval $[\frac{m}{6}, \frac{m}{3}]$ if we replace m by $m + 12$? Then we have to count the integers in the interval $[\frac{m}{6} + 2, \frac{m}{3} + 4]$; obviously this interval for $m + 12$ contains exactly 2 integers more than the corresponding interval for m, hence the number of integers has the same parity. The same thing happens for the other two intervals. By Gauss's Lemma, we conclude that $(\frac{12}{m}) = (\frac{12}{m+12})$; in particular, the Kronecker symbol $(\frac{12}{\cdot})$ has conductor dividing 12.

The fact that the conductor is not a proper divisor of 12 follows from

$$-1 = \left(\frac{12}{5}\right) \neq \left(\frac{12}{11}\right) = 1 \quad \text{and} \quad -1 = \left(\frac{12}{7}\right) \neq \left(\frac{12}{11}\right) = 1.$$

The proof of the general case[5] proceeds in the same way. We claim that if m is an odd natural number coprime to a, then the Jacobi symbol $(\frac{a}{m})$ only depends on the residue class of m modulo $4a$; in particular we claim that $(\frac{a}{m}) = (\frac{a}{m+4a})$. We may assume that a is positive: If a is negative, the claim follows from the observation $(\frac{a}{m}) = (\frac{-1}{m})(\frac{-a}{m})$ since the symbols on the right only depend on m modulo 4 and modulo $4|a|$, hence modulo $4|a|$.

Now consider the half system $A = \{1, 2, 3, \ldots, n\}$ modulo $m = 2n + 1$. The number of sign changes is equal to the number of integers ak lying in the intervals $(\frac{m}{2}, m)$, $(\frac{3m}{2}, 2m)$, …, until $(\frac{(2b-1)m}{2}, bm)$, where $b = \frac{a}{2}$ or $b = \frac{a-1}{2}$ according as a is even or odd.

Dividing through by a we see that this number is the same as the number of integers in the intervals

$$\left(\frac{m}{2a}, \frac{m}{a}\right), \quad \left(\frac{3m}{2a}, \frac{2m}{a}\right), \quad \ldots, \quad \left(\frac{(2b-1)m}{2a}, \frac{bm}{a}\right). \tag{3.8}$$

---

[5]This proof is lifted from Davenports beautiful book [27]; its basic idea goes back to the proof given by Arnold Scholz in [112].

If we replace m by $m+4a$, the number of integers in each of these intervals changes by an *even* number. This proves the weak modularity theorem that $(\frac{a}{m})$ only depends on m modulo $4a$.

It remains to prove the claim that $(\frac{a}{m}) = (\frac{a}{4a-m})$ for positive integers a and m. If we replace m by $a - m$, the intervals (3.8) become

$$\left(2 - \frac{m}{2a}, 4 - \frac{m}{a}\right), \ \left(6 - \frac{3m}{2a}, 8 - \frac{2m}{a}\right), \ \ldots, \ \left(4b - 2 - \frac{(2b-1)m}{2a}, 4b - \frac{bm}{a}\right).$$

Now the number of integers in the interval $(2 - \frac{m}{2a}, 4 - \frac{m}{a})$ has the same parity as the number of integers in the interval $(\frac{m}{2a}, \frac{m}{a})$. Indeed, if we subtract the numbers $2 - \frac{m}{2a}$ and $4 - \frac{m}{a}$ from 4 we see that the number of integers in $(2 - \frac{m}{2a}, 4 - \frac{m}{a})$ is equal to the number of integers in $(\frac{m}{a}, 2 + \frac{m}{2a})$. The union of this interval and $(\frac{m}{2a}, \frac{m}{a})$ is $(\frac{m}{2a}, 2 + \frac{m}{2a})$ minus the point $\frac{m}{a}$, so this union contains exactly two integers. Similar arguments show the same for the other intervals, and now the claim follows.

3.3.3 The Strong Modularity Theorem

The strong modularity theorem (which controls the conductor) is a consequence of the quadratic reciprocity law.

Theorem 3.20 (Strong Modularity Theorem) *Let Δ be the discriminant of a quadratic number fields. Then the Kronecker symbol $(\frac{\Delta}{\cdot})$ is modular with conductor $N = |\Delta|$.*

We first prove the claim for prime discriminants. We have already proved the claim if $\Delta \in \{-4, \pm 8\}$, so we may assume that Δ is an odd prime number. Since $(\frac{p}{\cdot})$, where $p \equiv 1 \bmod 4$ is prime, is defined modulo p, its conductor is either $N = p$ or $N = 1$. In the second case the symbols $(\frac{p}{m})$ would all have the same values, and by multiplicativity we conclude that we must have $(\frac{p}{m}) = 1$ for all natural numbers m coprime to p. If $p \equiv 5 \bmod 8$ we obtain the desired contradiction from the observation $(\frac{p}{2}) = -1$. Thus it remains to prove, for each prime $p \equiv 1 \bmod 8$, the existence of a natural number q with $(\frac{p}{q}) = -1$.

The existence of such primes (satisfying a few additional conditions) played a large role in the first proofs of the quadratic reciprocity law. Legendre's proof was incomplete since he did not succeed in proving the existence of such primes, and Gauss gave a highly ingenious proof of the existence of such a prime q (with the additional condition that $q < p$, which he needed for his induction proof to work) in his first proof of the quadratic reciprocity law.

Let us formulate the existence of such primes in the following form:

Theorem 3.21 *Let a be a nonzero integer. If $(\frac{a}{p}) = +1$ for all prime numbers $p \nmid a$, then a is a square number.*

Proof Assume that a is not a square number. We distinguish two cases:

- a is odd. Let n be an integer such that $(\frac{n}{|a|}) = 1$. One of n, $n + a$, $n + 2a$, and $n + 3a$ is an integer $\equiv 1 \bmod 4$. Replacing n by this number we have found a natural number $n \equiv 1 \bmod 4$ with $(\frac{n}{|a|}) = 1$. By quadratic reciprocity we have $(\frac{|a|}{n}) = -1$. Since $n \equiv 1 \bmod 4$, the first supplementary law implies $(\frac{a}{n}) = -1$. Thus there exists a prime number $p \mid n$ with $(\frac{a}{p}) = -1$.
- a is even. Then we may assume that $a = 2b$ for b odd. Choose an integer n with $(\frac{n}{|b|}) = -1$; adding multiples of b to n we can make sure that $n \equiv 1 \bmod 8$. As above, the quadratic reciprocity law and the first supplementary law implies $(\frac{b}{n}) = -1$, and since $n \equiv 1 \bmod 8$ we have $(\frac{2}{n}) = +1$, hence $(\frac{a}{n}) = (\frac{2b}{n}) = -1$. But then n must have a prime factor p such that $(\frac{a}{p}) = -1$.

$\square$

We call a discriminant $\Delta = \operatorname{disc} k$ of a quadratic number field k a *prime discriminant* if Δ is one of -4, ± 8, p, or $-q$ for primes $p \equiv 1 \bmod 4$ and $q \equiv 3 \bmod 4$. It is easy to see that each discriminant can be factored into prime discriminants:

Theorem 3.22 *Each discriminant of a quadratic number field can be written uniquely (up to order) as a product of prime discriminants.*

The proof is not difficult. First observe that either $\Delta \equiv 1 \bmod 4$ is odd, or Δ is divisible exactly by 4, or exactly by 8. In the second case, $\Delta = 4m$ for some $m \equiv 3 \bmod 4$, hence $\Delta = -4\Delta_1$ for some $\Delta_1 \equiv 1 \bmod 4$. In the last case we can always write $\Delta = \pm 8\Delta_1$ for some $\Delta_1 \equiv 1 \bmod 4$.

Since -4 and ± 8 are prime discriminants it is sufficient to prove that any squarefree odd integer $\Delta_1 \equiv 1 \bmod 4$ is the product of prime discriminants. To this end, write $\Delta_1 = p_1 \cdots p_r q_1 \cdots q_s$ for primes $p_j \equiv 1 \bmod 4$ and $q_j \equiv 3 \bmod 4$. Since $\Delta_1 \equiv 1 \bmod 4$, the number of primes factors $q_j \equiv 3 \bmod 4$ is even. But now

$$\Delta_1 = p_1 \cdots p_r(-q_1) \cdots (-q_s)$$

is a factorization into prime discriminants.

At this point we know that Kronecker symbols $(\frac{\Delta}{\cdot})$ for prime discriminants Δ are modular with conductor $N = |\Delta|$. Now let Δ be an arbitrary discriminant, and let $\Delta = \Delta_1 \cdots \Delta_r$ be its factorization into prime discriminants. For proving that the corresponding Kronecker character has conductor $N = |\Delta|$ we need to prove the following

Lemma 3.23 *Let $\kappa_1 = (\frac{\Delta_1}{\cdot})$ and $\kappa_2 = (\frac{\Delta_2}{\cdot})$ be Kronecker characters with coprime conductors N_1 and N_2, respectively. Then the Kronecker character $(\frac{\Delta}{\cdot})$, where $\Delta = \Delta_1\Delta_2$, has conductor $N = N_1 N_2$.*

It is easy to see that $(\frac{\Delta}{\cdot})$ is modular with defining modulus N. Let q be a prime number dividing N_1, write $N_1 = qn_1$, and assume that $(\frac{\Delta}{\cdot})$ is defined modulo $n = n_1 N_2$. Then

$$\left(\frac{\Delta}{a}\right) = \left(\frac{\Delta}{a+nk}\right)$$

for all $a > 0$ and $k \in \mathbb{N}$. But

$$\left(\frac{\Delta}{a+nk}\right) = \left(\frac{\Delta_1}{a+n_1 N_2 k}\right)\left(\frac{\Delta_2}{a+n_1 N_2 k}\right) = \left(\frac{\Delta_1}{a+n_1 N_2 k}\right)\left(\frac{\Delta_2}{a}\right),$$

$$\left(\frac{\Delta}{a+nk}\right) = \left(\frac{\Delta_1}{a}\right)\left(\frac{\Delta_2}{a}\right).$$

Thus we deduce that

$$\left(\frac{\Delta_1}{a+n_1 N_2 k}\right) = \left(\frac{\Delta_1}{a}\right).$$

This shows that κ_1 is defined modulo N_1 and modulo $n_1 N_2$; but then it is defined modulo $\gcd(N_1, n_1 N_2) = n_1$ contradicting our assumptions.

The strong modularity theorem now follows by induction on the number of prime discriminants dividing Δ.

3.4 $\mathbb{F}_p$-Rational Points on Curves

We will close this chapter by returning to the problem of counting points on algebraic curves such as Pell conics defined over the finite field $\mathbb{F}_p = \mathbb{Z}/p\mathbb{Z}$. We will express this number as a character sum and then evaluate such sums in a few simple cases (e.g., for Pell conics) as well as in a less trivial case, namely that of an elliptic curve.

Let K be a field and $f(x) \in K[x]$ a polynomial; we say that $C : y^2 = f(x)$ is a plane algebraic curve. A point $(x, y) \in K \times K$ satisfying this equation is called an affine point on C (if we think of C as living in the projective plane, then there might be additional points at infinity). If $\deg f = 2$, then the curve $C : y^2 = f(x)$ is called a conic. We claim

Lemma 3.24 *Consider the algebraic curve* $C : y^2 = f(x)$, *where* $f(x) \in \mathbb{Z}[x]$. *The number of affine* $\mathbb{F}_p$-*rational points on* C *is given by*

$$N_p(C) = \#C(\mathbb{F}_p) = p + \sum_{t=0}^{p-1} \left(\frac{f(t)}{p}\right).$$

For a given $x \in \mathbb{Z}/p\mathbb{Z}$, the congruence $y^2 \equiv f(x) \bmod p$ has 0, 1, or 2 solutions according as the Legendre symbol $(\frac{f(x)}{p})$ has the value $-1, 0$, and $+1$, respectively. Thus for a given integer x, the congruence $y^2 \equiv f(x) \bmod p$ has $1 + (\frac{f(x)}{p})$ solutions, hence

$$\#\mathcal{C}(\mathbb{F}_p) = \sum_{t=0}^{\infty} \left(1 + \left(\frac{f(t)}{p}\right)\right) = p + \sum_{t=0}^{\infty} \left(\frac{f(t)}{p}\right)$$

is the number of solutions of the congruence $y^2 \equiv f(x) \bmod p$.

Character sums of the form $\sum(\frac{f(x)}{p})$ are called *Jacobsthal sums* after Ernst Jacobsthal (1882–1965); the results in this section are all taken from his dissertation (see [67, 68]; for a different elementary evaluation see Monzingo [96]). L. von Schrutka [129] generalized Jacobsthal's results to primes $p \equiv 1 \bmod 3$ written in the form $p = a^2 + 3b^2$ (see also Chan et al. [17]). In his thesis [133], Widmer covered the cases above as well as primes of the form $p \equiv 1 \bmod 8$ written in the form $p = a^2 + 2b^2$. Hashimoto et al. [56] worked out the connection between Jacobsthal sums for primes $p = a^2 + 2b^2$ and the L-function of certain elliptic curves; for similar connections to elliptic curves see also [13, Section 6.4].

A simple observation that we will often use is the following:

Lemma 3.25 *For each odd prime number p We have*

$$\sum_{t=0}^{p} \left(\frac{t}{p}\right) = 0.$$

This is a consequence of the fact that there are as many quadratic residues as there are nonresidues modulo p. For a formal proof, let n be a quadratic nonresidue modulo p, and set $S = \sum(\frac{t}{p})$. Then $-S = (\frac{n}{p})S = \sum(\frac{nt}{p})$. But if t runs through a system of coprime residue classes modulo p, so does nt, hence the last sum is S. Now $-S = S$ implies that $S = 0$.

For polynomials f with degree 1 we have

Proposition 3.26 *Let $f(t) = at + b$ with $p \nmid a$. Then*

$$\sum_{t=0}^{p} \left(\frac{at+b}{p}\right) = 0.$$

This is clear since $at + b$ runs through a complete system of residue classes if t does.

Jacobsthal Sums for Quadratic Polynomials For quadratic polynomials $f(x) = ax^2 + bx + c$ we assume that $p \nmid a$ and that p is odd; completing the square we obtain $4af(x) = (2at + b)^2 - \Delta$ for $\Delta = b^2 - 4ac$ and

$$\sum_{t=0}^{p} \left(\frac{f(t)}{p}\right) = \left(\frac{a}{p}\right) \cdot \sum_{t=0}^{p} \left(\frac{t^2 - \Delta}{p}\right).$$

It is therefore sufficient to compute the character sum for polynomials of the form $f(t) = t^2 - D$. We now set

$$\psi(D) = \sum_{t=0}^{p-1} \left(\frac{t^2 - D}{p}\right).$$

Clearly $\psi(0) = p - 1$.

Lemma 3.27 *We have $\psi(a^2 D) = \psi(D)$ for all integers a with $p \nmid a$.*

This is easily seen to be true as

$$\psi(a^2 D) = \sum_{t=0}^{p-1} \left(\frac{t^2 - a^2 D}{p}\right) = \sum_{s=0}^{p-1} \left(\frac{a^2 s^2 - a^2 D}{p}\right) = \sum_{s=0}^{p-1} \left(\frac{s^2 - D}{p}\right) = \psi(D),$$

where we have used that s runs through a complete system of residues modulo p when $t = as$ does.

Next we show:

Lemma 3.28 *We have $\psi(1) = -1$.*

$$\psi(1) = \sum_{t=0}^{p-1} \left(\frac{t^2 - 1}{p}\right) = \sum_{t=0}^{p-1} \left(\frac{t-1}{p}\right)\left(\frac{t+1}{p}\right)$$

$$= \sum_{s=1}^{p-1} \left(\frac{s}{p}\right)\left(\frac{s+2}{p}\right) \qquad\qquad s = t - 1$$

$$= \sum_{s=1}^{p-1} \left(\frac{s}{p}\right)^{-1}\left(\frac{s+2}{p}\right) = \sum_{s=1}^{p-1} \left(\frac{1 + 2s^{-1}}{p}\right)$$

$$= \sum_{r=1}^{p-1} \left(\frac{1 + 2r}{p}\right) \qquad\qquad rs \equiv 1 \bmod p$$

$$= -1 + \sum_{r=1}^{p-1} \left(\frac{1 + 2r}{p}\right) = -1$$

which is what we wanted to prove.

When studying a family of objects it is often a good idea to consider them all at once; in the present case we can form the sum over all $\psi(D)$ and find

$$\sum_{D=1}^{p} \psi(D) = \sum_{D=0}^{p} \sum_{t=0}^{p-1} \left(\frac{t^2 - D}{p}\right) = \sum_{t=0}^{p-1} \sum_{D=0}^{p} \left(\frac{t^2 - D}{p}\right) = 0 \qquad (3.9)$$

by Proposition 3.26.

We will now compute this sum in a different way. We know that $\psi(D)$ only depends on $\left(\frac{D}{p}\right)$; thus if n denotes an arbitrary quadratic nonresidue modulo p, then we have

$$\sum_{D=1}^{p} \psi(D) = \psi(0) + \frac{p-1}{2} \cdot \big(\psi(1) + \psi(n)\big).$$

Since this sum is 0 and since $\psi(0) = p - 1$, we deduce that $\psi(1) + \psi(n) = -2$. Thus $\psi(1) = -1$ implies $\psi(n) = -1$, and we have shown

Proposition 3.29 *We have*

$$\psi(D) = \sum_{t=0}^{p-1} \left(\frac{t^2 - D}{p}\right) = \begin{cases} -1 & \text{if } p \nmid D, \\ p-1 & \text{if } p \mid D. \end{cases}$$

Jacobsthal Sums for a Cubic Polynomial The determination of Jacobsthal sums for cubic polynomials f is difficult. We will do this in the special case of the polynomial $f(x) = x^3 - x$. To this end we set

$$\phi_p(k) = \phi(k) = \sum_{t=1}^{p-1} \left(\frac{t}{p}\right)\left(\frac{t^2 - k}{p}\right). \qquad (3.10)$$

Lemma 3.30 *We have $\phi(a^2 k) = \left(\frac{a}{p}\right)\phi(k)$ for each a coprime to p.*

This is a simple calculation: Since at runs through a complete system of residue classes modulo p if t does, we have

$$\phi(a^2 k) = \sum_{t=1}^{p-1} \left(\frac{t}{p}\right)\left(\frac{t^2 - a^2 k}{p}\right) = \sum_{t=1}^{p-1} \left(\frac{at}{p}\right)\left(\frac{(at)^2 - a^2 k}{p}\right) = \left(\frac{a}{p}\right)\sum_{t=1}^{p-1} \left(\frac{t}{p}\right)\left(\frac{t^2 - k}{p}\right).$$

Now we claim

Theorem 3.31 *Let $p \equiv 1 \bmod 4$ be a prime number, and write $p = a^2 + 4b^2$. Then*

$$\phi(1) = \sum_{t=1}^{p-1} \left(\frac{t}{p}\right)\left(\frac{t^2 - 1}{p}\right) = 2a, \qquad (3.11)$$

where the sign of a is chosen in such a way that $a \equiv -\left(\frac{2}{p}\right) \bmod 4$.

In particular, the number of $\mathbb{F}_p$-rational points on the elliptic curve $y^2 = x^3 - x$ is $N_p = p + 1 - 2a$.

If we set $2x = \phi(r)$ and $2y = \phi(n)$, where r and n denote a quadratic residue and nonresidue modulo p, respectively, then the claim is that $x^2 + y^2 = p$. Since there are $\frac{p-1}{2}$ residues and $\frac{p-1}{2}$ nonresidues modulo p, we have

$$\sum_{k=1}^{p-1} \phi(k)^2 = 2(p-1)(x^2 + y^2).$$

We now compute $\sum \phi(k)^2$ directly. In our calculation we need the following

Lemma 3.32 *We have*

$$\sum_{k=1}^{p-1} \left(\frac{k}{p}\right)\left(\frac{k+b}{p}\right) = \begin{cases} -1 & \text{if } p \nmid b, \\ p-1 & \text{if } p \mid b. \end{cases}$$

This follows from Proposition 3.29 since

$$\sum_{k=1}^{p-1} \left(\frac{k}{p}\right)\left(\frac{k+b}{p}\right) = \sum_{k=1}^{p-1} \left(\frac{k^2+bk}{p}\right) = \sum_{k=1}^{p-1} \left(\frac{4k^2+4bk}{p}\right)$$

$$= \sum_{k=1}^{p-1} \left(\frac{(2k+b)^2 - b^2}{p}\right) = \sum_{t=1}^{p-1} \left(\frac{t^2 - b^2}{p}\right).$$

Now we have

$$\sum_{k=1}^{p-1} \phi(k)^2 = \sum_{k=1}^{p-1}\sum_{s=1}^{p-1} \left(\frac{s}{p}\right)\left(\frac{s^2-k}{p}\right) \sum_{t=1}^{p-1} \left(\frac{t}{p}\right)\left(\frac{t^2-k}{p}\right)$$

$$= \sum_{s,t=0}^{p-1} \left(\frac{st}{p}\right) \sum_{k=1}^{p-1} \left(\frac{s^2-k}{p}\right)\left(\frac{t^2-k}{p}\right).$$

Setting $s^2 - k = l$ and applying Lemma 3.32 we find

$$\sum_{k=1}^{p-1} \left(\frac{s^2-k}{p}\right)\left(\frac{t^2-k}{p}\right) = \sum_{l=1}^{p-1} \left(\frac{l}{p}\right)\left(\frac{l+t^2-s^2}{p}\right) = \begin{cases} -1 & \text{if } s \not\equiv \pm t, \\ p-1 & \text{if } s \equiv \pm t. \end{cases}$$

Thus

$$\sum_{k=1}^{p-1} \phi(k)^2 = (p-1)\left[\sum_{s=t}\left(\frac{t^2}{p}\right) + \sum_{s=-t}\left(\frac{-t^2}{p}\right)\right] - \sum_{s\neq\pm t}\left(\frac{st}{p}\right) = 2(p-1)p.$$

In fact, $\sum_{t=1}^{p-1}\left(\frac{t^2}{p}\right) = p-1$, so the sums in the brackets have value $2(p-1)$. Since

$$\sum_{s,t}\left(\frac{st}{p}\right) = \sum_{s=1}^{p-1}\left(\frac{s}{p}\right)\sum_{t=1}^{p-1}\left(\frac{t}{p}\right) = 0$$

we have

$$\sum_{s\neq\pm t}\left(\frac{t^2}{p}\right) = -2(p-1).$$

Thus the whole sum is $2(p-1)^2 + 2(p-1) = 2(p-1)p$ as claimed.

It remains to determine the sign of a. To this end we have to compute $\phi(1)$ modulo 4. Let R denote the number of residue classes t with $\left(\frac{t^3-t}{p}\right) = +1$ and N those with $\left(\frac{t^3-t}{p}\right) = +1$. Clearly $R + N = p - 3$ since the Legendre symbol vanishes for the residue classes $t \equiv \pm 1 \mod p$. The two residue classes $t = \pm i$, where $i^2 \equiv -1 \mod p$, give rise to the value $\left(\frac{t^3-t}{p}\right) = (-1)^{\frac{p-1}{4}}\left(\frac{-2}{p}\right) = +1$. The remaining $p-5$ residue classes can be divided into 4-tuples consisting of the residue classes $(r, -r, s, -s)$, where $rs \equiv 1 \mod p$. The residue classes in each 4-tuple clearly give rise to the same value $\left(\frac{t^3-t}{p}\right)$; thus $R \equiv 2 \mod 4$ and $N \equiv 0 \mod 4$.

Since $R + N = p - 3$ and $R - N = \phi(1)$, we find $\phi(1) + p - 3 = 2R \equiv 4 \mod 8$, hence $2a = \phi(1) \equiv 7 - p \mod 8$. This implies $a \equiv 1 \mod 4$ if $p \equiv 5 \mod 8$ and $a \equiv 3 \mod 4$ if $p \equiv 1 \mod 8$.

This result has a corollary, which is notoriously difficult to prove,[6] and which is going back to Gauss's work on biquadratic residues:

Corollary 3.33 *Let $p = 4n + 1$ be a prime number, and write $p = a^2 + b^2$ with $a \equiv 1 \mod 4$. Then*

$$a \equiv \frac{1}{2}\binom{2n}{n} \mod p.$$

---

[6]See the beautiful article [24] by Cosgrave and Dilcher for an introduction to such congruences.

This is an almost incredible congruence, but we can easily verify it for some small primes p:

p	a	$\binom{2n}{n}$	$\frac{1}{2}\binom{2n}{n}$ mod p
5	1	2	1
13	−3	20	−3
17	1	70	1
29	5	3432	5

The key to the proof is a useful congruence also going back to Gauss:

Lemma 3.34 *For each odd prime number p we have*

$$\sum_{x=1}^{p-1} x^m \equiv \begin{cases} 0 \bmod p & \text{if } (p-1) \nmid m, \\ -1 \bmod p & \text{if } (p-1) \mid m. \end{cases} \tag{3.12}$$

Set $S = \sum x^m$ and let g denote a primitive root modulo p. Then

$$g^m S = \sum_{x=1}^{p-1} (gx)^m \equiv S \bmod p,$$

since gx runs through a coprime system of residue classes modulo p when x does. Thus p divides $(g^m - 1)S$. Now $g^m \equiv 1 \bmod p$ if and only if m is a multiple of $p - 1$; thus if $(p - 1) \nmid m$, then $p \mid S$. If $(p - 1) \mid m$, on the other hand, then $x^{p-1} \equiv 1 \bmod m$ implies that

$$S = \sum_{x=1}^{p-1} x^m \equiv \sum_{x=1}^{p-1} 1 = p - 1 \equiv -1 \bmod p$$

as claimed.

We now apply this to the character sum (3.11); using $p = 4n + 1$ we find

$$\phi(1) = \sum_{t=1}^{p-1} \left(\frac{t^3 - t}{p}\right) \equiv \sum_{t=1}^{p-1} (t^3 - t)^{2n}$$

$$\equiv \sum_{t=1}^{p-1} \sum_{k=0}^{2n} \binom{2n}{k} t^{3k}(-t)^{2n-k} = \sum_{t=1}^{p-1} \sum_{k=0}^{2n} (-1)^k \binom{2n}{k} t^{2n+2k}$$

$$= \sum_{k=0}^{2n} (-1)^k \binom{2n}{k} \sum_{t=1}^{p-1} t^{2n+2k} \bmod p.$$

The only exponent $2n + 2k$ divisible by $p - 1 = 4n$ occurs for $k = n$; this shows that

$$\phi(1) \equiv -(-1)^n \binom{2n}{n} \bmod p.$$

Since $(-1)^n = (\frac{2}{p})$ this implies $a \equiv 1 \bmod 4$.

3.4.1   Another Proof of the Quadratic Reciprocity Law

As an application of the techniques introduced in this section we will give another proof of the quadratic reciprocity law. The idea is to count the number of $\mathbb{F}_p$-rational points on affine varieties

$$A_n : x_1^2 + x_2^2 + \ldots + x_n^2 = 1.$$

If we let $x_1, \ldots, x_n$ run through $\mathbb{F}_p$, we expect that $y = 1 - x_1^2 - \ldots - x_{n-1}^2$ is a square half the time, and if it is, then $y = x_n^2$ has (in general) two solutions. This suggests that we should expect about p^{n-1} points on A_n.

   Victor Amédée Lebesgue determined the number of $\mathbb{F}_p$-rational points on A_n in [73] and showed that this implies the quadratic reciprocity law. As in the proof of Lemma 3.24 it can be shown that

$$\#A_n(\mathbb{F}_p) = p^{n-1} + \sum \left(\frac{t_1 t_2 \cdots t_n}{p} \right),$$

where the sum is over all $t_1, \ldots t_n \in F_p$ with $t_1 + t_2 + \ldots + t_n = 1$. Eisenstein used such sums for proving the quadratic reciprocity law in [35]; a modern version of his proof can be found in [65].

   We will now present a version of this proof due to Wouter Castryck [16]. Instead of A_n, he considered

$$\mathcal{A}_n : x_1^2 - x_2^2 + x_3^2 - \ldots + x_n^2 = 1$$

for an odd integer n. Let $N_n(p)$ denote the number of $\mathbb{F}_p$-rational points on $\mathcal{A}_n$, i.e., the number of solutions of the congruence

$$x_1^2 - x_2^2 + x_3^2 - \ldots + x_n^2 \equiv 1 \bmod p$$

for an odd prime number p. Then $N_1(p) = 2$ since $x_1^2 \equiv 1$ mod p has exactly two solutions. Next $N_2(p) = p - 1$ since for solving the congruence

$$x_1^2 - x_2^2 = (x_1 - x_2)(x_1 + x_2) \equiv 1 \text{ mod } p$$

we can assign $x_1 - x_2$ an arbitrary nonzero value t (there are $p - 1$ choices), and then x_1 and x_2 are uniquely determined by the system of equations $x_1 - x_2 \equiv t$ and $x_1 + x_2 \equiv \frac{1}{t}$ mod p.

The quadratic reciprocity law will follow from counting $N_q(p)$ in two different ways; in fact, we will now prove

Proposition 3.35 *The number $N_n(p)$ of $\mathbb{F}_p$-rational points on $\mathcal{A}_n$ has the following properties:*

1. It satisfies the recursion

$$N_n = p^{n-2}(p - 1) + pN_{n-2}. \tag{3.13}$$

2. If n is odd, then $N_n(p)$ is given by

$$N_n = p^{n-1} + p^{\frac{n-1}{2}}. \tag{3.14}$$

3. For all integers $n \geq 1$ we have

$$N_n = p^{n-1} + \left(\frac{-1}{p}\right)^{\frac{n-1}{2}} \sum_{t_1 + \ldots + t_n = 1} \left(\frac{t_1 t_2 \cdots t_n}{p}\right). \tag{3.15}$$

Proof of (3.13) In the equation of $\mathcal{A}_n$, substitute $x_1 + x_2$ for x_1; then we obtain

$$x_1^2 + x_3^2 - \ldots + x_n^2 - 1 = -2x_1 x_2.$$

If $x_1 \neq 0$, then for each choice of $x_3, \ldots x_n$ there is a unique value x_2; the number of such points is $(p - 1)p^{n-2}$ since there are $p - 1$ choices for x_1 and p choices for each $x_3, \ldots, x_n$. If $x_1 = 0$, then there are p choices for x_2 if $x_3^2 - \ldots + x_n^2 = 1$ and none otherwise; the number of solutions in this case is pN_{n-2}. This proves (3.13).

Proof of (3.14) This equation holds for $n = 1$ since $N_1 = 1 + 1 = 2$. Assume that $N_{n-2} = p^{n-3} + p^{\frac{n-3}{2}}$ for some odd integer $n \geq 3$; then by (3.13) we have

$$N_n = p^{n-2}(p - 1) + pN_{n-2} = p^{n-2}(p - 1) + p(p^{n-3} + p^{\frac{n-3}{2}})$$

$$= p^{n-1} - p^{n-2} + p^{n-2} + p^{\frac{n-1}{2}} = p^{n-1} + p^{\frac{n-1}{2}},$$

and the claim follows by induction.

Proof of (3.15) If $N(x^2 = a)$ denotes the number of solutions of the equation $x^2 = a$ in $\mathbb{F}_p$, then

$$
N_n = \sum_{t_1+\ldots+t_n=1} N(x_1^2 = t_1) \cdot N(x_2^2 = -t_2) \cdots N(x_n^2 = t_n)
$$

$$
= \sum_{t_1+\ldots+t_n=1} \left(1 + \left(\frac{t_1}{p}\right)\right)\left(1 + \left(\frac{-t_2}{p}\right)\right) \cdots \left(1 + \left(\frac{t_1}{p}\right)\right)
$$

$$
= p^{n-1} + \left(\frac{-1}{p}\right)^{\frac{n-1}{2}} \sum_{t_1+\ldots+t_n=1} \left(\frac{t_1 t_2 \cdots t_n}{p}\right).
$$

Here we have used the fact that the sums $\sum(\frac{t_1}{p})$, $\sum(\frac{t_1 t_2}{p})$, $\ldots$, $\sum(\frac{t_1 t_2 \cdots t_{n-1}}{p})$, etc. vanish. For a proof it is, after reordering the t_i, sufficient to show that

$$
\sum_{t_1+\ldots+t_n=1} \left(\frac{t_1}{p}\right) \cdots \left(\frac{t_s}{p}\right) = 0
$$

when $1 \le s < n$. Clearly t_n is uniquely determined by $t_1, \ldots, t_{n-1}$, hence we can sum over all $t_1, \ldots t_{n-1}$ instead and get

$$
\sum_{t_1+\ldots+t_n=1} \left(\frac{t_1}{p}\right) \cdots \left(\frac{t_s}{p}\right) = \sum_{t_1,\ldots,t_{n-1}} \left(\frac{t_1}{p}\right) \cdots \left(\frac{t_s}{p}\right)
$$

$$
= p^{n-s-1} \sum_{t_1,\ldots,t_s} \left(\frac{t_1}{p}\right) \cdots \left(\frac{t_s}{p}\right)
$$

$$
= p^{n-s-1} \left(\sum_{t_1} \left(\frac{t_1}{p}\right)\right) \cdots \left(\sum_{t_s} \left(\frac{t_s}{p}\right)\right) = 0
$$

since all the character sums in the brackets vanish. This finishes the proof of (3.15).

A more conceptual proof of the vanishing of these character sums is the following: Let a denote a quadratic nonresidue modulo p and set $u_1 = at_1$, $u_2 = t_2$, $\ldots$, $u_{s-1} = t_{s-1}$ and u_n via $u_1 + \ldots + u_n = 1$. If S denotes the character sum above, then $aS = S$ as in the proof of Lemma 3.25, and this implies $S = 0$.

Proof of the Quadratic Reciprocity Law We know by (3.14) that the number of $\mathbb{F}_p$-rational points on $\mathcal{A}_q$, where q is an odd prime different from p, is

$$
N_q = p^{q-1} + p^{\frac{q-1}{2}} \equiv 1 + \left(\frac{p}{q}\right) \bmod q \tag{3.16}
$$

by Fermat's Little Theorem and Euler's criterion.

Now let us count this number in a different way. If $(t_1, \ldots, t_q)$ satisfies $t_1 + \ldots + t_q = 1$, then so does $(t_2, t_3, \ldots, t_q, t_1)$, and do the other shifts. There is a single element for which these shifts do not produce p distinct q-tuples, namely $(t, t, \ldots t)$ with $qt = 1$. This shows that the number of $(t_1, \ldots, t_q)$ with $t_1 + \ldots + t_q = 1$ is $\equiv 1 \bmod q$. Using (3.15) we find

$$N_q = p^{n-1} + \left(\frac{-1}{p}\right)^{\frac{q-1}{2}} \sum_{t_1 + \ldots + t_n = 1} \left(\frac{t_1 t_2 \cdots t_n}{p}\right)$$

$$\equiv 1 + (-1)^{\frac{p-1}{2} \cdot \frac{q-1}{2}} \left(\frac{t^q}{p}\right) = 1 + (-1)^{\frac{p-1}{2} \cdot \frac{q-1}{2}} \left(\frac{q}{p}\right) \bmod q,$$

where we have used Fermat's Little Theorem, the first supplementary law, the fact that distinct shifts give rise to q identical terms that vanish modulo q, and finally that $\left(\frac{t^q}{p}\right) = \left(\frac{t}{p}\right) = \left(\frac{q}{p}\right)$ since $t = \frac{1}{q}$.

Comparing this with (3.16) yields the congruence

$$\left(\frac{p}{q}\right) \equiv (-1)^{\frac{p-1}{2} \cdot \frac{q-1}{2}} \left(\frac{q}{p}\right) \bmod q,$$

which implies the quadratic reciprocity law since for odd primes q, a congruence of the form $\pm 1 \equiv \pm 1 \bmod q$ implies equality.

3.5 Terjanian's Theorem

The quadratic reciprocity law can often be used for proving that certain Diophantine equations do not have solutions in integers. A prominent example is the Fermat equation for even exponents:

Theorem 3.36 (Terjanian) *Let p be an odd prime number, and assume that $x^{2p} + y^{2p} = z^{2p}$ for integers x, y, and z. Then $2p \mid x$ or $2p \mid y$.*

As for the usual Fermat equation $x^p + y^p = z^p$, the "second" case where $p \mid xyz$ is much more difficult than the "first case"

Clearly we may assume that x, y, and z are pairwise coprime. Since x and y cannot both be odd (otherwise $z^{2p} \equiv 2 \bmod 4$), we may assume that x is even and y and z are odd. Now

$$x^{2p} = z^{2p} - y^{2p} = (z^2 - y^2) \cdot \frac{z^{2p} - y^{2p}}{z^2 - y^2}. \tag{3.17}$$

Set $Q_p(m, n) = \frac{m^p - n^p}{m - n}$; then $x^{2p} = (z^2 - y^2) Q_p(z^2, y^2)$. For coprime integers m and n we have

$$\gcd(Q_p(m, n), m - n) = \gcd(p, m - n). \tag{3.18}$$

For a proof, let $d = \gcd(m - n, Q_p(m, n))$. Since d divides $m - n$, we have $n \equiv m \bmod d$, hence

$$Q_p(m, n) = m^{p-1} + m^{p-2} n + \ldots + n^{p-1} \equiv p m^{p-1} \bmod d.$$

If q is a prime divisor of d and if $q \mid m$, then $q \mid n$ contradicting the assumption that m and n are coprime. Thus d and m are coprime, hence $d \mid p$ and thus $d = 1$ or $d = p$. This finishes the proof of (3.18).

By Eq. (3.18), there are only two possibilities:

(a) The two factors in (3.17) are coprime; then $Q_p(z^2, y^2)$ is a square.
(b) The prime p divides both factors; in particular $p \mid x^{2p}$ and therefore $p \mid x$. Since x was assumed to be even, we actually have $2p \mid x$.

Thus Theorem 3.36 follows if we can show that $Q_p(z^2, y^2)$ cannot be a square for odd integers z and y. Now there are many squares among the numbers $Q_p(m, n)$, for example:

$$\frac{3^5 - 1}{3 - 1} = 11^2, \quad \frac{5^3 - 3^3}{5 - 3} = 7^2, \quad \frac{7^4 - 1}{7 - 1} = 20^2, \quad 8^3 - 7^3 = 13^2.$$

The following theorem provides us with a large class of nonsquares of the form $Q_p(m, n)$:

Theorem 3.37 *Let p be an odd prime number, and let m and n be coprime integers $m \equiv n \equiv 1 \bmod 4$. Then $Q_p(m, n)$ is not a square number.*

Theorem 3.37 (and therefore Terjanian's Theorem 3.36) follows immediately from the following calculation of a Jacobi symbol:

Theorem 3.38 *If q is an odd integer, and if m and n are coprime natural numbers with $m \equiv n \equiv 1 \bmod 4$, then*

$$\left(\frac{Q_p(m, n)}{Q_q(m, n)} \right) = \left(\frac{p}{q} \right) \tag{3.19}$$

for all integers p coprime to q.

In fact, all we have to do is choose an odd prime number q with $\left(\frac{p}{q} \right) = -1$; by Theorem 3.38, $Q_p(m, n)$ is a quadratic nonresidue modulo $Q_q(m, n)$, hence cannot be a square.

For proving Theorem 3.38 we need a few elementary properties of the numbers $Q_p(m, n)$.

Lemma 3.39 *Let* $Q_p(m, n) = \frac{m^p - n^p}{m - n}$.

(a) If p is odd and $m \neq n$, then $Q_p(m, n)$ is positive.
(b) If $p = aq + r$, then

$$Q_p(m, n) = m^r Q_q(m^a, n^a) + n^{p-r} Q_r(m, n). \tag{3.20}$$

(c) If $p = aq - r$, then

$$Q_p(m, n) \equiv -n^{p-q} m^{q-r} Q_r(m, n) \bmod Q_q(m, n). \tag{3.21}$$

(d) If $\gcd(p, q) = \gcd(m, n) = 1$, then $\gcd(Q_p(m, n), Q_q(m, n,) = 1$.
(e) For positive integers m and n with $mn \equiv 1 \bmod 8$ and odd integers p we have

$$Q_p(m, n) \equiv p \bmod 8.$$

Proof

(a) Since $Q_p(-m, -n) = Q_p(m, n)$ for odd integers p we may assume that $m \geq 0$. Next $Q_p(0, n) = n^{p-1} > 0$ since $n \neq 0$ in this case. If $m > n \geq 0$, then both numerator and denominator of $Q_p(m, n)$ are positive, and if $0 < m < n$, then both numerator and denominator of $Q_p(m, n)$ are negative, hence $Q_p(m, n) > 0$.

(b) Using $p = aq + r$ we find

$$Q_p(m, n) = \frac{m^{aq+r} - n^{aq+r}}{m - n} = \frac{m^{aq+r} - m^r n^{aq} + m^r n^{aq} - n^{aq+r}}{m - n}$$

$$= m^r Q_q(m^a, n^a) + n^{p-r} Q_r(m, n)$$

as claimed.

(c) If $p = aq - r$, then we have[7]

$$Q_p(m, n) = \frac{m^p - n^p}{m - n} = \frac{m^{aq-r} - n^{aq-r}}{m - n}$$

$$= \frac{m^{aq-r} - m^{q-r} n^{aq-q}}{m - n} + \frac{m^{q-r} n^{aq-q} - n^{aq-r}}{m - n}$$

$$= m^{q-r} \frac{m^{aq-q} - n^{aq-q}}{m - n} + n^{aq-q-r} \frac{m^{q-r} n^r - n^r n^{q-r}}{m - n}$$

$$= m^{q-r} Q_{a-1}(m^q, n^q) + n^{aq-q-r} [-m^{q-r} Q_r(m, n) + Q_q(m, n)]$$

$$= m^{q-r} [Q_{a-1}(m^q, n^q) + n^{aq-q-r} Q_q(m, n)] - n^{aq-q-r} m^{q-r} Q_r(m, n).$$

---

[7]There is a misprint in Terjanian [121, Equation (2)].

Since

$$Q_{a-1}(m^q, n^q) = \frac{(m^q)^{a-1} - (n^q)^{a-1}}{m - n} = \frac{(m^q)^{a-1} - (n^q)^{a-1}}{m^q - n^q} \cdot \frac{m^q - n^q}{m - q}$$

$$\equiv 0 \bmod Q_q(m, n),$$

the claim now follows since the first factor is clearly an integer.

(d) Let $\ell > 1$ be a common prime divisor of $Q_p(m, n)$ and $Q_q(m, n)$ with $p+q \geq 1$ minimal. If $p = q$, then $p = q = 1$ since $\gcd(p, q) = 1$, and then $Q_1(m, n) = 1$: Contradiction. Since we may assume that $p > q$ we can write $\varepsilon p = aq + r$. Equation (3.20) and the fact that $Q_q(m, n)$ divides $Q_q(m^a, n^a)$ shows that d divides $n^{m-r} Q_r(m, n)$. If $\ell \mid n$, then $\ell \mid Q_p(m, n)$ implies $\ell \mid m$, which is impossible. Thus $\ell \mid Q_r(m, n)$. Thus $Q_p(m, n)$ and $Q_r(m, n)$ have a common divisor, and since $m + r < m + n$ this contradicts the minimality of $m + n$.

(e) Observe that $m^2 \equiv 1 \bmod 8$; now

$$Q_q(m, n) = m^{p-1} + m^{p-2}n + \ldots + n^{p-1}$$

$$\equiv 1 + mn + 1 + \ldots + mn + 1 \equiv p \bmod 8$$

since each of the p terms in this sum is $\equiv 1 \bmod 8$.

□

Next we give the

Proof of Theorem 3.38 If $p = q$, then $p = q = 1$ and $Q_p(m, n) = 1$. If there is a pair (p, q) for which the theorem fails, i.e., for which $(Q_p(m, n)/Q_q(m, n)) = -(p/q)$, then choose one such pair for which the sum $p + q$ is minimal. Then

$$\left(\frac{Q_q(m, n)}{Q_p(m, n)}\right) = (-1)^{\frac{p-1}{2} \cdot \frac{q-1}{2}} \left(\frac{Q_p(m, n)}{Q_q(m, n)}\right) = (-1)^{\frac{p-1}{2} \cdot \frac{q-1}{2}} \cdot \left[-\left(\frac{p}{q}\right)\right] = -\left(\frac{q}{p}\right);$$

in fact, the first step is an application of the quadratic reciprocity law together with the congruences $Q_p(m, n) \equiv p$ and $Q_q(m, n) \equiv q \bmod 4$ that we have proved in Lemma 3.39.(f), the second step is our assumption $(Q_p(m, n)/Q_q(m, n)) = -(p/q)$, and the last step is another application of the quadratic reciprocity law. This shows that the result also fails for the pair (q, p).

Thus we may assume that $p > q$. Write $p = aq \pm r$ with $0 \leq r < q$ and r odd. If $p = aq + r$, then by (3.20) we have

$$\left(\frac{Q_p(m, n)}{Q_q(m, n)}\right) = \left(\frac{n^{m-r} Q_r(m, n)}{Q_q(m, n)}\right) = \left(\frac{Q_r(m, n)}{Q_q(m, n)}\right)$$

since $m - r$ is even. But this implies

$$\left(\frac{Q_r(m, n)}{Q_q(m, n)}\right) = \left(\frac{Q_p(m, n)}{Q_q(m, n)}\right) = -\left(\frac{p}{q}\right) = -\left(\frac{aq + r}{q}\right) = -\left(\frac{r}{q}\right),$$

hence the result also fails for the pair (r, q). Since $r + q < p + q$, this contradicts the minimality of $m + n$.

If $p = aq - r$ for some odd natural number r, then by (3.21) we have

$$Q_p(m, n) \equiv -n^{p-q} m^{q-r} Q_r(m, n) \bmod Q_q(m, n);$$

since $p - q$ and $q - r$ are even, both n^{p-q} and m^{q-r} are squares, hence

$$\left(\frac{Q_p(m, n)}{Q_q(m, n)}\right) = \left(\frac{-Q_r(m, n)}{Q_q(m, n)}\right) = \left(\frac{-Q_r(m, n)}{Q_q(m, n)}\right).$$

Therefore

$$\left(\frac{-Q_r(m, n)}{Q_q(m, n)}\right) = \left(\frac{Q_p(m, n)}{Q_q(m, n)}\right) = -\left(\frac{p}{q}\right) = -\left(\frac{aq - r}{q}\right) = -\left(\frac{-r}{q}\right),$$

and now $Q_q(m, n) \equiv q \bmod 4$ implies

$$\left(\frac{Q_r(m, n)}{Q_q(m, n)}\right) = -\left(\frac{r}{q}\right),$$

so the pair (r, q) with $r + q < p + q$ is another pair for which the theorem fails, and this again contradicts the minimality of $p + q$. □

For an alternative proof of Theorem 3.38 see Exercises 3.14–3.16.

3.5.1 Summary

In this chapter we have

- explained the notions of Legendre, Jacobi, Kronecker, and Zolotarev symbols;
- shown the equivalence between the modularity of Kronecker symbols and the quadratic reciprocity law;
- given several proofs of the quadratic reciprocity law;
- determined the points on Pell conics over fields such as $\mathbb{Q}$ and $\mathbb{F}_p$.

We will return to the concept of modularity in Chap. 10.

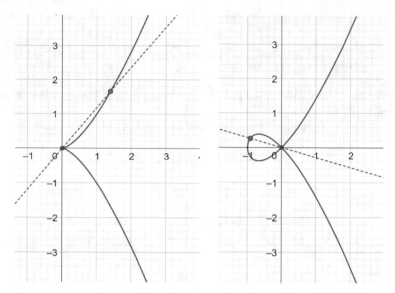

Fig. 3.1 The cubic curves $y^2 = x^3$ with a cusp and $y^2 = x^3 + x^2$ with a double point at the origin

3.6   Exercises

3.1. Parametrize the Pell conic $\mathcal{P} : x^2 + xy - my^2 = 1$ with discriminant $\Delta = 1 + 4m$. Show that the number of $\mathbb{F}_p$-rational points is given by $\#\mathcal{P}(\mathbb{F}_p) = p - \left(\frac{\Delta}{p}\right)$.

3.2. Curves of degree 3 can be parametrized only if they possess a singular point, i.e., a cusp as for the cubic $y^2 = x^3$ (Fig. 3.1, left graph) or a double point as for $y^2 = x^3 + x^2$ (Fig. 3.1, right graph). In these cases, the curves may be parametrized by lines through the singular point $(0, 0)$ (see [83]).

   Parametrize the cubic curves $y^2 = x^3$ and $y^2 = x^3 + x^2$. Also determine the number of $\mathbb{F}_p$-rational points on these curves.

3.3. We can think of the group $\mathbb{R}/\mathbb{Z}$ as the additive group of real numbers whose parts left of the decimal points are removed. For example, the sum of $\alpha = 0.7 + \mathbb{Z}$ and $\beta = 0.8 + \mathbb{Z}$ is $\alpha + \beta = 0.5 + \mathbb{Z}$.

   Show that the map $\lambda : \mathbb{R}/\mathbb{Z} \to S^1$ defined by $\lambda(t+\mathbb{Z}) = (\cos 2\pi t, \sin 2\pi t)$ from $\mathbb{R}/\mathbb{Z}$ to the unit circle S^1 is a group homomorphism, i.e., that

$$\lambda(s + t + \mathbb{Z}) = \lambda(s + \mathbb{Z}) + \lambda(t + \mathbb{Z}),$$

and that λ is a bijection.

   The element $t = \frac{1}{n} + \mathbb{Z}$ generates a cyclic subgroup of order n in $\mathbb{R}/\mathbb{Z}$. Deduce that $\lambda(t)$ generates a cyclic subgroup of order n on the unit circle. Show also that by identifying (x, y) with the complex number $x + yi$ this torsion group of order n consists exactly of the n-th roots of unity in $\mathbb{C}$.

3.4. Assume that $a = bc^2$ for nonzero integers a, b, c. Show that $(\frac{a}{p}) = (\frac{b}{p})$ for all primes $p \nmid c$.

3.5. Compute $(\frac{2}{15})$ and $(\frac{3}{35})$ using Gauss's Lemma.

3.6. Prove that $(\frac{-1}{m}) = (-1)^{\frac{m-1}{2}}$ for positive odd integers using the corresponding result for primes and the multiplicativity of the Jacobi symbol.

3.7. Show that $(a_1 a_2 \ldots a_n) = (a_1 a_2)(a_2 a_3) \cdots (a_{n-1} a_n)$.

3.8. Let π and ρ be permutations of the finite sets A and B, respectively. Show that they induce a permutation $\pi \times \rho$ on the product set $A \times B$, and that

$$\text{sign}(\pi \times \rho) = (\text{sign}\,\pi)^{\#B} \cdot (\text{sign}\,\rho)^{\#A}.$$

3.9. Show that Gauss's Lemma does not do what it is supposed to do when we restrict to coprime residue classes: Let $N = mn$ be the product of two coprime integers $m, n > 1$. Let $A = \{a_1, \ldots, a_{\phi(N)}\}$ denote a (coprime) half system modulo N. If a is an integer coprime to N and $a \cdot a_j \equiv (-1)^{s_j} a'_j$ for some $a'_j \in A$, then $(-1)^{\sum s_j} = 1$.

3.10. (Romanian Team Selection Test 2008) Let $m, n \geq 2$ be integers with $(2^m - 1) \mid (3^n - 1)$. Prove that n is even.

3.11. Show that $\phi(k) = 0$ (see 3.10) for all primes $p \equiv 3 \bmod 4$.

3.12. Show that Jacobsthal sums ϕ_m are multiplicative: $\phi_m(1)\phi_n(1) = \phi_{mn}(1)$ for coprime values of m and mn.

3.13. Let p be a prime number $\equiv 1 \bmod 4$. Show that the number of residue classes $t \bmod p$ with $1 \leq t \leq \frac{p-1}{2}$ for which $(\frac{t^3-t}{p}) = -1$ is a multiple of 4.

3.14. Prove that for odd coprime integers $m \equiv n \equiv 1 \bmod 4$ with $mn \equiv 1 \bmod 8$ we have

$$\left(\frac{Q_2(m, n)}{Q_q(m, n)}\right) = \left(\frac{m+n}{Q_q(m, n)}\right) = \left(\frac{2}{q}\right). \tag{3.22}$$

3.15. Prove the following result: If $r = 2t$ is even, then

$$\left(\frac{Q_r(m, n)}{Q_q(m, n)}\right) = \left(\frac{2}{q}\right)\left(\frac{Q_t(m^2, n^2)}{Q_q(m, n)}\right).$$

3.16. Prove Theorem 3.38 for odd natural numbers m, n with $mn \equiv 1 \bmod 8$ using induction on q.

3.17. Let RR denote the number of pairs of consecutive quadratic residues modulo an odd prime number p:

$$RR = \#\{(a, a+1) : 1 \leq a < p, \ (\tfrac{a}{p}) = (\tfrac{a+1}{p}) = +1\},$$

and define the numbers RN, NR, and NN correspondingly.

1. Compute these numbers for $p = 13$.
2. Show that $RR + RN + NR + NN = p - 2$.
3. Show that the number of pairs $(a, a + 1)$ with $(\frac{a}{p}) = (\frac{a+1}{p})$ is equal to
   $RR + NN = \frac{p-3}{2}$. Similarly, the number of $(a, a+1)$ with $(\frac{a}{p}) = -(\frac{a+1}{p})$
   is equal to $RN + NR = \frac{p-1}{2}$.
4. Let $p \equiv 1 \bmod 4$ be a prime number. Show that pairs of consecutive
   squares modulo p correspond to sets of point $(\pm x, \pm y)$ on the conic
   $\mathcal{H} : X^2 - Y^2 = 1$.
5. Show that the points on C_0 corresponding to a pair of consecutive
   quadratic residues form a subgroup of index 2 in C. Check that multiplication by 2 induces a map

$$C_0 \longrightarrow C : (X, Y) \mapsto (4X(X + 1), 2Y(2X + 1)).$$

   Deduce that if $(n, n + 1)$ is a pair of consecutive quadratic residues or
   nonresidues, then $(4n^2 + 4n, 4n^2 + 4n + 1)$ is a pair of quadratic residues
   and verify this directly. Derive formulas for RR and NN.
6. Show that

$$\left(1 + \left(\frac{a}{p}\right)\right)\left(1 + \left(\frac{a+1}{p}\right)\right) = \begin{cases} 4 & \text{if } (\frac{a}{p}) = (\frac{a+1}{p}) = +1, \\ 0 & \text{otherwise.} \end{cases}$$

Deduce that

$$4RR = \sum_{a=1}^{p-2} \left(1 + \left(\frac{a}{p}\right)\right)\left(1 + \left(\frac{a+1}{p}\right)\right). \tag{3.23}$$

Expand the product and show, using $\sum_{a=1}^{p-1} (\frac{a}{p}) = 0$, that

$$\sum_{a=1}^{p-2} \left(\frac{a}{p}\right) = -\left(\frac{-1}{p}\right) \quad \text{and} \quad \sum_{a=1}^{p-2} \left(\frac{a+1}{p}\right) = -1.$$

3.18. Show that, for primes $p \equiv 3 \bmod 4$, the quadratic residues modulo p form a
half system. Show that Gauss's Lemma holds trivially in this case.

3.19. Let p be an odd prime number. Show that the number N_n of $\mathbb{F}_p$-rational points
on $\mathcal{A}_n$ for even integers n is given by

$$N_n = p^{n-1} - p^{\frac{n-2}{2}}$$

for all $n \geq 2$.

3.20. Show that here exist infinitely many positive odd integers m and n with $m \equiv n \bmod 4$ such that $Q_4(m, n)$ is a square. Is it true that all such integers have a common divisor > 1?

3.21. The following proof of the quadratic reciprocity law based on Gauss's Lemma is due to Christian Zeller [135]. We will give it for $p = 5$ and $q = 23$.

- Write down the absolutely smallest remainders of $kp \bmod q$ and $hq \bmod p$:

k	1	2	3	4 5 6	7	8	9 10 11
$kp \bmod q$	5	10	−8	−3 2 7	−11	−6	−1 4 9

h	1 2
$hq \bmod p$	−2 1

- Let μ denote the number of negative remainders; then $(\frac{p}{q})(\frac{q}{p}) = (-1)^\mu$ by Gauss's Lemma; here $\mu = 6$.
- The negative remainders $-r = -1, -2$ with $0 < r < \frac{p}{2}$ occur exactly once in these tables (for $k = 9$ and $h = 1$);
- The other negative remainders $-r$ for $\frac{p}{2} < r < \frac{q}{2}$, with a few exceptions, come in pairs: The pairs $(k, k') = (3, 8)$ and $(4, 7)$ satisfy $k + k' = 11 = \frac{q-1}{2}$.
- The possible exceptions are the following: There exists a degenerate pair (k, k') with $k = k' = \frac{q-1}{4}$ if $q \equiv 1 \bmod 4$; in this case, the remainder is negative if $p \equiv 3 \bmod 4$. Observe that the value $k = \frac{q-1}{2}$ yields a positive remainder.

Now deduce the quadratic reciprocity law by counting the number of negative remainders. In our case, there are $\frac{p-1}{2}$ remainders $-r$ with $0 < r < \frac{p}{2}$ and an even number of negative remainders $-r$ with $\frac{p}{2} < r < \frac{q}{2}$, hence $(\frac{5}{23})(\frac{23}{5}) = +1$.

Chapter 4
Divisibility in Integral Domains

In this chapter we will study the notion of divisibility in general domains. We will restrict our attention to commutative domains R containing a unit[1] 1, i.e., an element with the property $1r = r$ for all $r \in R$. Recall that a ring R is called a *domain* if it does not contain any zero divisors, that is, if $ab = 0$ for elements $a, b \in R$ implies that $a = 0$ or $b = 0$. Subrings of fields are always domains, and every domain may be interpreted as a subring of its field of quotients (see Exercise 4.3). Our goal is the definition of units, primes, and irreducible elements and a first investigation of the question in which quadratic number rings the theorem of unique factorization holds.

4.1 Units, Primes, and Irreducible Elements

It is easy to transfer the notion of divisibility of integers in $\mathbb{Z}$ to arbitrary domains R: Given $a, b \in R$, we say that b divides a if there is a $c \in R$ such that $a = bc$, and we write $b \mid a$ in this case. More generally we write $a \equiv b \bmod mR$ if $m \mid (a - b)$ in R. Congruences in R have the usual properties; we leave the proofs as exercises (see Exercise 4.9).

Proposition 4.1 *Let R be a domain; for all $a, b, c, d, m, n \in R$, we have*

(a) $a \equiv b \bmod m$, $c \equiv d \bmod m \Longrightarrow a + c \equiv b + d \bmod m$;
(b) $a \equiv b \bmod m$, $c \equiv d \bmod m \Longrightarrow ac \equiv bd \bmod m$; and
(c) $n \mid m$ *und* $a \equiv b \bmod m \Longrightarrow a \equiv b \bmod n$.

The properties (a) and (b) are equivalent to the statement that $a \equiv b \bmod m$ implies $f(a) \equiv f(b) \bmod m$ for all polynomials $f \in \mathbb{Z}[x]$. The following result shows

---

[1] The standard example of a domain without 1 is the ring of even integers.

© The Author(s), under exclusive license to Springer Nature Switzerland AG 2021
F. Lemmermeyer, *Quadratic Number Fields*, Springer Undergraduate
Mathematics Series, https://doi.org/10.1007/978-3-030-78652-6_4

that certain congruences in $\mathcal{O}_k$ imply congruences in $\mathbb{Z}$; it allows us to work in the bigger ring $\mathcal{O}_k$ and then pull back results from there to the ring of ordinary integers.

Proposition 4.2 *Let $a, b, m \in \mathbb{Z}$. Then $a \equiv b \bmod m$ in $\mathcal{O}_k$ if and only if $a \equiv b \bmod m$ in $\mathbb{Z}$.*

Proof The congruence $a \equiv b \bmod m$ in $\mathcal{O}_k$ is equivalent to $a - b = m\gamma$ for some $\gamma \in \mathcal{O}_k$. Since $\gamma = \frac{a-b}{m}$, we have $\gamma \in \mathcal{O}_k \cap \mathbb{Q}$, and now Proposition 2.4 shows that $\gamma \in \mathbb{Z}$, and hence $a \equiv b \bmod m\mathbb{Z}$ in $\mathbb{Z}$. The converse is trivial.   □

The following result is also useful for computing with quadratic irrationalities; the simple proof is given in Exercise 4.10.

Proposition 4.3 *Let $\{1, \omega\}$ be an integral basis of a quadratic number field, and let $m \in \mathbb{Z}$ be an integer. Then $m \mid (a + b\omega)$ in $\mathcal{O}_k$ if and only if $m \mid a$ and $m \mid b$.*

Elements of a domain R that divide 1 are called *units* of R. The set $R^\times$ of all units forms a group with respect to multiplication; it is called the *unit group* of R. Examples of unit groups of some well-known rings are the following:

R	$\mathbb{Z}$	$\mathbb{Z}[x]$	$\mathbb{Q}[x]$	$\mathbb{Z}[\sqrt{-2}\,]$	$\mathbb{Z}[i]$	$\mathbb{Z}[\sqrt{2}\,]$
$R^\times$	$\{\pm 1\}$	$\{\pm 1\}$	$\mathbb{Q}^\times$	$\{\pm 1\}$	$\{\pm 1, \pm i\}$	$\{\pm(1 + \sqrt{2})^n\}$

The computation of units in number fields is often challenging; checking whether a given element is a unit is rather easy:

Proposition 4.4 *An element $\varepsilon \in \mathcal{O}_k$ is a unit if and only if $N\varepsilon = \pm 1$. If we write $\varepsilon = \frac{t+u\sqrt{m}}{2}$ for integers $t \equiv u \bmod 2$, then ε is a unit if and only if $t^2 - mu^2 = \pm 4$.*

Proof Let $\varepsilon \in \mathcal{O}_k$ be a unit; then $\varepsilon\eta = 1$ for some $\eta \in \mathcal{O}_k$, and taking the norm yields $N\varepsilon N\eta = N(1) = 1$. Since $N\varepsilon$ and $N\eta$ are integers whose product is 1, we either have $N\varepsilon = N\eta = 1$ or $N\varepsilon = N\eta = -1$. Conversely, $N\varepsilon = \pm 1$ for some $\varepsilon \in \mathcal{O}_k$ means $\pm\varepsilon\varepsilon' = 1$, and hence ε is a unit.

If $\varepsilon = \frac{t+u\sqrt{m}}{2}$ is a unit, then clearly $t^2 - mu^2 = \pm 4$. If conversely $t^2 - mu^2 = \pm 4$ and $m \equiv 2, 3 \bmod 4$, then it follows that t and u both must be even, and hence $\varepsilon = \frac{t}{2} + \frac{u}{2}\sqrt{m} \in \mathbb{Z}[\sqrt{m}\,]$. If $m \equiv 1 \bmod 4$, on the other hand, then $t \equiv u \bmod 2$. In both cases, ε is a unit in $\mathcal{O}_k$.   □

It follows that the norm yields a group homomorphism $E_k \longrightarrow E_{\mathbb{Q}} = \{\pm 1\}$, where $E_k = \mathcal{O}_k^\times$ and $E_{\mathbb{Q}} = \mathbb{Z}^\times$ are the unit groups of $\mathcal{O}_k$ and $\mathbb{Z}$.

The unit groups in complex quadratic number fields can be described explicitly.

Theorem 4.5 *Let $m < 0$ be squarefree, $k = \mathbb{Q}(\sqrt{m}\,)$, and $R = \mathcal{O}_k$ the ring of integers in k. Then*

$$R^\times = \begin{cases} \langle i \rangle & \text{if } m = -1; \\ \langle -\rho \rangle & \text{if } m = -3; \\ \langle -1 \rangle & \text{otherwise.} \end{cases}$$

Here $i = \sqrt{-1}$ denotes a primitive fourth and $\rho = \frac{1}{2}(-1 + \sqrt{-3})$ a primitive cube root of unity.

Proof Assume first that $m \equiv 1, 2 \bmod 4$, and let $\varepsilon = a + b\sqrt{-m}$ be a unit. Then $1 = N\varepsilon = a^2 + mb^2$ (the case $N\varepsilon = -1$ cannot occur since $m > 0$). For $m > 1$, this implies $a = \pm 1$ and $b = 0$, and hence $\varepsilon = \pm 1$ (and of course ± 1 are units). If $m = 1$, there are four possibilities, namely $a = \pm 1, b = 0$ and $a = 0, b = \pm 1$. All these units are powers of $i = \sqrt{-1}$.

If $m \equiv 3 \bmod 4$, we set $\varepsilon = \frac{1}{2}(a + b\sqrt{-m})$ for integers a, b and find $4 = a^2 + mb^2$ as a necessary and sufficient condition for ε to be a unit. For $m > 3$, there are again only the trivial solutions corresponding to $\varepsilon = \pm 1$; if $m = 3$, then we obtain the units

$$\pm 1, \quad \pm \frac{-1 + \sqrt{-3}}{2}, \quad \pm \frac{1 + \sqrt{-3}}{2}.$$

Setting $\rho = \frac{-1 + \sqrt{-3}}{2}$ (this is a cube root of unity since $\rho^3 = 1$), we find that E_k is generated by $-\rho$ (a primitive sixth root of unity). $\square$

The determination of the unit group of rings of integers in real quadratic number fields boils down to solving the *Pell equation* $t^2 - mu^2 = \pm 4$; we will prove in Chap. 7 below that this equation has integral solutions whenever $m \geq 2$ is not a square. At this point we only observe that $\varepsilon = 1 + \sqrt{2}$ is a unit with infinite order in $\mathbb{Z}[\sqrt{2}]$ (see Theorem 2.7): If we had $(1 + \sqrt{2})^n = \pm 1$ for some $n \geq 1$, then taking absolute values (after identifying $\sqrt{2}$ with the positive real square root of 2), we obtain $1 = |\pm 1| = |1 + \sqrt{2}|^n > 1$, and similarly $1 = |\pm 1| = |1 + \sqrt{2}|^n = |1 - \sqrt{2}|^{-n} < 1$ if $n \leq -1$. In particular, $\mathbb{Z}[\sqrt{2}]$ has infinitely many units.

John Pell (1611–1685) was an English mathematician. His name got attached to the Pell equation through a mistake by Euler, who apparently confused him with Lord William Brouncker. It was Brouncker who developed a method for solving such equations in integers in connection with Fermat's challenge in 1657 for the English mathematicians. The proof that Brouncker's method always leads to a solution was given much later by Lagrange.

A method for solving the Pell equation similar to Brouncker's had already been developed by Indian mathematicians, in particular Brahmagupta (ca. 598–670) and Bhaskara II (1114–1185); their contributions (see Plofker [104]) became known in Europe only during the nineteenth century. We will present a method for solving the Pell equation in Chap. 7.

Elements $a, b \in R$ are called *associated*, if there is a unit $e \in R^\times$ such that $a = be$; we write $a \sim b$ and verify easily that this defines an equivalence relation on R.

Irreducible and Prime Elements An element $a \in R \setminus R^\times$ is called *irreducible* if a has only trivial divisors, that is, units and associates. More exactly: a is irreducible in R if $a = bc$ implies that b or c is a unit. An element $p \in R \setminus R^\times$ is called *prime* if

$p \mid ab$ implies that $p \mid a$ or $p \mid b$. Observe that units are by definition neither prime nor irreducible.

Proposition 4.6 *Prime elements are irreducible.*

Proof Let a be prime. If we could factor a, there would exist $b, c \in R \setminus R^\times$ with $a = bc$. Now $a \mid bc$; if $a \mid b$, i.e., $b = ad$ for some $d \in R$, then $a = acd$, hence $1 = cd$, and c is a unit in contradiction to our assumption.   □

A simple criterion for the primality of an element in a ring is the following:

Proposition 4.7 *An element $p \in R$ is prime if and only if the residue class ring R/pR of the residue classes modulo p is a domain.*

The proof is simple. The residue class ring modulo p does not have a zero divisor if $ab \equiv 0 \bmod p$ implies that $a \equiv 0 \bmod p$ or $b \equiv 0 \bmod p$. But this is just a version of the definition of a prime element, which states that an element is prime if $p \mid ab$ implies that $p \mid a$ or $p \mid b$.

4.1.1 Elements with Prime Norm Are Prime

We have already seen that elements $\pi \in \mathcal{O}_k$ for which $p = |N\pi|$ is a rational prime are always irreducible. As a matter of fact, such elements are always prime. This will follow easily from the theory of ideals that we will develop later; here we will give a direct proof based on Proposition 4.7.

Proposition 4.8 *If k is a quadratic number field with ring of integers $\mathcal{O}_k$, then each $\pi \in \mathcal{O}_k$ with prime norm is prime.*

This is easy to see if $\mathcal{O}_k$ is a unique factorization domain (see the next section): Elements with prime norm are irreducible, and in unique factorization domains, irreducible elements are prime. In order to prove this for general rings $\mathcal{O}_k$, we show that the residue class ring $\mathcal{O}_k/\pi\mathcal{O}_k$ does not have zero divisors. In fact, we will show that $\mathcal{O}_k/\pi\mathcal{O}_k \simeq \mathbb{F}_p = \mathbb{Z}/p\mathbb{Z}$ is isomorphic to the field with p elements.

To this end, let $\{1, \omega\}$ be an integral basis of $\mathcal{O}_k$; then $\pi = a + b\omega$ for integers $a, b \in \mathbb{Z}$. We claim that b is not divisible by π (and thus not divisible by $p = |\pi\pi'|$). In fact, $\pi \mid b$ implies $\pi \mid a$ since $a = \pi - b\omega$, and taking norms, we find $p \mid a^2$ and $p \mid b^2$. Since p is prime, this implies that $p \mid a$ and $p \mid b$. But then $\pi = a + b\omega$ would be divisible by p, and hence π' would be a unit: a contradiction.

Thus there exists an integer $c \in \mathbb{Z}$ with $bc \equiv 1 \bmod p$, and in particular, we have $bc \equiv 1 \bmod \pi\mathcal{O}_k$. We find $b\omega \equiv -a \bmod \pi$, after multiplying through by c, thus $\omega \equiv -ac \bmod \pi\mathcal{O}_k$. If any $\gamma = r + s\omega \in \mathcal{O}_k$ is given, then we find $\gamma \equiv r - sac \bmod \pi\mathcal{O}_k$, and thus modulo π every element is congruent to an ordinary integer. Reducing this number modulo p (and p is a multiple of π), we find that γ is congruent to one of the numbers $0, 1, 2, \ldots, p - 1$ modulo π.

Now it is easy to show that there are no zero divisors in the ring of residue classes: If we had $\alpha\beta \equiv 0 \bmod \pi$ and if $A, B \in \{0, 1, \ldots, p - 1\}$ are integers with $\alpha \equiv A \bmod \pi\mathcal{O}_k$ and $\beta \equiv B \bmod \pi\mathcal{O}_k$, then $\pi \mid AB$; taking norms yields $p \mid A^2B^2$, and hence $p \mid A$ or $p \mid B$. Thus $A = 0$ or $B = 0$, and therefore $\alpha \equiv A = 0 \bmod \pi$ or $\beta \equiv B = 0 \bmod \pi$.

Proposition 4.9 *Let p be an odd prime number and $\mathcal{O}_k$ the ring of integers in $k = \mathbb{Q}(\sqrt{m})$. Then p is prime in $\mathcal{O}_k$ if and only if the congruence $x^2 \equiv m \bmod p$ is not solvable.*

Proof If $x^2 \equiv m \bmod p$ is solvable, then $p \mid (x+\sqrt{m})(x-\sqrt{m})$, but $p \nmid (x\pm\sqrt{m})$. Thus p is not prime.

Now we show that p remains prime in $\mathcal{O}_k$ if $(\frac{m}{p}) = -1$. This case is not covered by Proposition 4.8 since here $N(p) = p^2$ is not prime. The idea for proving the result is the same as in the proof of Proposition 4.8: We show that the residue classes modulo p in $\mathcal{O}_k$ form a field.

We will give the proof in the case where $\mathcal{O}_k = \mathbb{Z}[\sqrt{m}]$. Here the residue classes modulo p in $\mathcal{O}_k$ are represented by the p^2 elements $a + b\sqrt{m}$ with $0 \leq a, b < p$; clearly every $\alpha \in \mathcal{O}_k$ is congruent modulo p to one of these elements, and they are pairwise distinct. These residue classes form a ring, and we want to show that they form a field. This will follow if we can write down an inverse for each residue class $a + b\sqrt{m} \bmod p$ different from $0 \bmod p$. Now

$$\frac{1}{a + b\sqrt{m}} = \frac{a - b\sqrt{m}}{a^2 - mb^2},$$

and the denominator is $\equiv 0 \bmod p$ if and only if a and b are divisible by p (otherwise m would be a quadratic residue modulo p). Since $0 \leq a, b < p$, this implies $a = b = 0$. In fact, $a^2 \equiv mb^2 \bmod p$ implies either (if $b \neq 0$) that $(\frac{a}{b})^2 \equiv m \bmod p$, and then $x^2 \equiv m \bmod p$ is solvable and m is a quadratic residue modulo p, or (if $b = 0$) that $a^2 \equiv 0 \bmod p$ and hence $a = 0$. Thus for each nonzero residue class $a + b\sqrt{m} \bmod p$, the inverse is given by $\frac{a-b\sqrt{m}}{a^2-mb^2} \bmod p$.

In the case $m \equiv 1 \bmod 4$, the residue classes modulo p are represented by elements $a + b\omega$ with $0 \leq a, b < p$; the rest of the proof is left to the readers as an exercise. $\square$

For $p = 2$, there is a corresponding criterion that may be proved in a similar manner.

Proposition 4.10 *The element $p = 2$ is prime in the ring of integers $\mathcal{O}_k$ of the quadratic number field $k = \mathbb{Q}(\sqrt{m})$ if and only if $m \equiv 5 \bmod 8$.*

We leave the proof as an exercise for the readers.

4.2   Unique Factorization Domains

A domain in which the theorem of unique factorization holds is called a unique factorization domain (UFD). More exactly, we demand

UFD–1. Each non-unit $\neq 0$ is a product of finitely many irreducible elements.
UFD–2. Irreducible elements are prime.

There are domains in which UFD–1 fails: In the domain A that is obtained by adjoining all 2^n-th roots of 2 to $\mathbb{Z}$, namely $A = \mathbb{Z}[\sqrt{2}, \sqrt[4]{2}, \sqrt[8]{2}, \ldots]$, the element 2 cannot be written as a finite product of irreducible elements since

$$2 = \sqrt{2}\sqrt{2} = \sqrt[4]{2}\,\sqrt[4]{2}\,\sqrt[4]{2}\,\sqrt[4]{2} = \ldots.$$

The defining property of unique factorization domains is that the factorization guaranteed by UFD–1 should be unique:

UFD–3. Let $a \in R \setminus \{0\}$ and $a = ep_1 \cdots p_s = e'q_1 \cdots q_t$, where $e, e' \in R^\times$ are units and where the p_j and q_j are irreducible elements in R. Then $s = t$, and we can rearrange the q_j in such a way that $p_i \sim q_i$ for $i = 1, \ldots, s$.

Clearly, UFD–3 holds in any unique factorization domain.

Proposition 4.11 *Conditions UFD–2 and UFD–3 are equivalent in every domain R in which UFD–1 holds.*

Proof UFD–2 $\Longrightarrow$ UFD–3: Since the p_i are irreducible, they are prime by assumption. In particular, p_1 divides one of the factors q_j, say q_1. Since q_1 is irreducible, we must have $p_1 \sim q_1$. Since R is a domain, p_1 may be canceled, and we obtain $e_1 p_2 \cdots p_s = e'_1 q_2 \cdots q_t$. Induction now yields the claim.

UFD–3 $\Longrightarrow$ UFD–2: Let a be irreducible and $a \mid xy$, where $x, y \in R$. Then there exists an element $b \in R$ with $ab = xy$. Because of UFD–3, the decomposition into irreducible elements is unique up to order and units; thus an associate of a must occur in the factorization of x or y, and we find $a \mid x$ or $a \mid y$. Thus a is prime.   □

Since $1 + \sqrt{-5}$ is irreducible in $R = \mathbb{Z}[\sqrt{-5}]$, but not prime, R is not a unique factorization domain. This fact also proves that the theorem of unique factorization in $\mathbb{Z}$, which often seems obvious to beginners in number theory, requires a proof.

We call an element d in some domain a common divisor of elements $a, b \in R$ if $d \mid a$ and $d \mid b$. How should we choose a "greatest" common divisor among these common divisors? In the ordinary integers, we can choose the greatest divisor with respect to the absolute value, but this is not a suitable definition for general domains R. What we want is a definition of the greatest common divisor in terms of divisibility alone: We call $d \in R$ a *greatest common divisor* of $a, b \in R$ and write $d \sim \gcd(a, b)$ if d has the following properties:

GCD–1. d is a common divisor of a and b, i.e., $d \mid a$ and $d \mid b$.

GCD–2. Every common divisor of a and b divides d, i.e., if $c \mid a$ and $c \mid b$ for some $c \in R$, then $c \mid d$.

Again we would like to emphasize the fact that this definition is well suited for building a theory of greatest common divisors but cannot easily be used for finding a greatest common divisor of two elements in some domain.

In unique factorization domains, the greatest common divisor of two elements can be written down explicitly. In fact, if $a = u \prod p^{\alpha_p}$ and $b = v \prod p^{\beta_p}$ are the prime factorizations of a and b (with units $u, v \in R^{\times}$), then we can easily show that $d = \prod p^{\min(\alpha_p, \beta_p)}$ is a greatest common divisor of a and b. One has to remark that even in the case of the ordinary integers, finding the prime factorization of two (large) integers can be very difficult.

Two elements a and b of some unique factorization domain R are called *coprime* (or relatively prime) if their greatest common divisor is a unit. Observe that we demand that R be a unique factorization domain. In fact, in domains without unique factorization, a greatest common divisor need not exist, and if it does, it need not have the properties we expect from a greatest common divisor, such as $\gcd(a, b)^2 = \gcd(a^2, b^2)$.

Proposition 4.12 *If R is a unique factorization domain, if $a, b \in R$ are coprime, and if $ab = ex^n$ $(n \geq 2)$ for some unit $e \in R^{\times}$ and some $x \in R$, then there exist units $e_1, e_2 \in R^{\times}$ and elements $c, d \in R$ such that $a = e_1 c^n$ and $b = e_2 d^n$, where $cd = x$ and $e_1 e_2 = e$.*

Proof We prove this by induction on the number of prime factors of a. If a is a unit, then the claim follows with $c = 1$, $d = x$, $e_1 = a$, and $e_2 = ea^{-1}$.

Assume that the claim is true for all $a \in R$ with at most t different prime factors, and let $p \in R$ be a prime with $p \mid a$. Assume that $p^h \parallel a$ (we write $p^h \parallel a$ if $p^h \mid a$ and $p^{h+1} \nmid a$, i.e., if p^h is the largest power of p that divides a). Since $p^h \parallel x^n$ (here we use the fact that a and b are coprime), we must have $h = nk$ for some $k \in \mathbb{N}$ and $p^k \parallel x$. Thus $a = p^h a_1$, $x = p^k x_1$ and $a_1 b = ex_1^n$. By induction assumption, we have $a_1 = e_1 c^n$ and $b = e_2 d^n$, and now the claim follows since $a = e_1(cp^k)^n$. $\square$

Corollary 4.13 *If R is a unique factorization domain, if $\gcd(a, b) = p$ for elements $a, b, p \in R$, where p is prime, and if $ab = ex^n$ $(n \geq 2)$ for some $e \in R^{\times}$ and $x \in R$, then there exist units $e_1, e_2 \in R^{\times}$ and $c, d \in R$ with $a = e_1 pc^n$ and $b = e_2 p^{n-1} d^n$ (after switching a and b, if necessary).*

Proof Exercise 4.28. $\square$

4.3 Principal Ideal Domains

Principal ideal domains will play a minor role in this chapter, mainly as a link in the chain of inclusions

Euclidean Domains $\subset$ Principal Ideal Domains $\subset$ Unique Factorization Domains that we will use for constructing unique factorization domains. Both inclusions are proper; for rings of integers in quadratic number fields (and in fact of general number fields), the second inclusion is in fact an equality.

First we will have to explain the notion of a principal ideal domain. To this end, consider a domain R; a subring I of R is called an *ideal* of R if $I \cdot R \subseteq I$. Thus an ideal is a subset of a domain that is closed with respect to addition ($I + I \subseteq I$) as well as with respect to multiplication by arbitrary elements of the domain R.

Observe that I is a subring of R if the weaker condition $I \cdot I \subset I$ is satisfied. In the domain $R = \mathbb{Z}$, it can be shown that *each* subring is an ideal. The following example shows that this is not true for general domains: The set

$$M = \mathbb{Z} + 2\sqrt{m}\,\mathbb{Z} = \{a + 2b\sqrt{m} : a, b \in \mathbb{Z}\}$$

is a subring of $\mathbb{Z}[\sqrt{m}\,]$, but not an ideal. This is because $MR = R$; in fact, $1 \in M$ implies that each element of R is contained in MR. Since $\sqrt{m} \in R \setminus M$, the subring M is not an ideal.

It is very easy to write down examples of ideals. If we are given elements $a_1, \ldots, a_n \in R$, then the set of all R-linear combinations

$$I = (a_1, \ldots, a_n) := \{a_1 r_1 + \ldots + a_n r_n : r_j \in R\}$$

of these elements is an ideal called the ideal generated by $a_1, \ldots, a_n$. Clearly I is closed with respect to addition; thus it remains to verify that $IR \subseteq I$. But this is easy: Since $a = a_1 r_1 + \ldots + a_n r_n \in I$, clearly $ar = a_1(r_1 r) + \ldots + a_n(r_n r)$ is an element of I.

In our proofs we have to consider ideals generated by infinitely many elements $a_1, a_2, \ldots$ These ideals $I = (a_1, a_2, \ldots)$ are by definition the set of all *finite* R-linear combinations of the elements $a_i \in I$.

Remark In fields $R = K$, there are only two different ideals, namely the zero ideal (0) and the unit ideal $(1) = R$.

Ideals generated by a single element a are called *principal ideals*. These have the form $I = (a) = \{ar : r \in R\}$; occasionally, we will write $I = aR$. Principal ideals (a) consist of all multiples of a.

The transition from elements to principal ideals consists essentially in disregarding units.

Lemma 4.14 *For $a, b \in R$, the following assertions are equivalent:*

1. $(a) = (b)$;
2. *There is a unit $e \in R^\times$ with $a = be$.*

The proof is a simple exercise.

A domain in which each ideal is principal is called a *principal ideal domain* (PID). Clearly, the ring $\mathbb{Z}$ of ordinary integers is a PID; in fact, the ideal $(a_1, \ldots, a_n)$ is generated by the greatest common divisor $d = \gcd(a_1, \ldots, a_n)$. Not every unique factorization domain is a principal ideal domain; the best known example is the domain $\mathbb{C}[x, y]$ of polynomials in two variables with complex coefficients; here, (x, y) is not principal, as is easily seen.

Remark The fact that $\mathbb{C}[x, y]$ is a unique factorization domain follows from a well-known theorem in algebra: If R is a UFD, then so is the polynomial ring $R[y]$. Since $R = \mathbb{C}[x]$ is a UFD (this ring is even Euclidean—see Sect. 4.4), the claim follows.

Now we prove that principal ideal domains have unique factorization.

Theorem 4.15 *Principal ideal domains are unique factorization domains.*

Proof Assume that UFD–1 is not satisfied. Then there is an $a_1 \in R$ that cannot be written as a product of irreducible elements (in particular, a_1 is not irreducible). Thus, $a_1 = a_2 b_2$ (for non-units $a_2, b_2 \in R \setminus R^\times$), where one of the factors, say a_2, is not a product of irreducible elements. Thus, $a_2 = a_3 b_3$, etc., and we obtain a chain of elements $a_1, a_2, a_3 \ldots \in R$ with $a_2 \mid a_1$, $a_3 \mid a_2$, $\ldots$, where a_i and a_{i+1} are not associated.

Now consider the ideal $I = (a_1, a_2, \ldots)$ generated by the a_i. By assumption, there is an element $a \in R$ with $I = (a)$, and thus there exist $m \in \mathbb{N}$ and $r_i \in R$ such that $a = r_1 a_1 + \ldots + r_m a_m$. Since $a_m \mid a_{m-1} \mid \cdots \mid a_1$, we have $a_m \mid a$. Since $a_{m+1} \in (a)$, there is an element $r \in R$ such that $a_m = ar$, i.e., with $a \mid a_{m+1}$. By construction of the a_i, we have $a_{m+1} \mid a_m$, and hence a_m and a_{m+1} are associated in contradiction to the construction of the a_i.

Now we show that irreducible elements are prime (UFD–2). To this end, let $a \in R$ be irreducible, and let $x, y \in R$ be given with $a \mid xy$ and $a \nmid x$; then we have to show that $a \mid y$. Now $(a, x) = (d)$ for some $d \in R$; thus $d \mid a$ and $d \mid x$. If we had $d \sim a$, it would follow that $a \mid x$ in contradiction to our assumption. Since a is irreducible, d must be a unit. Thus $d^{-1} \in R$, and therefore $1 = d^{-1}d \in (d) = (a, x)$, i.e., there exist $m, n \in R$ with $1 = ma + nx$. Multiplication by y yields $y = may + nxy$, and since $a \mid xy$, we find $a \mid y$. This is what we wanted to show. $\square$

An important property of principal ideal domains is the fact that they are Bézout domains:[2] A domain R is called a *Bézout domain* if for all $a, b \in R$ there exists a $d \sim \gcd(a, b)$ such that $d = ar + bs$ is an R-linear combination of a and b. Principal ideal domains are always Bézout domains: Given $a, b \in R$, we form the ideal $I = (a, b)$; since R is a principal ideal domain, there is an element $d \in R$ with $(a, b) = (d)$. We claim that $d \sim \gcd(a, b)$. In fact, since $a \in (d)$, there is a $t \in R$ with $a = dt$; this shows that $d \mid a$, and similarly we find that $d \mid b$, and hence d is a common divisor of a and b. On the other hand, $d \in (a, b)$ implies that there are elements $r, s \in R$ with $d = ar + bs$; if e is any common divisor of a and b, then e

---

[2]Étienne Bézout (1730–1783) was a French mathematician, an author of textbooks. Bézout proved the existence of Bézout elements for polynomial rings; in the case of integers, they already occurred in the work of Bachet.

divides $ar + bs = d$, and hence d is a greatest common divisor of a and b. Observe that we have proved the Bézout property en passant.

4.4 Euclidean Domains

In his *Lectures on number theory* [31, p. 20], Dirichlet (actually we do not know how much of this is due to Dedekind) discusses the foundations of elementary number theory and then writes the following:

> It is now clear that the whole structure rests on a single foundation, namely the algorithm for finding the greatest common divisor of two numbers. [...] any analogous theory, for which there is a similar algorithm for the greatest common divisor, must also have consequences analogous to those in our theory.

In order to show that some domain R is a unique factorization domain, we will at first use the Euclidean algorithm. A function $f : R \longrightarrow \mathbb{N}_0$ is called a *Euclidean function* if it has the following properties:

EA–1. $f(a) = 0$ if and only if $a = 0$.
EA–2. For all $a \in R$ and $b \in R \setminus \{0\}$, there exists a $c \in R$ such that $f(a - bc) < f(b)$.

If there exists a Euclidean function on R, then R is called a *Euclidean domain*.

The ring of integers $\mathbb{Z}$ is Euclidean with respect to the absolute value $| \cdot |$. Other examples of Euclidean domains will be given in the Exercises section. The first domain $R \neq \mathbb{Z}$ that was shown to be Euclidean was the ring $\mathbb{Q}[X]$ of polynomials with rational (or real) coefficients. The existence of a Euclidean algorithm in this domain was proved by the Dutch mathematician Simon Stevin (1548–1620). Stevin wrote almost a dozen textbooks and helped to popularize the decimal system in Europe.

Theorem 4.16 *Euclidean domains are principal ideal domains.*

Proof Let f be a Euclidean function on R, and let $A \subseteq R$ be an ideal in R. Among the elements in $A \setminus \{0\}$, there is one, say a, for which f is minimal (in fact, the values of f are natural numbers). We claim that $A = (a)$. Since $a \in A$, we clearly have $(a) \subseteq A$; it remains to prove the reverse inclusion. To this end, take an arbitrary $b \in A$; by EA–2, there is a $q \in R$ with $f(b - aq) < f(a)$; since $f(a)$ was chosen minimal on $A \setminus \{0\}$, we have $f(b - aq) = 0$, and EA–2 implies that $b = aq$. Thus $b \in (a)$, and since $b \in A$ was arbitrary, we even have $A \subseteq (a)$. $\square$

In particular, Euclidean domains have the Bézout property, i.e., given an ideal (a, b), an element $d \sim \gcd(a, b)$ can be written as $d = ar + bs$ with $r, s \in R$. The advantage of working in a Euclidean ring is that given $a, b \in R$, we can compute the greatest common divisor $d \sim \gcd(a, b)$ as well as the Bézout elements r and s using the Euclidean algorithm.

To this end, take elements $a, b \in R \setminus \{0\}$; applying the Euclidean algorithm, we find $q_0, r_1 \in R$ with $a - bq_0 = r_1$ and $f(r_1) < f(b)$. Similarly, there exist $q_1, r_2 \in R$ with $b - r_1 q_1 = r_2$ and $f(r_2) < f(r_1)$ (unless we already have $r_1 = 0$; in this case, $a = bq_1$ and $d = b = 0a + 1b$, so everything is trivial). Continuing in this way, we find a chain of equations

$$
\begin{aligned}
a - bq_0 &= r_1 & f(r_1) &< f(b), \\
b - r_1 q_1 &= r_2 & f(r_2) &< f(r_1), \\
r_1 - r_2 q_2 &= r_3 & f(r_3) &< f(r_2), \\
&\ \ \vdots & &\ \ \vdots \\
r_{n-2} - r_{n-1} q_{n-1} &= r_n & f(r_n) &< f(r_{n-1}) \\
r_{n-1} - r_n q_n &= r_{n+1} & f(r_{n+1}) &< f(r_n).
\end{aligned}
$$

Now the natural numbers $f(r_j)$ cannot become arbitrarily small; thus there exists an index $n \in \mathbb{N}$ with $r_{n+1} = 0$. We then claim that $r_n \sim \gcd(a, b)$. In fact, it follows from the last row that $r_n \mid r_{n-1}$, and then the next to last row gives $r_n \mid r_{n-2}$, and in this way we climb the ladder until we reach $r_n \mid r_1$, $r_n \mid b$ and $r_n \mid a$. Thus r_n is a common divisor of a and b.

Conversely, if d is any common divisor of a and b, then the first row tells us that $d \mid r_1$, the second $d \mid r_2$, etc., and eventually we reach $d \mid r_n$. In other words, r_n is a greatest common divisor.

It may be said that the definition of the greatest common divisor is chosen in such a way that the proof of this fundamental result on the Euclidean algorithm becomes essentially trivial.

We obtain the Bézout elements $r, s \in R$ as follows: We start with $r_n = r_{n-2} - r_{n-1} q_{n-1}$ and replace the r_j with the maximal index by the linear combination in the preceding row, in our case r_{n-1} by $r_{n-1} = r_{n-3} - r_{n-2} q_{n-2}$. Now we have written r_n as a linear combination of r_{n-2} and r_{n-3}. Next we replace r_{n-2} by $r_{n-2} = r_{n-4} - r_{n-3} q_{n-3}$, etc., until we finally have written r_n as an R-linear combination of a and b.

4.4.1   Summary

We have defined the following notions in quadratic number rings:

- divisibility and congruences,
- units and associate elements, and
- primes and irreducible elements.

Among the important results, we have obtained are the following:

- Primes are irreducible; the converse holds in unique factorization domains.
- We have the inclusions
  Unique Factorization Domains $\supset$ Principal ideal domains $\supset$ Euclidean domains.

Moreover we know that in unique factorization domains, there exist greatest common divisors $d = \gcd(a, b)$; in principal ideal domains, there exist Bézout elements: We can write the greatest common divisor as a $\mathbb{Z}$-linear combination of a and b: $d = am + bn$. Finally, in Euclidean domains, we have an algorithm for computing greatest common divisors as well as Bézout elements.

4.5 Exercises

4.1. In the ring $R = \mathbb{Z}[x]$ of polynomials, show that $x \mid f(x)$ for some $f \in R$ if and only if $f(0) = 0$. Show more generally that $(x - a) \mid f(x)$ if and only if $f(a) = 0$.

   Show that these properties continue to hold in polynomial rings $R = K[x]$ over fields K. What about polynomial rings over domains or arbitrary rings?

4.2. Show that (1.12) is also a counterexample to the Four Numbers Theorem in $\mathbb{Z}[\sqrt{-5}\,]$, whereas (1.11) is compatible with the Four Numbers Theorem in $\mathbb{Z}[\sqrt{-2}\,]$.

4.3. Let R be a domain. Consider the set S of pairs (p, q) and define an equivalence relation on S by $(p, q) \sim (r, s)$ if and only if $ps = qr$. On the set K of equivalence classes, define addition and multiplication via

   - $(p, q) + (r, s) = (ps + qr, qs)$;
   - $(p, q) \cdot (r, s) = (pr, qs)$.

   Show that this is well defined and that it makes K into a field with neutral elements $(0, 1)$ for addition and $(1, 1)$ for multiplication.

   Show that the map $\iota : R \longrightarrow K : r \to (r, 1)$ is an injective ring homomorphism. The field K is called the quotient field of R, and we may regard R as a subring of K via the embedding ι.

4.4. Let $R \subseteq S$ be domains, and let $a, b, m \in R$. Does $a \equiv b \bmod m$ in R imply the same congruence in S? Is the converse true?

4.5. Each fraction in $\mathbb{Q}$ can be reduced to lowest terms in a unique way; in $\mathbb{Z}[\sqrt{-5}\,]$, on the other hand, $\frac{1+\sqrt{-5}}{2} = \frac{3}{1-\sqrt{-5}}$, and both fractions are reduced to lowest terms. Find more such examples.

4.6. Let $\alpha, \beta \in \mathcal{O}_k$; show that $\alpha \mid N\alpha$. If moreover $\alpha \mid \beta$, then $N\alpha \mid N\beta$ (even in $\mathbb{Z}$).

4.7. Show that if $\sqrt{-2} \mid y$ in $\mathbb{Z}[\sqrt{-2}]$ for some $y \in \mathbb{Z}$, then $2 \mid y$.

Show more generally that $\sqrt{m} \mid y$, where m is squarefree, always implies that $m \mid y$.

Find a counterexample to the claim that $\alpha \mid y$ always implies $N\alpha \mid y$.

4.8. Show that $a + bi \equiv a + b \mod (1 + i)$ in $\mathbb{Z}[i]$.

4.9. Prove Proposition 4.1.

4.10. Prove Proposition 4.3.

4.11. Show that $a \mid b$ in $\mathbb{Z}$ implies $a \mid b$ in the ring of integers $\mathcal{O}_k$ in a quadratic number field k.

4.12. Show that the set of units $R^\times$ in some ring R is a group with respect to multiplication.

4.13. Show that if $R = K$ is a field, then $K^\times = K \setminus \{0\}$.

4.14. If R is a domain and $R[X]$ the ring of polynomials in one variable X with coefficients from R, then $R[X]^\times = R^\times$, that is, the units in this polynomial ring are all constant.

Show, on the other hand, that the polynomial $2X + 1$ in $(\mathbb{Z}/4\mathbb{Z})[X]$ is a unit.

4.15. Show that the unit groups of the domains $R = \mathbb{Z}[\sqrt{m}]$ for $m < -1$ are given by $R^\times = \{-1, +1\}$.

4.16. Let $\mathcal{O}_k$ be the ring of integers in a quadratic number field k, and let $E_k = \mathcal{O}_k^\times$ be its unit group. Show that E_k is a Gal $(k/\mathbb{Q})$-module (see Exercise 2.16).

4.17. Show: If R is a domain containing $\mathbb{Z}$, and if π is prime in R, then the smallest natural number divisible by π in R is a prime number.

4.18. Show that $N\alpha = 1$ for $\alpha = \frac{1+2i}{1-2i} \in \mathbb{Q}(i)$, but that α is not a unit in $\mathbb{Z}[i]$. Construct infinitely many such examples.

4.19. Show that $\mathbb{Z}$ is Euclidean with respect to the absolute value.

4.20. Show that the polynomial ring $K[x]$, where K is a field, is Euclidean with respect to $f(a) = 2^{\deg a}$, where $\deg a$ denotes the degree of $a \in K[x]$, and where we have set $\deg 0 = -\infty$ in order to have $2^{\deg 0} = 2^{-\infty} = 0$.

4.21. Discuss the examples $2 \cdot 3 = -\sqrt{-6} \cdot \sqrt{-6}$ in $\mathbb{Z}[\sqrt{-6}]$, $2 \cdot 3 = \sqrt{6} \cdot \sqrt{6}$ in $\mathbb{Z}[\sqrt{6}]$, and $2 \cdot 7 = (2 + \sqrt{-10})(2 - \sqrt{-10})$ in $\mathbb{Z}[\sqrt{10}]$ as in (1.12).

4.22. Consider the quadratic number field $k = \mathbb{Q}(\sqrt{m})$; which of the rational prime numbers $p \in \{2, 3, 5\}$ in $\mathcal{O}_k$ with $m \in \{-5, -3, -2, -1, 2, 3, 5\}$ are irreducible and which are not?

4.23. Show that elements $\pi \in \mathcal{O}_k$ are irreducible if $N\pi$ is a rational prime.

4.24. Let R be a unique factorization domain. Show:

a. $\gcd(a^2, b^2) = (\gcd(a, b))^2$ for all $a, b \in R$.
b. If $\gcd(a, b) = 1$, then $\gcd(a^2, b) = 1$.
c. $\gcd(a + b, b) = \gcd(a, b)$.
d. $\gcd(ra, rb) = r \gcd(a, b)$.

4.25. Show that the elements $a = 1 + \sqrt{-5}$ and $b = 1 - \sqrt{-5}$ do not have a common divisor except ± 1, but that 2 is a common divisor of a^2 and b^2.

4.26. Let S be the domain you obtain by adjoining the element $\omega = \frac{1}{2}(1 + \sqrt{-5})$ to $R = \mathbb{Z}[\sqrt{-5}]$. Show that $S = R[\frac{1}{2}]$ and $S \cap \mathbb{Q} = \mathbb{Z}[\frac{1}{2}]$.

Show moreover that the decomposition (1.12) is not an example for nonunique factorization into irreducible elements because $3 = \frac{1}{2}(1 - \sqrt{-5})(1+\sqrt{-5})$ is a factorization of 3 into the unit $\frac{1}{2}$ and the two irreducible (and even prime) elements $1 \pm \sqrt{-5}$. Explain the equation $3 \cdot 3 = (2 - \sqrt{-5})(2 + \sqrt{-5})$ by giving a factorization into irreducible elements.

4.27. Solve the Diophantine equation $x^2 + 5y^2 = z^2$ by setting $x + y\sqrt{-5} = (r + s\sqrt{-5})^2$ as Euler did, and show that the resulting parametrization $x = r^2 - 5s^2$, $y = 2rs$ does not yield all integral solutions of the equation.

Use the domain $S = \mathbb{Z}[\sqrt{-5}, \frac{1}{2}]$ from the preceding exercise for obtaining a complete parametrization of the solutions.

4.28. Prove Corollary 4.13. Hint: Try to obtain $a = pa_1$ and $b = p^{n-1}b_1$, and then apply Proposition 4.12 to a_1 and b_1.

4.29. Determine all integral points on the elliptic curve $4y^2 = x^3 + 1$, i.e., all pairs $(x, y) \in \mathbb{Z} \times \mathbb{Z}$ satisfying this equation.

4.30. Find all ring homomorphisms κ from $\mathbb{Z}[\sqrt{-5}]$ to $\mathbb{Z}/2\mathbb{Z}$, $\mathbb{Z}/3\mathbb{Z}$ and $\mathbb{Z}/5\mathbb{Z}$, and determine their kernels.

4.31. Show that the even integers $2\mathbb{Z}$ form an ideal in $\mathbb{Z}$. More generally, the sets $m\mathbb{Z}$ for arbitrary $m \in \mathbb{Z}$ are ideals in $\mathbb{Z}$.

4.32. Let (a) and (b) be principal ideals in some domain R. Show that $a \mid b$ if and only if $(a) \supseteq (b)$. Show moreover that this implies the equivalence of the following assertions:

  a. $(a) = (b)$;
  b. $a \mid b$ and $b \mid a$;
  c. $a = be$ for some unit $e \in R^{\times}$.

4.33. Show that the set

$$T = \left\{ \begin{pmatrix} a & b \\ 0 & d \end{pmatrix} : a, b, d \in \mathbb{Z} \right\}$$

is a subring of $R = M_2(\mathbb{Z})$, the ring of all 2×2-matrices with entries from $\mathbb{Z}$ (this ring is neither commutative nor a domain since it contains zero divisors), but that T is not an ideal in R. Hint: Consider the product of the identity matrix with a lower triangular matrix such as $\begin{pmatrix} 1 & 0 \\ 1 & 1 \end{pmatrix}$.

4.34. Let $R \subseteq S$ be domains. Show that $I \cap R$ is an ideal in R if I is an ideal in S.

4.35. If I is a nonzero ideal in the ring of integers $\mathcal{O}_k$ of a quadratic number field k, then I contains a natural number $\neq 0$. (Hint: Take the norm). Show that, on the other hand, the ideal (X) in the polynomial rings $\mathbb{Z}[X]$ and $\mathbb{Q}[X]$ does not contain any natural number $\neq 0$.

4.36. Show that the polynomial ring $\mathbb{Z}[x]$ admits a lot more homomorphisms into simpler rings than the rings of integers $\mathcal{O}_k$; show in particular that the reductions π_p modulo p and π_x modulo x yield the following commutative diagram:

$$\begin{array}{ccc}
\mathbb{Z}[x] & \xrightarrow{\ \pi_p\ } & \mathbb{F}_p[x] \\
\ \downarrow{\scriptstyle \pi_x} & & \ \downarrow{\scriptstyle \pi_p} \\
\mathbb{Z} & \xrightarrow{\ \pi_p\ } & \mathbb{F}_p
\end{array}$$

4.37. Let k be a quadratic number field. Show that $\mathbb{Z}$ is a subring of $\mathcal{O}_k$, but not an ideal in $\mathcal{O}_k$.

4.38. Show that the set $2\mathbb{Z} + \sqrt{2}\,\mathbb{Z}$ is an ideal in $\mathbb{Z}[\sqrt{2}\,]$ consisting of the multiples of $\sqrt{2}$. Show moreover that $\mathbb{Z} + 2\sqrt{2}\,\mathbb{Z}$ is a subring of $\mathbb{Z}[\sqrt{2}\,]$, but not an ideal.

4.39. An order $\mathcal{O}$ in some quadratic number field is a subring of $\mathcal{O}_k$ that properly contains $\mathbb{Z}$. Consider the set $\mathcal{F} = \{f \in \mathbb{Z} : f\omega \in \mathcal{O} \text{ for all } \omega \in \mathcal{O}_k\}$. Show that $\mathcal{F}$ is an ideal in $\mathbb{Z}$; the generator $f > 0$ of this ideal $\mathcal{F} = (f)$ is called the *conductor* of the order $\mathcal{O}$. Show that the maximal order $\mathcal{O}_k$ has conductor 1.

4.40. Show that $\gcd(2, x) = 1$ in the unique factorization domain $\mathbb{Z}[x]$ and that there do not exist associated Bézout elements, i.e., that there do not exist polynomials $p, q \in \mathbb{Z}[x]$ with $2p(x) + xq(x) = 1$.

Is $(2, x)$ a principal ideal in $\mathbb{Z}[x]$ or in $\mathbb{Q}[x]$?

4.41. Find ideals in $\mathbb{Z}[\sqrt{-6}\,]$, $\mathbb{Z}[\sqrt{-10}\,]$, and $\mathbb{Z}[\sqrt{10}\,]$ that are not principal.

4.42. Let R be the domain of all algebraic integers. Show that 2 does not possess a factorization into irreducible elements. Also show that the ideal $(2, \sqrt{2}, \sqrt[4]{2}, \sqrt[8]{2}, \ldots)$ is not principal in R and that it is not even finitely generated (this means that it is not generated by finitely many elements, i.e., it does not have the form $(a_1, \ldots, a_n)$ for suitable elements $a_j \in R$).

4.43. Let R be a domain containing $\mathbb{Z}$ (for example, $R = \mathcal{O}_k$). Show that if $a, b \in \mathbb{Z}$ are coprime in $\mathbb{Z}$, then they are also coprime in R. (Hint: Bézout).

4.44. Compute the Bézout elements for $\gcd(21, 15)$ in $\mathbb{Z}$.

4.45. For $n \geq 3$, compute the greatest common divisor of the polynomials $x^n + x^2 - 2$ and $x^2 - 1$ in $\mathbb{Z}[x]$ (the result will depend on n). How can the result that $x - 1$ is always a common divisor be verified in advance?

4.46. Let $\alpha, \beta \in \mathcal{O}_k$ and $(N\alpha, N\beta) = 1$ in $\mathbb{Z}$. Then $\gcd(\alpha, \beta) \sim 1$ in $\mathcal{O}_k$ even if $\mathcal{O}_k$ is not a unique factorization domain.

4.47. Bézout elements can be used for inverting residue classes. Assume for example that a and m are coprime integers; show how to find the inverse of the residue class $a \bmod m$ in $(\mathbb{Z}/m\mathbb{Z})^\times$ (i.e., the element $b \in \mathbb{Z}$ such that $ab \equiv 1 \bmod m$). Compute $\frac{1}{2} \bmod 21$ and $\frac{1}{5} \bmod 33$.

4.48. Study the equation $y^2 = x^3 + 9$ in integers.

4.49. Use the factorization $(y - k)(y + k) = x^3$ to deduce results on the integral solutions of the Diophantine equation $y^2 = x^3 + k^2$ for a fixed integer k. This is more of an open problem than an exercise. Do not despair if you cannot find a complete solution (and look for an error if you do).

4.50. For integers k, study the Diophantine equation $y^2 = x^3 - k^2$. You should be able to prove that this equation is solvable for $k = b(3a^2 - b^2)$ or $k = 2(a^3 + 3a^2b - 3ab^2 - b^3)$. For $k = 88$, there are two different representations $k = b(3a^2 - b^2)$, and hence there are at least two solutions of the equation also $y^2 = x^3 - k^2$ in this case. Can the number of solutions become arbitrarily large?

4.51. Solve the Diophantine equation $(1 + 8i)x + (5 + 4i)y = 1$ in $\mathbb{Z}[i]$.

Chapter 5
Arithmetic in Some Quadratic Number Fields

Although already Euler had used numbers of the form $a + b\sqrt{-2}$ for solving the Diophantine equation $y^2 + 2 = x^3$ in integers, it was Gauss who laid the foundations for the arithmetic of quadratic number rings such as $\mathbb{Z}[i]$ by defining prime elements and units and proving unique factorization for the first number ring strictly larger than the ordinary integers. He did so in his second memoir on biquadratic residues published in 1831.

In this chapter we will discuss a few examples of quadratic number rings. For the example of the ring of integers in $\mathbb{Q}(\sqrt{5})$, we refer to the dissertation by Dodd [32]; more examples may be found in Sommer's book [118] and in the still excellent introduction to number theory by Hardy and Wright [53].

5.1 The Gaussian Integers

We will start our journey through various quadratic number rings with the ring $\mathbb{Z}[i]$ of Gaussian integers.

5.1.1 $\mathbb{Z}[i]$ Is Norm-Euclidean

Consider the domain $R = \mathbb{Z}[i]$; we want to show that the norm is a Euclidean function in R. To this end, we have to find, for each $\alpha \in R$ and each $\beta \in R \setminus \{0\}$, an element $\gamma \in R$ such that

$$N(\alpha - \beta\gamma) < N(\beta). \tag{5.1}$$

F. Lemmermeyer, *Quadratic Number Fields*, Springer Undergraduate
Mathematics Series, https://doi.org/10.1007/978-3-030-78652-6_5

Since we are dealing with infinitely many pairs (α, β), this looks difficult. But if we divide (5.1) by $N(\beta)$ using the multiplicativity of the norm, we see that it is sufficient to find some $\gamma \in R$ for each $\xi = \alpha/\beta \in k = \mathbb{Q}(i)$ such that

$$N(\xi - \gamma) < 1. \tag{5.2}$$

If we can find such a $\gamma \in R$ for some ξ, then we can solve this inequality for any $\zeta \in k$ that differs from ξ by an integer $\eta \in R$ since

$$N((\xi - \eta) - (\gamma - \eta)) = N(\xi - \gamma) < 1.$$

Thus it is sufficient to consider only those $\xi \in k$ that have the form $\xi = x + yi$ with $|x|, |y| \leq \frac{1}{2}$. We claim that for all such ξ, a single element γ suffices, namely $\gamma = 0$. In fact, we have

$$N(\xi - \gamma) = N(\xi) = x^2 + y^2 \leq \frac{1}{4} + \frac{1}{4} = \frac{1}{2} < 1.$$

Thus $\mathbb{Z}[i]$ is Euclidean with respect to the norm, and in particular it is a unique factorization domain.

In the plane of Gaussian numbers, the elements $\xi \in \mathbb{Q}(i)$ with $N(\xi) \leq \frac{1}{2}$ lie inside a circle with radius $1/\sqrt{2}$ around the origin. If we place a circle with this radius around each lattice point, i.e., around each point $a + bi$ with $a, b \in \mathbb{Z}$ (see Fig. 5.1), then these circles cover the whole plane. This implies that for each $\xi \in \mathbb{Q}(i)$, there is a $\gamma \in \mathbb{Z}[i]$ satisfying (5.2), and in fact we may even demand that $N(\xi - \gamma) \leq \frac{1}{2}$.

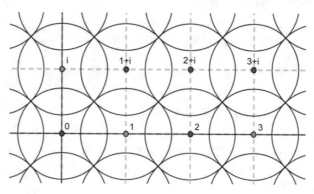

Fig. 5.1 The covering of $\mathbb{Z}[i]$ with circles of radius $\frac{1}{\sqrt{2}}$

This proof that $\mathbb{Z}[i]$ is Euclidean is constructive: We can use it for finding the greatest common divisor of two Gaussian integers. In order to compute, e.g., $\gcd(1 + 12i, 7 + 4i)$, we find the nearest integer to

$$\frac{1 + 12i}{7 + 4i} = \frac{(1 + 12i)(7 - 4i)}{(7 + 4i)(7 - 4i)} = \frac{55 + 80i}{65} = \frac{11 + 16i}{13},$$

which is $1+i$, and we obtain, as the first step in the Euclidean algorithm, the equation

$$1 + 12i = (1 + i)(7 + 4i) + (-2 + i).$$

The next step consists of the observation that

$$\frac{7 + 4i}{-2 + i} = \frac{(7 + 4i)(2 + i)}{(-2 + i)(2 + i)} = -2 - 3i.$$

Thus the Euclidean algorithm produces the chain of equations

$$1 + 12i = (1 + i)(7 + 4i) + (-2 + i),$$
$$7 + 4i = (-2 - 3i)(-2 + i),$$

which implies that $\gcd(1 + 12i, 7 + 4i) \sim -2 + i \sim 1 + 2i$.

Prime Elements and Associates Now that we know that $R = \mathbb{Z}[i]$ is a Euclidean domain and therefore has the unique factorization property, we would like to determine the primes in R. We start with an observation valid for all quadratic number fields.

Proposition 5.1 *Let $\mathcal{O}_k$ be the ring of integers in some quadratic number field k. Then for each prime $\pi \in \mathcal{O}_k$, there is a unique prime number $p \in \mathbb{N}$ with $\pi \mid p$. In particular, $N\pi = \pm p$ or $N\pi = \pm p^2$.*

Proof Since $\pi \mid N\pi$, we see that π divides the prime factor p of $N\pi \in \mathbb{Z}$; if we also had $\pi \mid q$ for a different prime $q \neq p$, then π would divide $\gcd(p, q) = 1$, which is impossible since π is not a unit. The second claim follows easily from $\pi \mid p$ by taking the norm. In fact, we find $N\pi \mid p^2$ in $\mathbb{Z}$, and since $N\pi \neq \pm 1$ (otherwise π would be a unit), we are left with the possibilities $N\pi = \pm p$ and $N\pi = \pm p^2$. $\square$

Thus there exist the following possibilities:

(1) p is prime in $\mathcal{O}_k$; then $Np = p^2$, and we say that p is inert in k.
(2) p is not prime in $\mathcal{O}_k$, but irreducible; this can only happen if $\mathcal{O}_k$ is not a unique factorization domain.
(3) p is reducible $\mathcal{O}_k$.

Let us have a close look at the third case. Here $p = \alpha\beta$ for non-units $\alpha, \beta \in R$. It follows from $N\alpha N\beta = p^2$ that $N\alpha = N\beta = \pm p$ (if one factor had norm 1, it would be a unit); finally, $\pm p = N\alpha = \alpha\alpha'$ shows that we must have $\beta = \pm\alpha'$. Thus if we write π instead of α, then $\pm p = \pi\pi'$, where π and π' are primes with norm $\pm p$. The only question remaining is whether π and π' are distinct prime elements or whether they are associated.

Below we will discuss this question in general; here we are content with studying the prime elements in $\mathbb{Z}[i]$. We claim:

Proposition 5.2 *Let $p \in \mathbb{N}$ be a rational prime number. Then, there are the following possibilities:*

1. *$p = 2$: Then, p is reducible in $\mathbb{Z}[i]$. In fact, $2 = i(1 - i)^2$, and $\pi = 1 - i$ is, up to associates, the only prime element dividing 2.*
2. *$p \equiv 3 \bmod 4$: Then, p is inert, i.e., it remains prime in $\mathbb{Z}[i]$ and has norm p^2.*
3. *$p \equiv 1 \bmod 4$: Then, $p = \pi\pi'$ for prime elements $\pi = a + bi$ and $\pi' = a - bi$ in $\mathbb{Z}[i]$. Here the primes π and π' are not associated.*

Proof The first claim is easily verified. For proving the second claim, we assume that $p \equiv 3 \bmod 4$ is not prime in $\mathbb{Z}[i]$; since this is a unique factorization domain, p must be reducible, and hence $\pm p = N\pi$ for some prime $\pi = a + bi$. Clearly, the positive sign must hold, and then $p = a^2 + b^2$, which is never $\equiv 3 \bmod 4$ since squares are always $\equiv 0, 1 \bmod 4$: a contradiction.

Now assume that $p \equiv 1 \bmod 4$. By Euler's criterion, -1 is a quadratic residue modulo p, and hence there is an $x \in \mathbb{N}$ with $x^2 \equiv -1 \bmod p$ (this also follows easily from the existence of a primitive root g modulo p: since $g^{(p-1)/2} \equiv -1 \bmod p$, the congruence class $x \equiv g^{(p-1)/4} \bmod p$ solves the congruence $x^2 \equiv -1 \bmod p$). This implies that $x^2 + 1 = (x + i)(x - i)$ is divisible by p. Since none of the two factors is divisible by p in $\mathbb{Z}[i]$, p cannot be prime in $\mathbb{Z}[i]$, and since this ring is a unique factorization domain, p must be reducible. Thus $p = \pi\pi'$ for some element $\pi = a + bi$. If $\pi \sim \pi'$, then $\pi'/\pi = \pi'^2/p = (a^2 - b^2 + 2abi)/p$ must be integral, and hence $p \mid a^2 - b^2$ and $p \mid ab$. The second condition yields $p \mid a$ or $p \mid b$, which implies that we have $p \mid a$ and $p \mid b$, hence $p \mid \pi$. Taking norms now gives a contradiction. $\square$

As a corollary we obtain the following:

Corollary 5.3 (Two-Squares Theorem by Fermat and Euler) *Each prime number of the form $4n + 1$, where $n \in \mathbb{N}$, is a sum of two square numbers.*

We also remark that the Euclidean algorithm provides us with a method for computing the representation of a prime $p = 4n + 1$ as a sum of two squares from a solution of the congruence $x^2 \equiv -1 \bmod p$. In fact, all we have to do is compute $\gcd(x + i, p) = a + bi$ in $\mathbb{Z}[i]$ because then $p = a^2 + b^2$.

Another consequence of unique factorization in $\mathbb{Z}[i]$ is:

Corollary 5.4 (Euler's Factorization Theorem) *If x and y are coprime and $m = x^2 + y^2$, then to each factorization $m = p_1 \cdots p_t$ into primes there corresponds a factorization $\mu = x + yi = \pi_1 \cdots \pi_t$ such that $p_j = N\pi_j$ for $1 \leq j \leq t$.*

Quadratic Residues We will now briefly look at Fermat's Little Theorem in the domain $\mathbb{Z}[i]$ of Gaussian integers. It is easily checked that, for odd prime numbers p, we have

$$(a + bi)^p \equiv \begin{cases} a + bi \bmod p, & \text{if } p \equiv 1 \bmod 4, \\ a - bi \bmod p, & \text{if } p \equiv 3 \bmod 4. \end{cases} \tag{5.3}$$

This immediately implies:

Theorem 5.5 (Fermat's Little Theorem) *Let $\pi \in \mathbb{Z}[i]$ be prime. Then all elements $\alpha \in \mathbb{Z}[i]$ not divisible by π satisfy the congruence*

$$\alpha^{N\pi - 1} \equiv 1 \bmod \pi.$$

For the proof of the second claim, we observe that $(a + bi)^p \equiv a - bi \bmod \pi$, and hence $(a + bi)^{p^2} \equiv a + bi \bmod p$. If $\alpha = a + bi$ is not divisible by p, we are allowed to cancel α in this congruence.

We also have, in analogy with the ordinary integers, the following proposition.

Proposition 5.6 (Euler's Criterion) *If $\pi \in \mathbb{Z}[i]$ is prime with odd norm and if $\alpha \in \mathbb{Z}[i]$ not divisible by π, then the following assertions are equivalent:*

1. *α is a quadratic residue modulo π, i.e., the congruence $\alpha \equiv \xi^2 \bmod \pi$ is solvable with $\xi \in \mathbb{Z}[i]$;*
2. *The congruence $\alpha^{(N\pi - 1)/2} \equiv 1 \bmod \pi$ holds.*

This result allows us to define the quadratic residue symbol $\left[\frac{\alpha}{\pi}\right]$ with values in $\{\pm 1\}$ by the congruence

$$\alpha^{(N\pi - 1)/2} \equiv \left[\frac{\alpha}{\pi}\right] \bmod \pi.$$

Dirichlet has shown how to derive the quadratic reciprocity law in $\mathbb{Z}[i]$ first formulated by Gauss from the reciprocity law in ordinary integers by a few simple calculations (see [77]).

Theorem 5.7 (Quadratic Reciprocity Law) *If π and λ are non-associated primes with odd norm in $\mathbb{Z}[i]$, and if $\pi \equiv \lambda \equiv 1 \bmod 2$, then*

$$\left[\frac{\lambda}{\pi}\right] = \left[\frac{\pi}{\lambda}\right],$$

The generalization of the quadratic reciprocity law to general quadratic number fields is quite technical but can be done in a similar way. The generalization to arbitrary number fields, on the other hand, requires much deeper means and leads, as Hilbert has shown, directly to class field theory; see for example the last chapter in Hecke's introduction to algebraic number theory [60].

5.1.2 Fermat's Last Theorem in Quadratic Number Fields

Certain Fermat equations do have solutions in quadratic number fields; we can easily verify, for example, that

$$\left(\frac{1+\sqrt{-7}}{2}\right)^4 + \left(\frac{1-\sqrt{-7}}{2}\right)^4 = 1^4 \qquad\qquad \text{in} \quad \mathbb{Z}[\tfrac{1+\sqrt{-7}}{2}],$$

$$\left(\frac{9+\sqrt{-31}}{2}\right)^3 + \left(\frac{9-\sqrt{-31}}{2}\right)^3 = (-3)^3 \qquad \text{in} \quad \mathbb{Z}[\tfrac{1+\sqrt{-31}}{2}],$$

$$(5 - 9\sqrt{5})^3 + (12\sqrt{5})^3 = (5 + 9\sqrt{5})^3 \qquad \text{in} \quad \mathbb{Z}[\sqrt{5}].$$

On the other hand, the cubic and the quintic Fermat equations $x^3 + y^3 = z^3$ and $x^5 + y^5 = z^5$ do not have nontrivial solutions in $\mathbb{Z}[\tfrac{1+\sqrt{-3}}{2}]$ and $\mathbb{Z}[\tfrac{1+\sqrt{5}}{2}]$, respectively; we will prove these claims below. In this section we will solve (following Hilbert [61]) the Fermat equation with exponent 4.

Theorem 5.8 *The equation $\alpha^4 + \beta^4 = \gamma^2$ has only trivial solutions in $\mathbb{Z}[i]$.*

If $a + bi \in \mathbb{Z}[i]$ has odd norm, then a and b have different parity, and hence $a^2 - b^2 \equiv \pm 1 \bmod 4$ and $2ab \equiv 0 \bmod 4$; this implies $(a+bi)^2 = a^2 - b^2 + 2abi \equiv \pm 1 \bmod 4$. If the elements $\alpha, \beta \in \mathbb{Z}[i]$ have odd norm, then $\alpha^4 + \beta^4 \equiv 2 \bmod 4$, but 2 is not a square modulo 4 (the only squares modulo 4 are 0, $2i$, and ± 1). Thus we may assume that β has even norm.

We will now show that if the equation $\alpha^4 + \varepsilon\lambda^{4n}\beta^4 = \gamma^2$ is solvable, where $\varepsilon \in \{\pm 1, \pm i\}$ is a unit, $\lambda = 1 + i$, and where β is not divisible by $1 + i$, then

(1) $n \geq 2$;
(2) there exist $\alpha_1, \beta_1, \gamma_1 \in \mathbb{Z}[i]$ and a unit ε_1 with

$$\alpha_1^4 + \varepsilon_1\lambda^{4(n-1)}\beta_1^4 = \gamma_1^2,$$

where β_1 is not divisible by $1 + i$.

By applying (2) sufficiently often, we will find a solution in which the exponent of λ vanishes, and this contradicts (1).

For proving the first claim, we assume that

$$\alpha^4 + \varepsilon\lambda^4\beta^4 = \gamma^2$$

with $\lambda \nmid \beta$. Then $\alpha^4 \equiv 1 \bmod 8$ implies $1 + \varepsilon\lambda^4 \equiv \gamma^2 \bmod \lambda^6$. Clearly, $\gamma^2 \not\equiv -1 \bmod 4$, and hence we must have $\gamma \equiv 1 \bmod 2$. But then $\gamma^2 - 1 = (\gamma - 1)(\gamma + 1)$ is divisible by $\lambda^2 \cdot \lambda^3 = \lambda^5$, and this implies $\lambda^5 \mid \varepsilon\lambda^4\beta^4$. Thus $\lambda \mid \beta$, and this proves the claim.

Now assume that

$$\varepsilon \lambda^{4n} \beta^4 = \gamma^2 - \alpha^4 = (\gamma - \alpha^2)(\gamma + \alpha^2)$$

for some $\beta \neq 0$. Since $\gamma^2 \equiv 1 \bmod 4$, we have $\gamma \equiv \alpha^2 \equiv 1 \bmod 2$, and hence both factors on the right are divisible by λ^2. Since any common divisor divides their sum $2\gamma^2$, these factors have greatest common divisor λ^2. Replacing γ by $-\gamma$ if necessary, we thus have

$$\gamma + \beta^2 = \eta \lambda^{4n-2} \zeta^4, \quad \gamma - \beta^2 = \eta' \lambda^2 \xi^4$$

for units η and η'. Subtracting these equations, we obtain

$$2\beta^2 = \eta \lambda^{4n-2} \zeta^4 - \eta' \lambda^2 \xi^4.$$

Dividing through by 2, we get, using $\lambda^2 = 2i$,

$$\beta^2 = \eta i \lambda^{4(n-1)} \zeta^4 - \eta' i \xi^4.$$

Since $\beta^2 \equiv 1 \bmod 2$, we deduce that $\eta' i \equiv 1 \bmod 2$, and hence $\eta' i = \pm 1$. Dividing through by $-1 = i^2$ if necessary, we now have

$$\eta_1 \lambda^{4(n-1)} \zeta^4 + \xi^4 = \beta^2.$$

This proves our claims.

5.2 The Eisenstein Integers

The domain $\mathbb{Z}[\rho]$, where $\rho = \frac{-1+\sqrt{-3}}{2}$ is a primitive cube root of unity, is called the ring of Eisenstein integers. Gotthold Eisenstein (1823–1852) died at a very young age (as did his contemporaries Galois, Abel, and Riemann). He is best known for a geometric interpretation of a key lemma in Gauss's third proof of the quadratic reciprocity law,[1] his irreducibility criterion (which actually goes back to Theodor Schönemann (1812–1868); see [26]), and for the Eisenstein series in the theory of modular forms. Eisenstein used the domain $\mathbb{Z}[\rho]$ in his proof of the cubic reciprocity law.

$\mathbb{Z}[\rho]$ **Is Norm-Euclidean** For verifying the criterion given by Eq. (5.2), we have to show that for each $\xi = x + y\sqrt{-3} \in k = \mathbb{Q}(\sqrt{-3})$, there exists a $\gamma = \frac{1}{2}(a + b\sqrt{-3}) \in \mathcal{O}_k$ (here we have $a \equiv b \bmod 2$) with $N(\xi - \gamma) < 1$. Now $\xi - \gamma =$

---

[1] See the beautiful article [72].

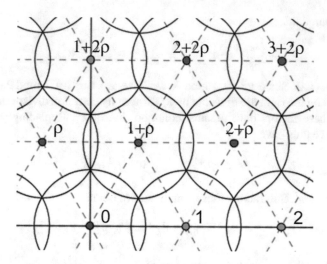

Fig. 5.2 The ring of Eisenstein integers is norm-Euclidean

$\frac{1}{2}((2x - a) + (2y - b)\sqrt{-3})$; clearly, we can choose $b \in \mathbb{Z}$ in such a way that $|2y - b| \leq \frac{1}{2}$. Next we have to determine an integer $a \in \mathbb{Z}$ with $a \equiv b \bmod 2$ in such a way that $|2x - a|$ becomes small. By choosing a from the integers $\equiv b \bmod 2$, we can make $|2x - a| \leq 1$ (the nearest integer with given parity has at most distance 1 from $2x$). But then $N(\xi - \gamma) \leq \frac{1}{4}(1 + \frac{3}{4}) = \frac{7}{16} < 1$.

In the diagram in Fig. 5.2, each number $x + y\sqrt{-3}$ corresponds to the point $(x, y\sqrt{3})$ in $\mathbb{R}^2$; the domain $\mathbb{Z}[\rho]$ then corresponds to a 2-dimensional lattice, and we have drawn a circle with radius $1/\sqrt{3}$ around each lattice point. These circles cover the whole plane, and hence the constant $\frac{7}{16}$ in the proof above may be improved to $\frac{1}{3}$, and we can always find an integral γ such that $N(\xi - \gamma) \leq \frac{1}{3}$. The figure also shows that this bound is best possible, since for smaller values, we lose for example the point corresponding to $\xi = \frac{1}{3}\sqrt{-3}$.

Prime Elements and Associates Since the domain $R = \mathbb{Z}[\rho]$ contains only six units, namely ± 1, $\pm \rho$, and $\pm \rho^2$, each nonzero element has six associates. If we write $\alpha = a + b\rho$, then we find

$$\begin{aligned}
\alpha &= a + b\rho & -\alpha &= -a - b\rho \\
\alpha\rho &= -b + (a - b)\rho & -\alpha\rho &= b + (b - a)\rho\,. \\
\alpha\rho^2 &= b - a - a\rho & -\alpha\rho^2 &= a - b + a\rho
\end{aligned}$$

Now $\sqrt{-3} = \rho - \rho^2$ is a prime element with $3 = -(\sqrt{-3})^2$, whereas $\lambda^2 = -3\rho$ for the element $\lambda = 1 - \rho$.

When is an element $\alpha = a + b\rho$ not divisible by λ? Since $\alpha = a + b\rho = a + b - b(1 - \rho) \equiv a + b \bmod \lambda$, this is true if and only if $a + b \not\equiv 0 \bmod 3$. In this case one of the three numbers a, b, or $a - b$ is divisible by 3, and the list above shows that there is an associate of α whose coefficient of ρ is divisible by 3.

Proposition 5.9 *If $\alpha \in \mathbb{Z}[\rho]$ is not divisible by $\sqrt{-3}$, then there is a $t \in \{0, 1, 2\}$ such that $\rho^t \alpha = a + b\rho$ with $b \equiv 0$ mod 3.*

If $\alpha = a + b\rho$ and if b is a multiple of 3, then 2α can be written in the form $2\alpha = L + 3M\sqrt{-3}$. In particular, $4N\alpha = L^2 + 27M^2$.

On the other hand, it is clear that at least one of the three numbers a, b, and $a - b$ is even; the same argument as above then shows the following:

Proposition 5.10 *For each $\alpha \in \mathbb{Z}[\rho]$, there is a $t \in \{0, 1, 2\}$ such that $\rho^t \alpha = a + b\rho$ with $b \equiv 0$ mod 2. In other words, α has an associate of the form $c + d\sqrt{-3}$ with $c, d \in \mathbb{Z}$.*

The determination of prime elements in $\mathbb{Z}[\rho]$ proceeds exactly as for $\mathbb{Z}[i]$, so it will be sufficient to record the result and leave the details for the readers. We will, however, at least explain how to show that the congruence $x^2 \equiv -3$ mod p is solvable for primes $p \equiv 1$ mod 3. To this end, we set $r = g^{(p-1)/3}$, where g is a primitive root modulo p. Clearly, $r^3 \equiv 1$ mod 3, i.e., r is a primitive cube root of unity in $\mathbb{Z}/3\mathbb{Z}$. If ρ denotes a primitive cube root of unity in $\mathbb{C}$, then we know how to construct a square root of -3: we simply write $\lambda = 2\rho + 1$ and obtain $\lambda^2 = 1 + 4\rho + 4\rho^2 = -3$. This suggests that we set $x = 2r + 1$; then $x^2 = 1 + 4r + 4r^2 = -3 + 4(1 + r + r^2)$. It is therefore sufficient to show that $S = 1 + r + r^2 \equiv 0$ mod p. But it follows from $rS = r + r^2 + r^3 \equiv r + r^2 + 1 = S$ mod p that $p \mid (r - 1)S$. Since $r \not\equiv 1$ mod p, we conclude that S is divisible by p as claimed.

Proposition 5.11 *The domain $\mathbb{Z}[\rho]$ is Euclidean and thus a unique factorization domain. The prime elements in this domain are the following:*

1. *$\lambda = 1 - \rho = \sqrt{-3}\rho^2$ is the prime dividing 3;*
2. *the inert primes $q \equiv 2$ mod 3;*
3. *the elements π and π' with $\pi\pi' = p$, where p is a prime number $\equiv 1$ mod 3.*

As a corollary, we obtain Fermat's Little Theorem

$$\alpha^{N\pi - 1} \equiv 1 \text{ mod } \pi$$

for all $\alpha \in \mathbb{Z}[\rho]$ not divisible by the prime π. In particular, we observe that if $\pi = 2$; then

$$\alpha^3 \equiv 1 \text{ mod } 2$$

for all $\alpha \in \mathbb{Z}[\rho]$ not divisible by 2.

5.2.1   The Cubic Fermat Equation $x^3 + y^3 + z^3 = 0$

Euler has given a proof of Fermat's claim that the Diophantine equation

$$x^3 + y^3 + z^3 = 0 \tag{5.4}$$

has only trivial solutions (those with $xyz = 0$) in integers. In this proof he has used properties of numbers of the form $c^2 + 3d^2$ for which he did not give complete proofs.[2] The first proof of Fermat's Last Theorem for cubes using the arithmetic of $\mathbb{Z}[\rho]$ was given by Gauss, who showed the stronger result that Eq. (5.4) does not have nontrivial solutions in the larger ring $\mathbb{Z}[\rho]$. In the following, we will give a rigorous version of Euler's proof using the methods of Gauss. The idea behind the proof goes back to Fermat, who called his method *infinite descent*.

For proving that a Diophantine equation in many variables $x, y, z, \ldots$ does not have a solution in integers, assume you do have such a solution $(x, y, z, \ldots)$, and then show that for each such solution there is a *smaller* solution $(u, v, w, \ldots)$ (smaller in the sense that e.g. $|u| < |x|$). Since natural numbers cannot decrease indefinitely, this results in a contradiction. For a simple application of this technique, see Exercise 5.7.

Theorem 5.12 *The Diophantine equation $x^3 + y^3 + z^3 = 0$ has only trivial solutions in integers.*

Instead of proving this theorem for $x, y, z \in \mathbb{Z}$, we show (as Gauss) that the cubic Fermat equation does not have a solution in the domain $\mathbb{Z}[\rho]$. This follows immediately by setting $\alpha = x^3$, $\beta = y^3$, and $\gamma = z^3$ in the following theorem, whose smooth formulation I learned from Paul Monsky[3].

Theorem 5.13 *Let $\alpha, \beta, \gamma \in \mathbb{Z}[\rho] \setminus \{0\}$. If $\alpha + \beta + \gamma = 0$ and $\alpha\beta\gamma = \mu^3$ for some $\mu \in \mathbb{Z}[\rho]$, then $\alpha\beta\gamma = 0$ or $\alpha\beta\gamma = 1$.*

Proof We may assume that α, β, and γ are pairwise coprime. Among all counterexamples, we choose one for which $N(\alpha\beta\gamma)$ is minimal. Then there exist $A_1, B_1, C_1 \in \mathbb{Z}[\rho]$ with

$$\alpha = \rho^a A_1^3, \quad \beta = \rho^b B_1^3, \quad \gamma = \rho^c C_1^3.$$

If (α, β, γ) is a solution, then so is $(\alpha/\rho^a, \beta/\rho^a, \gamma/\rho^a)$; thus we may assume that $a = 0$.

---

[2] The gap is the one that we have pointed out in Chap. 1, namely the missing proof for the decomposition theorem for numbers of the form $x^2 + 3y^2$: If $c^2 + 3d^2 = r^3$, then there exist integers p and q with $c = p(p^2 - 9q^2)$ and $d = 3q(p^2 - q^2)$; see [10].
[3] See https://mathoverflow.net/questions/39561.

Since $\alpha\beta\gamma = \mu^3$, we must have $b + c \equiv 0 \bmod 3$, and there remain the possibilities $(b, c) = (0, 0), (1, 2), (2, 1)$. Switching the roles of β and γ if necessary, we may assume that $(b, c) = (0, 0)$ or $(b, c) = (1, 2)$.

(a) $(b, c) = (0, 0)$. In this case we have $a = b = c = 0$, and hence

$$\alpha = A_1^3, \quad \beta = B_1^3, \quad \gamma = C_1^3.$$

We set $\alpha_1 = B_1 + C_1, \beta_1 = \rho B_1 + \rho^2 C_1$ and $\gamma_1 = \rho^2 B_1 + \rho C_1$. Then

- $\alpha_1 + \beta_1 + \gamma_1 = B_1(1 + \rho + \rho^2) + C_1(1 + \rho + \rho^2) = 0$;
- $\alpha_1 \beta_1 \gamma_1 = B_1^3 + C_1^3 = \beta + \gamma = -\alpha = (-A_1)^3$;
- $\beta_1 + \gamma_1 = (B_1 + C_1)(\rho + \rho^2) = -(B_1 + C_1) \neq 0$ since $\beta + \gamma = -\alpha \neq 0$.
- $N(\alpha_1 \beta_1 \gamma_1) = N(\alpha) \mid N(\alpha\beta\gamma)$; in particular, we have $N(\alpha_1 \beta_1 \gamma_1) \leq N(\alpha\beta\gamma)$, and if we had equality here, we would have $N(\beta) = N(\gamma) = 1$. Thus B_1 and C_1 are units, and hence $\beta, \gamma = \pm 1$. This is only possible if (α, β, γ) is, up to a permutation, equal to $(0, 1, -1)$, which contradicts our assumption.

(b) $(b, c) = (1, 2)$. In this case we have

$$\alpha = A_1^3, \quad \beta = \rho B_1^3, \quad \gamma = \rho^2 C_1^3.$$

We set $\alpha_1 = B_1 + C_1, \beta_1 = \rho\alpha$, and $\gamma_1 = \rho^2\alpha$. Then,

- $\alpha_1 + \beta_1 + \gamma_1 = B_1(1 + \rho + \rho^2) + C_1(1 + \rho + \rho^2) = 0$;
- $\alpha_1 \beta_1 \gamma_1 = (B_1 + C_1)^3$;
- $\beta_1 + \gamma_1 = (B_1 + C_1)(\rho + \rho^2) = -(B_1 + C_1) \neq 0$ since $\beta + \gamma = -\alpha \neq 0$.
- $N(\alpha_1 \beta_1 \gamma_1) = N(\alpha) \mid N(\alpha\beta\gamma)$; in particular, we have $N(\alpha_1 \beta_1 \gamma_1) \leq N(\alpha\beta\gamma)$, and if we had equality here, we would have $N(\beta) = N(\gamma) = 1$.
  Thus B_1 and C_1 are units, and hence $\beta = \pm\rho$ and $\gamma = \pm\rho^2$. This is only possible if (α, β, γ) is, up to sign, equal to $(1, \rho, \rho^2)$.

Thus $(\alpha_1, \beta_1, \gamma_1)$ is a solution with $N(\alpha_1) < N(\alpha)$ contrary to our assumption that the solution we started with is one for which $N\alpha$ is minimal. This contradiction completes the proof. $\square$

With the same method, we can show that the equation $x^3 + y^3 = 3z^3$ has only trivial solutions; similarly, the only integral solutions of $x^3 + y^3 = 2z^3$ with $xyz \neq 0$ are (x, x, x). Both results are due to Adrien-Marie Legendre (1752–1833), who first stated the quadratic reciprocity law the way we know it, and whose main claim to fame are his textbook in number theory and his contributions to the theory of elliptic functions (or rather elliptic integrals). Legendre also claimed that the equation $x^3 + y^3 = az^3$ has only trivial solutions for $a = 3, 4, 5, 6, 8, \ldots$ Théophile Pépin later observed that $17^3 + 37^3 = 6 \cdot 21^3$. Finally, Trygve Nagell has shown that for $a > 2$, the equation $x^3 + y^3 = az^3$ either has no nontrivial solution or infinitely many primitive solutions. A solution $(x, y, z) \in \mathbb{Z}^3$ is called primitive if x, y, z are pairwise coprime. Any solution $(x, y, z) \in \mathbb{Z}^3 \setminus \{(0, 0, 0)\}$ gives rise to infinitely many non-primitive solutions (kx, ky, kz) for $k \in \mathbb{Z}$.

Another result can be found in the book [14] on elliptic curves by J.W.S.Cassels. There you can find a sketch of a proof that the equation $x^3 + y^3 = q_1 q_2 z^3$, where $q_1 \equiv 2 \bmod 9$ and $q_2 \equiv 5 \bmod 9$ are prime, has only trivial solutions. It is natural to ask how this is connected to elliptic curves. In fact, the cubic Fermat equation $x^3 + y^3 = z^3$ is an elliptic curve: Dividing through by z and setting $r = x/z$, $s = y/z$, we obtain $r^3 + s^3 = 1$; with $r = u + v$ and $s = u - v$, we get $2u^3 + 6uv^2 = 1$, and hence $2 + 6(v/u)^2 = 1/u^3$. Multiplying through by 6^3 and setting $Y = 36v/u$, $X = 6/u$, we obtain the equation of an elliptic curve $Y^2 = X^3 - 432$ in the well-known Weierstrass form.

5.3 The Lucas–Lehmer Test

It is known since Euclid that there is no largest prime number; nevertheless, there usually is a *largest known* prime number, mainly because there is no simple formula for computing arbitrarily large primes. Fermat once believed to have found such a formula: He conjectured that all numbers of the form $F_n = 2^{2^n} + 1$ are prime (and in fact he almost believed to have a proof). Euler later showed that $F_5 = 2^{32} + 1 = 641 \cdot 6700417$ is composite, and in fact, no other Fermat prime beyond F_4 has been discovered until now. For quite a few years now, the largest known prime number always has been a prime number of the form $2^p - 1$, where p is prime; such numbers are called Mersenne numbers.

Marin Mersenne (1588–1648) was a French priest. He corresponded with most mathematicians and many scientists of his time, in particular with Fermat. He is known for his conjecture that $p = 2, 3, 5, 7, 13, 17, 19, 31, 67, 127$, and 257 (these are all numbers that differ at most by 3 from a power of 2) are the only prime numbers ≤ 257 for which $2^p - 1$ is prime. Later it was shown that the Mersenne numbers for $p = 67$ and $p = 257$ are composite and that $p = 61, 89$, and 107 give rise to primes. The smallest Mersenne number not completely factored today is M_{1207}; this number has the prime factors $131\,071, 228\,479, 48\,544\,121$, and $212\,885\,833$, and the remaining factor is a composite number with 337 digits.

It is easy to show that $2^p - 1$ is composite if p is. This follows from the fact that $2^a - 1$ always divides $2^{ab} - 1$ since

$$x^{ab} - 1 = (x^a - 1)(x^{ab-a} + x^{ab-2a} + \ldots + x^a + 1).$$

The reason why the largest known prime is usually a Mersenne prime is that there is a very effective primality test for such numbers developed by Édouard Lucas (1842–1891) and Derrick Lehmer (1905–1991). In fact, the number $M_p = 2^p - 1$ (where $p \geq 3$) is prime if and only if $S_{p-1} \equiv 0 \bmod M_p$, where the sequence S_n is defined recursively by $S_1 = 4$ and $S_{n+1} = S_n^2 - 2$. Using this test, Lucas was able to show that $2^{127} - 1$ is prime.

Example Let $p = 5$; then $M_5 = 31$, and we find

$$S_1 = 4$$

$$S_2 = 14$$

$$S_3 = 194 \equiv 8 \bmod 31$$

$$S_4 \equiv 62 \equiv 0 \bmod 31,$$

which shows that M_5 is prime.

The reason why this test works is related to the fact that $M_p + 1$ has a simple prime factorization (it is a power of 2). It is intimately connected to the arithmetic of the quadratic number field $\mathbb{Q}(\sqrt{3})$. At first sight, this number field has nothing to do with the Lucas–Lehmer test; looking more carefully at the situation, we observe the following lemma, which is easily proved by induction.

Lemma 5.14 Let $\omega = 2 + \sqrt{3}$ be the fundamental unit of $\mathbb{Z}[\sqrt{3}]$ and $\overline{\omega} = 2 - \sqrt{3}$ its conjugate. Then $S_{n+1} = \omega^{2^n} + \overline{\omega}^{2^n}$ for all $n \geq 0$.

This lemma connects the Lucas–Lehmer test with the arithmetic of $\mathbb{Z}[\sqrt{3}]$. At the heart of the matter lies the group structure of the unit group, whose geometric interpretation we have presented in Chap. 2. This interpretation has the advantage that it is analogous to similar theorems based on the group structure of $(\mathbb{Z}/p\mathbb{Z})^\times$ or the group of rational points on elliptic curves. We next present the results we need for understanding the Lucas–Lehmer test.

5.3.1 The Arithmetic in $\mathbb{Z}[\sqrt{3}]$

We begin by showing that $R = \mathbb{Z}[\sqrt{3}]$ is norm-Euclidean. To this end, let $\xi = x + y\sqrt{3} \in k = \mathbb{Q}(\sqrt{3})$ be given, and choose $\alpha = a + b\sqrt{3} \in R$ in such a way that $|x - a| < \frac{1}{2}$ and $|y - b| \leq \frac{1}{2}$. Then,

$$|N(\xi - \alpha)| = |(x - a)^2 - 3(y - b)^2| \leq \frac{3}{4} \qquad \text{since}$$

$$(x - a)^2 - 3(y - b)^2 \leq (x - a)^2 \leq \frac{1}{4} \qquad \text{and}$$

$$(x - a)^2 - 3(y - b)^2 \geq -3(y - b)^2 \geq -\frac{3}{4}.$$

In particular, R is norm-Euclidean.

The geometric interpretation of the Euclidean algorithm is not as simple as for the ring of Gaussian integers. We embed the domain $\mathbb{Z}[\sqrt{3}]$ into the Euclidean plane via the map

$$x + y\sqrt{3} \mapsto (x + y\sqrt{3}, x - y\sqrt{3}).$$

The number 1 then corresponds to the point $(1, 1)$, and $\sqrt{3}$ is mapped to $(\sqrt{3}, -\sqrt{3})$. The norm of $x + y\sqrt{3}$ is then the product of the coordinates of the points, and hence the elements (ξ, η) with norm 1 lie on the hyperbola $\xi\eta = 1$.

If we move the fundamental domain between the images of the elements 0, 1, $\sqrt{3}$, and $1 + \sqrt{3}$ to that whose corners are the images of $\frac{1}{2}$, $-\frac{1}{2}$, $-\frac{1}{2} + \sqrt{3}$, and $\frac{1}{2} + \sqrt{3}$, then we see that this region completely lies inside the hyperbolas $\xi\eta = 1$ and $\xi\eta = -1$, which makes clear again that $\mathbb{Z}[\sqrt{3}]$ is norm-Euclidean (see Fig. 5.3).

Proposition 5.15 *Let q be a rational prime number that is inert in R. Then R/qR is a finite field with q^2 elements.*

Proof Clearly, the residue class ring modulo qR contains at most q^2 elements since each integral element $a + b\sqrt{3}$ is congruent modulo qR to one of the elements of $\{r + s\sqrt{3} : 0 \le r, s \le q - 1\}$. Moreover, it is easily seen that the elements of this set are pairwise incongruent modulo q, which implies that the residue class ring does indeed have q^2 elements. Finally, R/qR does not have any zero divisors: $\alpha\beta \equiv 0 \bmod qR$ implies, since q is prime, that $\alpha \equiv 0 \bmod qR$ or $\beta \equiv 0 \bmod qR$.

It is therefore sufficient to show that finite rings without zero divisors are fields. All we have to do is show the existence of inverses. Assume therefore that A is a finite domain and that $a \ne 0$. Since A is finite, the sequence $a, a^2, \ldots, a^m$ must contain two equal elements, say $a^i = a^j$ for some $i < j$. Since A is a domain, we may cancel i, which gives us $a^{j-i} = 1$. But then a^{j-i-1} is an inverse of a. $\qquad\square$

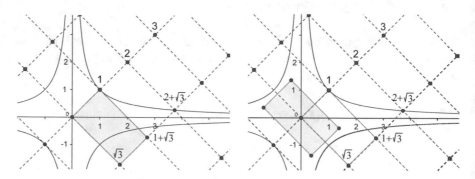

Fig. 5.3 The domain $\mathbb{Z}[\sqrt{3}]$ is norm-Euclidean

It follows easily from the quadratic reciprocity law that

$$\left(\frac{3}{p}\right) = \begin{cases} +1, & \text{if } p \equiv \pm 1 \bmod 12, \\ -1, & \text{if } p \equiv \pm 5 \bmod 12. \end{cases}$$

The prime elements of $\mathbb{Z}[\sqrt{3}\,]$ can now be determined as for $\mathbb{Z}[i]$ and $\mathbb{Z}[\rho]$.

Proposition 5.16 *The following elements are, up to associates, the only prime elements in $\mathbb{Z}[\sqrt{3}\,]$:*

1. $1 + \sqrt{3}$ *is the prime dividing* 2 *since* $2\varepsilon = (1 + \sqrt{3}\,)^2$ *with* $\varepsilon = 2 + \sqrt{3}$.
2. $\sqrt{3}$ *is the prime divisor of* 3 *since* $\sqrt{3}^2 = 3$.
3. *The prime numbers* $q \equiv \pm 5 \bmod 12$ *are inert.*
4. *The prime numbers* $p \equiv \pm 1 \bmod 12$ *split into two distinct prime elements* π *and* π'; *in particular, every prime* $p \equiv \pm 1 \bmod 12$ *can be written in the form* $\pm p = x^2 - 3y^2$.

Fermat's Little Theorem also holds in $\mathbb{Z}[\sqrt{3}\,]$; as in other quadratic number rings, it is a consequence of the following general and elementary observation:

Proposition 5.17 *Let* $p \nmid 4m$ *be prime and* $k = \mathbb{Q}(\sqrt{m})$; *then for all* $\alpha \in \mathcal{O}_k$, *we have*

$$\alpha^p \equiv \begin{cases} \alpha \\ \overline{\alpha} \end{cases} \bmod p, \ \ \textit{if } \left(\frac{m}{p}\right) = \begin{cases} +1 \\ -1. \end{cases}$$

Proof Write $\alpha = \frac{1}{2}(a + b\sqrt{m})$ with $a, b \in \mathbb{Z}$; then the fact that the binomial coefficients $\binom{p}{t}$ are divisible by p for each $1 \leq t \leq p - 1$ implies that

$$(2\alpha)^p \equiv a^p + b^p \sqrt{m}^p \equiv a + \left(\frac{m}{p}\right) b\sqrt{m} \bmod p$$

since $a^p \equiv a \bmod p$ and $\sqrt{m}^p = m^{(p-1)/2}\sqrt{m}$. The claim now follows from $2^p \equiv 2 \bmod p$. □

5.3.2 The Lucas–Lehmer Test

Assume first that $q = M_p = 2^p - 1$ is prime; we want to show that M_p passes the Lucas–Lehmer test, i.e., that S_{p-1} is divisible by M_p. To this end, we observe that $M_p \equiv 7 \bmod 8$ since $p \geq 3$ is odd and that $M_p \equiv 1 \bmod 3$; this shows that $M_p \equiv 7 \bmod 24$. We claim that M_p is irreducible in $\mathbb{Z}[\sqrt{3}\,]$.

In fact, if we had $M_p = \pi\pi'$ for an element $\pi = a + b\sqrt{3}$, then we would have $a^2 - 3b^2 = N\pi = \pm M_p$; since $M_p \equiv 1 \bmod 3$ and $a^2 - 3b^2 \equiv 0, 1 \bmod 3$ only the positive sign can hold. But then $a^2 - 3b^2 \equiv a^2 + b^2 \equiv 0, 1 \bmod 4$ in contradiction to $M_p \equiv 7 \bmod 8$.

Since $R = \mathbb{Z}[\sqrt{3}\,]$ is a unique factorization domain, M_p is not only irreducible but prime in R. If $q \geq 5$ is any prime element in R, then

$$(a + b\sqrt{3})^q \equiv a + (\tfrac{3}{q})b\sqrt{3} \equiv a - b\sqrt{3} \bmod qR$$

since $(\tfrac{3}{q}) = -1$ for $q \equiv 7 \bmod 24$. For $a = 2$ and $b = 1$, this yields $\omega^q \equiv \overline{\omega} \bmod qR$ (this congruence is analogous to the congruence (5.3) in $\mathbb{Z}[i]$), and thus $\omega^{q+1} \equiv \omega\overline{\omega} = 1 \bmod qR$. Since R/qR is a field, the element 1 has at most two (and in fact exactly two) square roots, namely 1 and -1. In particular, $\omega^{(q+1)/2} \equiv \pm 1$. We claim that the positive sign holds.

To this end, we observe that $2\omega = 4 + 2\sqrt{3} = (1 + \sqrt{3})^2$ is a square; thus we find $2^{(q+1)/2}\omega^{(q+1)/2} = (1 + \sqrt{3})^{q+1}$. The binomial expansion of this expression shows that $(1 + \sqrt{3})^q \equiv 1 + \sqrt{3}^q = 1 + 3^{(q-1)/2}\sqrt{3} \bmod q$. By Euler's criterion, we have $3^{(q-1)/2} = -(-3)^{(q-1)/2} \equiv -1 \bmod q$, and hence $(1 + \sqrt{3})^{q+1} \equiv (1 + \sqrt{3})(1 - \sqrt{3}) = -2 \bmod qR$. Since $2^{(q+1)/2} = 2 \cdot 2^{(q-1)/2} \equiv 2 \bmod q$, we finally find

$$\omega^{(q+1)/2} = 2^{-(q+1)/2}(1 + \sqrt{3})^{q+1} \equiv -1 \bmod qR$$

as claimed. Using $\omega\overline{\omega} = 1$, we now obtain

$$S_{p-1} = \omega^{(q+1)/4} + \overline{\omega}^{(q+1)/4} = \omega^{(q+1)/4}\big(1 + \omega^{-(q+1)/2}\big) \equiv 0 \bmod qR.$$

Assume conversely that $S_{p-1} \equiv 0 \bmod q$; then $\omega^{(q+1)/2} \equiv -1 \bmod qR$. Since $\frac{q+1}{2} = 2^{p-1}$ is a power of 2, $\frac{q+1}{2}$ must be the smallest exponent $n > 0$ for which $\omega^n \equiv -1 \bmod qR$. On the other hand, for each prime divisor $\ell \mid q$, the same congruence $\omega^{(q+1)/2} \equiv -1 \bmod \ell R$ holds, and again the exponent $\frac{q+1}{2}$ is minimal. Now either $\omega^{\ell+1} \equiv 1 \bmod \ell R$ or $\omega^{\ell-1} \equiv 1 \bmod \ell R$ by Proposition 5.17, i.e., we have $\ell - 1 \geq 2\frac{q+1}{2} = q + 1$ or $\ell + 1 \geq 2q + 1$. The first case is impossible, the second shows that $\ell \geq q$, and hence all divisors of q are $\geq q$. But then q is prime.

5.4 Fermat's Last Theorem for the Exponent 5

Dirichlet's first mathematical result was his proof [29] that the quintic Fermat equation $x^5 + y^5 = z^5$ has only the trivial solutions with $xyz = 0$ in integers. Legendre completed Dirichlet's proof before Dirichlet did but did not deal properly with the issues of unique factorization and units. In the following, we will give a streamlined version of Dirichlet's proof using ideas due to J. Plemelj [103] and L. Tschakaloff [125].

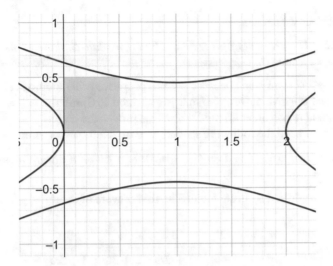

Fig. 5.4 The ring $\mathbb{Z}[\omega]$ is norm-Euclidean: The square defined by $0 \le x, y \le \frac{1}{2}$ lies inside the region cut out by the hyperbolas $(x-1)^2 - 5y^2 = 1$ and $(x-1)^2 - 5y^2 = -1$

First we observe that the ring of integers in $\mathbb{Q}(\sqrt{5})$ is $\mathbb{Z}[\omega]$ with $\omega = \frac{1+\sqrt{5}}{2}$.

Proposition 5.18 *The ring $\mathbb{Z}[\omega]$ is Euclidean with respect to the absolute value of the norm. In particular, it is a unique factorization domain.*

As before we need to show that for every $\xi \in \mathbb{Q}(\sqrt{5})$, there is an element $\gamma \in \mathbb{Z}[\omega]$ with $|N(\xi - \gamma)| < 1$. By symmetry, it is enough to prove the existence of γ for $\xi = x + y\sqrt{5}$ with $0 \le x, y \le \frac{1}{2}$. But for such ξ, the value $\gamma = 1$ always works with the single exception $(\frac{1}{2}, \frac{1}{2})$, for which we can pick $\gamma = 0$ (see Fig. 5.4).

We also see that -1 and ω are units in $\mathbb{Z}[\omega]$; it will follow from the results in the next chapter that every unit is, up to sign, a power of ω.

Assume now that $x^5 + y^5 = z^5$ for nonzero integers. Classical proofs of Fermat's Last Theorem for exponents ≥ 5 are usually divided into the first and the second case. The first case, where one of x, y, or z is divisible by p, is a lot easier to prove than the second case.

Observe that $x^5 \equiv \pm 1, \pm 7 \mod 25$ if $5 \nmid x$. This shows that the only solutions of $x^5 + y^5 \equiv z^5 \mod 25$ in rational integers are those in which $5 \mid xyz$.

Now we have to prove something similar in $\mathbb{Z}[\omega]$. In the following, we write $\lambda = \sqrt{5}$ for the unique ramified prime element. Then $x \equiv \pm 1, \pm 2 \mod \lambda$ for any x coprime to λ, and this implies (binomial expansion) that $x^5 \equiv \pm 1, \pm 32 \equiv \pm 7 \mod \lambda^3$. By the same reasoning as above, we obtain the following lemma.

Lemma 5.19 *If $x^5 + y^5 = z^5$ in $\mathbb{Z}[\omega]$, then $\lambda \mid xyz$.*

The next fact we need is:

Lemma 5.20 *For every nonzero residue class a modulo 5, there exists a unit $\eta \in \mathbb{Z}[\omega]$ with $\eta \equiv a$ mod 5.*

Since $\omega \equiv -2$ mod λ and because -2 is a primitive root modulo 5, $\omega^4 = 2+3\omega$ is the smallest positive power of ω that is $\equiv 1$ mod λ. Similarly, $\omega^{10} \equiv -1$ mod 5 then shows that ω has order 20 modulo 5. This proves the claim.

We will also need the following special case of "Kummer's Lemma":

Lemma 5.21 *If $\eta \in \mathbb{Z}[\omega]$ is a unit congruent to a rational integer modulo 5, then ω is a fifth power.*

This can be verified by brute force: The first powers of ω are

n	1	2	3	4	5
ω^n	ω	$1+\omega$	$1+2\omega$	$2+3\omega$	$3+5\omega$

Thus $\omega^5 \equiv 3$ mod 5 is the smallest positive power congruent to a rational integer modulo 5, and the exponents of all other such powers are divisible by 5.

Now consider the equation $x^5 + y^5 = z^5$. We may assume without loss of generality that $\lambda \mid z$; in fact, if $\lambda \mid y$, for example, then write the equation in the form $x^5 + (-z)^5 = (-y)^5$.

Assume therefore that $x^5 + y^5 = z^5$ and $\lambda \mid z$. The equation

$$(x+y)[(x+y)^4 - 5xy(x+y)^2 + 5x^2y^2] = z^5$$

shows that $\lambda \mid (x+y)$; therefore, the expression in the square brackets is divisible exactly by λ^2, and hence $x + y$ must be divisible by λ^3. Since any common divisor divides $5x^2y^2$, we conclude that the greatest common divisor of both factors is $\lambda^2 = 5$.

Next we multiply $x^5 + y^5 = z^5$ with a suitable power of ω^5 to make sure that $x \equiv 1$ mod 5 and therefore $y \equiv -1$ mod 5. Next

$$x^5 + y^5 = (x+y)(x^4 - x^3y + x^2y^2 - xy^3 + y^4)$$
$$= (x+y)[(x+y)^4 - 5xy(x^2 + xy + y^2)]$$
$$= (x+y)[(x+y)^4 - 5xy(x+y)^2 + 5x^2y^2]$$
$$= (x+y)[\lambda xy - \omega(x+y)^2][\lambda xy + \omega'(x+y)^2].$$

By unique factorization, there exist $\alpha, \beta, \gamma \in \mathbb{Z}[\omega]$ and units ε_j with

$$x + y = \varepsilon_1 \lambda^3 \gamma^5,$$

$$\lambda xy - \omega(x + y)^2 = \varepsilon_2 \lambda \alpha^5,$$

$$\lambda xy + \omega'(x + y)^2 = \varepsilon_3 \lambda \beta^5.$$

Dividing the second equation by λ, we get

$$xy - \omega \frac{x+y}{\lambda} = \varepsilon_2 \alpha^5.$$

Since $xy \equiv -1 \bmod \lambda^2$, we find that ε_1 is congruent to a rational integer modulo 5 and therefore is a fifth power; the same argument shows that ε_3 is a fifth power, and since the product of the three units is 1, so is ε_1. We may therefore assume that $\varepsilon_j = 1$ and obtain

$$x + y = \lambda^3 \gamma^5, \tag{5.5}$$

$$\lambda xy - \omega(x + y)^2 = \lambda \alpha^5, \tag{5.6}$$

$$\lambda xy + \omega'(x + y)^2 = \lambda \beta^5. \tag{5.7}$$

Subtracting (5.7) from (5.6) and using $\omega + \omega' = 1$, we obtain

$$\lambda \alpha^5 + \lambda(-\beta)^5 = \lambda(x + y)^2 = \lambda^6 \gamma^{10} = \lambda(\lambda \gamma^2)^5$$

and, after dividing through by λ,

$$\alpha^5 + (-\beta)^5 = (\lambda \gamma^2)^5 \tag{5.8}$$

with $\alpha\beta\gamma = z$.

Thus we have obtained a new solution to the quintic Fermat equation in which γ has fewer *distinct* prime factors than z except when α and β are units (observe that they are coprime to λ). But this is impossible, as we will show now.

In fact, dividing (5.8) through by α^5, we find an equation $1 \pm \omega^{5k} = \gamma^5$ for some γ divisible by λ. Since $\omega^{5k} \equiv \pm 1 \bmod 5$, the integer k is even. Dividing our equation by its conjugate and using

$$\frac{1 + \omega^{5k}}{1 + \omega'^{5k}} = \omega^{5k} \frac{1 + \omega^{5k}}{\omega^{5k} + (-1)^{5k}} = \omega^{5k},$$

we obtain

$$\omega^{5k} = \pm \left(\frac{\gamma}{\gamma'} \right)^5,$$

and this shows that $\gamma' = \pm\omega^k\gamma$. Thus γ and γ' have the same prime factorization, and hence γ is a product of a power of the ramified prime element λ and a rational integer. Since $\lambda^2 = 5$, we have $\gamma = a$ or $\gamma = \lambda a$ for a rational integer a.

Since $1 \pm \omega^{5k} = a^5$ immediately implies $k = 0$ and $a = 0$, we must have $1 \pm \omega^{5k} = \lambda^{5j}a^5$ for some odd integer j. Taking the trace of $1\pm(F_{5k-1}+F_{5k}\omega) = \lambda^{5j}a^5$ yields $0 = 2\pm(2F_{5k-1}+F_{5m}) = 2\pm(F_{5k-2}+3F_{5k-1})$, hence $F_{5k-2}+3F_{5k-1} = \pm 2$, and this is impossible.

5.5   Euclidean Number Fields

There are only a few norm-Euclidean quadratic number fields. The norm-Euclidean complex quadratic number fields are $\mathbb{Q}(\sqrt{m})$ for $m = -1, -2, -3, -7, -11$. It can in fact be shown that the other complex quadratic number fields not only are not norm-Euclidean, but that they also do not admit any Euclidean function at all.[4] Nevertheless, the rings of integers in the quadratic number fields with $m = -19$, $-43, -67, -163$ are unique factorization domains, as we shall see below.

In the real quadratic case, the situation is less clear. The classification of all norm-Euclidean real quadratic number fields was completed in the 1950s. Here is the result:

Theorem 5.22 *The rings of integers in* $\mathbb{Q}(\sqrt{m})$, $m > 0$, *are norm-Euclidean exactly for*

$$m = 2, 3, 5, 6, 7, 11, 13, 17, 19, 21, 29, 33, 37, 41, 57, 73.$$

The full proof of this result is very technical; we now present a clever idea for proving that several rings with small discriminant are norm-Euclidean due to Oppenheim [101]. More geometric proofs can be found in [34].

Assume first that $K = \mathbb{Q}(\sqrt{m})$ with $m \equiv 2, 3 \bmod 4$. The ring $\mathcal{O}_K$ is norm-Euclidean if for every $(x, y) \in \mathbb{Q} \times \mathbb{Q}$, there exists a pair of integers (a, b) such that $|N(x + y\sqrt{m} - (a + b\sqrt{m}))| < 1$, i.e., with

$$|(x - a)^2 - m(y - b)^2| < 1. \tag{5.9}$$

We now assume that $\mathcal{O}_K$ is *not* norm-Euclidean. We will show that this implies $m \geq 8$, and then it will follow that the rings with $m \leq 7$ *are* norm-Euclidean.

---

[4]This result is due to Theodore Motzkin (1908–1970) [98]. It can be proved quite easily and has played a big role in recent years; see Lemmermeyer [76].

We may also assume that $0 \le x, y \le \frac{1}{2}$. Since $\mathcal{O}_K$ is not norm-Euclidean, there exists a pair (x, y) such that one of the inequalities

$$P(a, b): \qquad\qquad (x - a)^2 \ge 1 + m(y - b)^2 \qquad\qquad (5.10)$$

$$N(a, b): \qquad\qquad m(y - b)^2 \ge 1 + (x - a)^2 \qquad\qquad (5.11)$$

is true for all pairs of integers (a, b). We will consider the following set of inequalities:

(a, b)	$P(a, b)$	$N(a, b)$
$(0, 0)$	$x^2 \ge 1 + my^2$	$my^2 \ge 1 + x^2$
$(1, 0)$	$(x - 1)^2 \ge 1 + my^2$	$my^2 \ge 1 + (x - 1)^2$
$(-1, 0)$	$(x + 1)^2 \ge 1 + my^2$	$my^2 \ge 1 + (x + 1)^2$

Now clearly $P(0, 0)$ is false; therefore, $N(0, 0)$ must be true. But then $P(1, 0)$ is false, and hence $N(1, 0)$ must hold. Next $P(-1, 0)$ and $N(1, 0)$ imply $(1 + x)^2 \ge 2 + (1 - x)^2$, hence $4x \ge 2$, $x = \frac{1}{2}$, and $my^2 = \frac{5}{4}$, and hence y is irrational. Thus $P(-1, 0)$ is false, and $N(-1, 0)$ must hold. But now $my^2 \ge 1 + (1 + x)^2 \ge 2$ implies $m \ge 8$.

Therefore the Euclidean algorithm holds for all $m < 8$, i.e., for $m = 2, 3, 5, 6, 7$. The very same proof works for fields with odd discriminant if we replace the inequality (5.9) by

$$|(x - \tfrac{b}{2} - a)^2 - \tfrac{m}{4}(y - b)^2| < 1,$$

and the result is that $\mathbb{Z}[\frac{1+\sqrt{m}}{2}]$ is norm-Euclidean if $m < 32$, i.e., for $m = 5, 13, 17, 21, 29$.

For real quadratic number fields, there are fields that are Euclidean but not norm-Euclidean, and in fact it is expected (and can be proved by assuming the truth of the Generalized Riemann Hypothesis) that all number fields whose ring of integers is a unique factorization domain are Euclidean for a suitable function (with the exception of complex quadratic fields). The first number field that was known not to be norm-Euclidean and was shown to be Euclidean is $\mathbb{Q}(\sqrt{69})$; nowadays, many examples are known; see, e.g., Harper [54, 55].

The fact that the domains listed in Theorem 5.22 are norm-Euclidean can nowadays be done by computer (as we have seen, it is possible to do this by hand for small values of the discriminant). The proof that the other fields are not norm-Euclidean is much more technical. The heart of the proof is an article by Davenport, which uses the language of quadratic forms; see [49].

5.5.1 Dedekind–Hasse Criterion

It is possible to show that a quadratic number ring has unique factorization by a criterion going back to Dedekind and Hasse.[5] Helmut Hasse (1898–1979) was one of the leading number theorists in the first half of the twentieth century. The Local–Global Principle for quadratic forms and for norm equations in cyclic extensions of number fields, explicit reciprocity laws or the Riemann conjecture for elliptic curves were results that have more or less defined the progress in number theory in the 1930s.

The Dedekind–Hasse criterion is a weakening of the existence of a Euclidean algorithm.

Theorem 5.23 *Let k be a quadratic number field; then $\mathcal{O}_k$ is a unique factorization domain if and only if for all $\alpha, \beta \in \mathcal{O}_k \setminus \{0\}$ with $\beta \nmid \alpha$, there exist elements $\gamma, \delta \in \mathcal{O}_k$ with $0 < |N(\alpha\gamma - \beta\delta)| < |N\beta|$.*

Using this criterion, we can, with some effort, prove the following theorem that we will obtain later (in Theorem 6.17) using the theory of ideals:

Theorem 5.24 *Let k be a quadratic number field with discriminant Δ; set*

$$
M_k = \begin{cases} \sqrt{\Delta/5} & \text{if } \Delta > 0, \\ \sqrt{-\Delta/3} & \text{if } \Delta < 0. \end{cases}
$$

Then $\mathcal{O}_k$ is a unique factorization domain if and only if for all prime numbers $p < M_k$ with $(\Delta/p) \neq -1$, there exist elements $\pi \in \mathcal{O}_k$ with $|N\pi| = p$.

Using this result, we can quickly verify that there are nine complex quadratic number fields whose ring of integers has unique factorization, namely those with discriminants

$$
\Delta = -3, -4, -7, -8, -11, -19, -43, -67, -163.
$$

It is a very deep result first conjectured by Gauss[6] and proved independently by Heegner, Stark, and Baker that there are no other complex quadratic fields with unique factorization.

---

[5]The Dedekind–Hasse criterion was published by Helmut Hasse [57]. Emmy Noether later found this criterion among Dedekind's papers when she edited his collected works [28]; see also [90, Anm. 1, S. 60].

[6]Gauss formulated this conjecture for class numbers of binary quadratic forms with even middle coefficients.

5.6 Quadratic Unique Factorization Domains

For most quadratic number fields, we can often decide that the ring of integers does not have unique factorization just by looking at its discriminant. We will later explain our partial results here by more advanced techniques such as the ambiguous class number formula.

In this section we will prove that the ring of integers in a quadratic number field cannot have unique factorization if its discriminant Δ has more than two distinct prime factors. The proof we will give is due to Laszlo Rédei [106]; in Chap. 9 we will derive this observation as a corollary of the much more general theorem on the structure of the ideal class group.

Theorem 5.25 *Let $k = \mathbb{Q}(\sqrt{m})$ be a quadratic number field whose ring of integers $\mathcal{O}_k$ is a unique factorization domain, and let $\Delta = \operatorname{disc} k$ denote its discriminant,*

If $\Delta < 0$, then $\Delta = -4, -8$ or $\Delta = -q$ for some positive prime number $q \equiv 3 \bmod 4$.

If $\Delta > 0$, then Δ is either a prime discriminant or the product of two negative prime discriminants:

$$
m = \begin{cases} p, & p \equiv 1, 3 \bmod 4 \text{ prime}, \\ 2q, & q \equiv 3 \bmod 4 \text{ prime}, \\ pq, & p \equiv q \equiv 3 \bmod 4 \text{ prime}. \end{cases}
$$

For the proof, we need the following:

Lemma 5.26 *Let $\mathcal{O}_k$ be a unique factorization domain. If $p \mid \Delta$ is a prime factor of the discriminant, then there is a unit $\eta \in E_k$ with $\sqrt{p\eta} \in \mathcal{O}_K$.*

Proof Let $k = \mathbb{Q}(\sqrt{m})$ for some squarefree integer m. If $p \mid m$, then we set $\alpha = \gcd(p, \sqrt{m})$ and find $\alpha^2 = (p^2, m) = p$, that is, $\alpha^2 = p\eta$ for a unit $\eta \in E_k$. If $p \nmid m$, then we must have $p = 2$ and $m \equiv 3 \bmod 4$; in this case, we set $\alpha = \gcd(2, 1 + \sqrt{m})$ and obtain $\alpha^2 = \gcd(4, 1 + m + 2\sqrt{m}) = \gcd(4, 2\sqrt{m}) = 2$, and now we conclude as above that $\alpha^2 = 2\eta$ for some unit $\eta \in E_k$. $\qquad\square$

Now we are ready to prove Theorem 5.25. If $\Delta = -3$ or $\Delta = -4$, then unique factorization holds. If $\Delta < -4$, then $E_k = \{-1, +1\}$. By Lemma 5.26, for each prime $p \mid \Delta$, the field k contains the element $\sqrt{-p}$. Thus there can be at most one such prime p, and Δ must be a prime discriminant.

In the case $\Delta > 0$, we start by showing that if $\mathcal{O}_k$ has unique factorization, then $\Delta = \operatorname{disc} k$ is either a prime discriminant or the product of two negative prime discriminants. Assume therefore that unique factorization holds, and write $p\varepsilon_p = \alpha_p^2$ for each prime $p \mid \Delta$. Clearly, both ε_p and ε_p' are positive, and hence $N\varepsilon_p = +1$.

If there is a prime $p \mid \Delta$ for which ε_p is a square, then $p\varepsilon = \alpha^2$ implies that p is a square in k. Thus in this case, there is only one such prime, and Δ must be a prime discriminant.

Assume therefore that ε_p is not a square for all $p \mid \Delta$. Since every positive unit is a power of the fundamental unit ε, it follows that $p\varepsilon$ is a square for each prime $p \mid \Delta$. If p and q are such primes, then $pq\varepsilon^2$ and therefore also pq is a square in k, and hence $k = \mathbb{Q}(\sqrt{pq})$.

Thus it remains to show that we cannot have $p \equiv q \equiv 1 \bmod 4$ or $p \equiv 1 \bmod 4$ and $q = 2$. In both cases, $pq = a^2 + b^2$ would be a sum of two squares, and we may assume that a is odd. We set $\rho = b + \sqrt{pq}$ and $\omega = \gcd(a, \rho)$. Observe that either b is odd and pq is even, or b is even and pq is odd; this implies that $\gcd(\rho, 2) = 1$. Next

$$\gcd(\rho, \rho') = \gcd(\rho, \rho + \rho') = \gcd(\rho, 2b) = \gcd(\rho, b) = 1$$

since this gcd divides $\gcd(\rho\rho', b) = \gcd(-a^2, b) = 1$. This implies

$$\omega^2 = \gcd(a^2, \rho^2) = \gcd(\rho\rho', \rho^2) = \rho \gcd(\rho, \rho') \sim \rho;$$

observe that the gcd has the usual properties since $\mathcal{O}_k$ has unique factorization. Thus ω^2/ρ is a unit with negative norm: a contradiction.

5.6.1   Euler's Polynomial

In his letter from September 28, 1743, Goldbach pointed out to Euler that the quadratic polynomial $f(x) = x^2 + 19x - 19$ represents many prime numbers for small values of x (the first composite number is $f(19) = 19 \cdot 37$). Some time later, Euler gave the polynomial $n(x) = x^2 - x + 41$, which represents prime numbers for $0 \le x \le 40$. The discriminant of Euler's polynomial is $\Delta = -163$, and the ring of integers of the quadratic number field $\mathbb{Q}(\sqrt{-163})$ is, as we will show in the next chapter, a unique factorization domain. Georg Frobenius [42] and Juri Rabinowitsch [105] have discovered that this is no coincidence, and today the mathematical literature knows hundreds if not thousands of publications (see, e.g., [7, 120] and Paulo Ribenboim's book [109], to mention but three) that deal with this topic. We claim:

Theorem 5.27 *The ring of integers in the quadratic number field $\mathbb{Q}(\sqrt{-p})$, where $p \equiv 3 \bmod 8$ is a prime number, is a unique factorization domain if and only if the polynomial $n(x) = x^2 + x + \frac{p+1}{4}$ attains prime values for $0 \le x < \frac{p-3}{4}$.*

We start with the following remark:

Lemma 5.28 *If k is a complex quadratic number field with discriminant $\Delta \le -11$ for which the ring of integers is a unique factorization domain, then $\Delta = -p$ is prime and we have $p \equiv 3 \bmod 8$.*

Proof We already know that $\Delta = -p$ must be a prime discriminant, and thus we have $p \equiv 3 \bmod 4$. If we had $p \equiv 7 \bmod 8$, then 2 in $\mathcal{O}_k$ cannot prime since $2 \mid$

$\frac{1+p}{4} = \frac{1-\sqrt{-p}}{2} \cdot \frac{1+\sqrt{-p}}{2}$ divides a product without dividing one of the factors. Since $\mathcal{O}_k$ is a unique factorization domain, 2 must be reducible, and this is only possible if $\mathcal{O}_k$ contains an element of norm 2. Since $N(\frac{x+y\sqrt{-p}}{2}) = \frac{x^2+py^2}{4} \geq \frac{x^2+11y^2}{4} > 2$, this is impossible. $\square$

If there is a composite number occurring among the values of the polynomial $n(x)$ for $0 \leq x < \frac{p-3}{4}$, then this number is divisible by a prime number q. This prime q is odd since $n(x) = x^2 + x + \frac{p+1}{4} \equiv x^2 + x + 1 \equiv 1 \bmod 2$ for all integers x. Thus $n(x) = x^2 + x + \frac{p+1}{4} = aq$, where we may assume that $q^2 \leq x^2 + x + \frac{p+1}{4}$ since each composite integer N contains a prime divisor $\leq \sqrt{N}$. But then

$$4q^2 \leq 4x^2 + 4x + p + 1 = (2x+1)^2 + p \leq \left(\frac{p-1}{2}\right)^2 + p = \left(\frac{p+1}{2}\right)^2,$$

and thus $q \leq \frac{p+1}{4}$. Observe that $(2x+1)^2 + p = aq$ implies that $(\frac{-p}{q}) = +1$. Since $q \mid \frac{2x+1-\sqrt{-p}}{2} \cdot \frac{2x+1+\sqrt{-p}}{2}$, the prime q is not prime in $\mathcal{O}_k$, hence reducible, and there exists an element with norm q in $\mathcal{O}_k$. On the other hand, we have

$$N\left(\frac{x+y\sqrt{-p}}{2}\right) = \frac{x^2+py^2}{4} \geq \frac{1+p}{4}$$

unless the norm is a square (and thus $y = 0$). This contradiction proves the claim.

For proving the converse, namely that $\mathcal{O}_k$ is a unique factorization domain if the values of $n(x) = x^2 + x + \frac{p+1}{4}$ are prime numbers for $0 \leq x < \frac{p-3}{4}$, we use Theorem 5.24. We have to show that for all primes $q < \sqrt{p/3}$ that are not inert, there exist elements $\pi \in \mathcal{O}_k$ with $|N\pi| = q$. Since $p \equiv 3 \bmod 8$, we have $q > 2$.

Assume therefore that $-p \equiv a^2 \bmod q$; we may assume moreover that $a = 2x + 1$ is odd and find that q divides $(2x+1)^2 + p = 4x^2 + 4x + p + 1 = 4n(x)$ and therefore $n(x)$. Changing x modulo q does not change this divisibility by q, and hence there is an integer x with $0 \leq x < q$ and $q \mid n(x)$. Since $q < \sqrt{p/3} < \frac{p-3}{4}$ for $p \geq 11$, we have $q \mid n(x)$; on the other hand, $n(x)$ is prime by assumption. Thus $n(x) = q$, and therefore q is the norm of an element in $\mathcal{O}_k$. This completes the proof.

5.6.2 Summary

In this chapter we have discussed a few minor applications of the theory of quadratic number fields. In particular, we have shown that $\mathbb{Z}[i]$ and $\mathbb{Z}[\rho]$ are norm-Euclidean domains and that the decomposition of primes p in these rings is connected with the representations of p in the form $x^2 + y^2$ and $x^2 + 3y^2$, respectively.

5.7 Exercises

5.1. Determine $\gcd(26-29i, 13+4i)$ using the Euclidean algorithm in $\mathbb{Z}[i]$. Verify the result using the prime factorization of these numbers. Also determine the corresponding Bézout elements.

5.2. Let $p = a^2 + b^2$ be an odd prime number. Show that a and b can be computed from a solution of the congruence $x^2 \equiv -1 \bmod p$ by applying the Euclidean algorithm to the numbers p and $x + i$ in $\mathbb{Z}[i]$.

This consequence of the fact that $\mathbb{Z}[i]$ is Euclidean can of course be generalized. How would you prove that each positive prime number $p \equiv 1, 3 \bmod 8$ can be written in the form $p = c^2 + 2d^2$?

5.3. Show that both $\{0, \pm 1, \pm i\}$ and $\{0, 1, 2, 3, 4\}$ are complete systems of residues modulo $1 + 2i$ in $\mathbb{Z}[i]$.

5.4. Show that the associates of $a + bi \in \mathbb{Z}[i]$ are given by $\pm(a+bi), \pm(-b+ai)$.

5.5. Show that for $\alpha \in \mathbb{Z}[i]$, the following assertions are equivalent:

1. $(1+i) \nmid \alpha$.
2. $N\alpha$ is odd.
3. $N\alpha \equiv 1 \bmod 4$.
4. α has an associate of the form $a + bi$ with $a - 1 \equiv b \equiv 0 \bmod 2$.
5. α has an associate congruent to $1 \bmod (2 + 2i)$.

5.6. Solve the Pythagorean equation $x^2 + y^2 = z^2$ by factoring the left side and using the arithmetic of $\mathbb{Z}[i]$.

5.7. Use infinite descent to show that the equation $x^3 + 3y^3 + 9z^3 = 0$ has only the trivial solution $(0, 0, 0)$ and generalize.

5.8. Compute the quadratic residue symbols $\left[\frac{1+2i}{3+2i}\right]$, $\left[\frac{1+4i}{3+2i}\right]$ and $\left[\frac{1+2i}{1+4i}\right]$.

5.9. In the following, we prove the quadratic reciprocity law in $\mathbb{Z}[i]$ using an idea of Dirichlet.

1. Show by comparing the definitions of the quadratic residue symbols in $\mathbb{Z}$ and $\mathbb{Z}[i]$ that for primes $\pi \in \mathbb{Z}[i]$ with odd prime norm p and $a \in \mathbb{Z}$ we always have $\left[\frac{a}{\pi}\right] = \left(\frac{a}{p}\right)$.

Show next that $\left[\frac{a}{q}\right] = 1$ for all $a \in \mathbb{Z}$ not divisible by $q \equiv 3 \bmod 4$.

2. If $\pi = a + bi \equiv 1 \bmod 2$ is prime, then $\left[\frac{a}{a+bi}\right] = 1$.
3. Let $\pi = a + bi$ and $\lambda = c + di$ be primes $\equiv 1 \bmod 2$ with norms $N\pi = p$ and $\lambda = q$. Use the congruences $ci \equiv d \bmod (c + di)$ and $\left(\frac{c}{q}\right) = 1$ for proving $\left[\frac{a+bi}{c+di}\right] = \left(\frac{ac+bd}{q}\right)$.
4. Use the quadratic reciprocity law in $\mathbb{Z}$ for verifying that $\left(\frac{ac+bd}{pq}\right) = 1$.
5. Prove the quadratic reciprocity law in $\mathbb{Z}[i]$.

5.10. Show using Theorem 5.24 that $\mathcal{O}_k$ is a unique factorization domain for $\Delta = -19, -43, -67, -163$.

5.11. Show that the ring

$$S = \mathbb{Z}[\sqrt{-17}, \tfrac{1}{2}] = \{2^{-n}(a + b\sqrt{-17}) : a, b, n \in \mathbb{Z}\}$$

is Euclidean with respect to the norm N_u defined by taking the maximal odd factor of the usual norm $N(x + y\sqrt{-17}) = x^2 + 17y^2$.

5.12. Show that the ring

$$R = \mathbb{Z}[\sqrt{-5}, \tfrac{1}{2}] = \{2^{-n}(a + b\sqrt{-5}) : a, b, n \in \mathbb{Z}\}$$

is Euclidean with respect to the norm N_u defined by taking the maximal odd factor of the usual norm $N(x + y\sqrt{-5}) = x^2 + 5y^2$; for example, $N_u(1 + \sqrt{-5}) = 3$.

Show that the unit group of R is $R^\times = \langle -1, 2 \rangle$ and thus is isomorphic as an abelian group to $\mathbb{Z}/2\mathbb{Z} \times \mathbb{Z}$.

The domains in this exercise are called rings of S-integers (in our case we had $S = \{2\}$). These domains are occasionally used if one would like to apply theorems that hold for principal ideal domains. The price one has to pay is a larger unit group, which usually outweighs the advantage of having unique factorization in almost all number theoretic problems.

5.13. Find the prime factorizations of 7, 13, and 19 in $\mathbb{Z}[\rho]$.

5.14. Show that for each $\alpha \in \mathbb{Z}[\rho]$, we have $\alpha^3 \equiv 0, 1 \bmod 2$.

5.15. The integral solutions of the equation $y^2 = x^3 + 24$ are $(1, 5)$, $(-2, 4)$, $(10, 32)$, and $(8\,158, 736\,844)$. How close to this result can you come by factoring $y^2 - 24 = x^3$ in the quadratic number field $\mathbb{Q}(\sqrt{6})$?

5.16. (See [87]) Consider Goldbach's polynomial $f(x) = x^2 + 19x - 19$, and show that it represents infinitely many composite integers by verifying the identity

$$f(x^2 + 20x - 19) = f(x) \cdot f(x + 1).$$

Show similarly that, for Euler's polynomial $n(x) = x^2 - x + 41$, we have

$$n(x^2 + 41) = n(x) \cdot n(x + 1).$$

Show more generally that $f((x + f(x)))$ is, for any polynomial f with integral coefficients and degree n, the product of $f(x)$ and another polynomial with integral coefficients.

5.17. Show that the only integral solutions of $y^2 + 1 = 2x^3$ are $(x, y) = (1, \pm 1)$.

5.18. In his proof of the first case of Fermat's Last Theorem for the exponent 5, Gauss considered the equation $x^5 + y^5 + z^5 = 0$ and set $y + z = a, z + x = b$ and $x + y = c$. Show that this implies

$$(a + b + c)^5 = 80abc(a^2 + b^2 + c^2).$$

Verify this identity and derive the first case of Fermat's Last Theorem for the exponent 5.

5.19. (Werebrussow) Let $\phi(x, y) = x^2 + xy - y^2$. Verify the identity

$$x^5 + y^5 = (x + y)\phi(x^2 - xy + y^2, x^2 - 2xy + y^2).$$

5.20. Generalize the congruence (5.3) to general quadratic number fields.

5.21. The congruence (5.3) implies divisibility results for recurring series of order 2. We will be content with explaining a few results for Fibonacci numbers; general results were obtained by Siebeck [116].

Recall Binet's formula (2.5) for the Fibonacci numbers:

$$U_n = \frac{\omega^n - \omega'^n}{\omega - \omega'},$$

where $\omega = \frac{1+\sqrt{5}}{2}$ and $\omega' = \frac{1-\sqrt{5}}{2}$; in particular, $\omega - \omega' = \sqrt{5}$.

Now prove the following congruences: For primes p with $\left(\frac{5}{p}\right) = +1$, we have

$$U_{p-1} \equiv 0, \qquad U_p \equiv 1, \qquad U_{p+1} \equiv 1 \bmod p \quad \text{if} \quad \left(\frac{5}{p}\right) = +1,$$

$$U_{p-1} \equiv 1, \qquad U_p \equiv -1, \qquad U_{p+1} \equiv 0 \bmod p \quad \text{if} \quad \left(\frac{5}{p}\right) = -1.$$

Chapter 6
Ideals in Quadratic Number Fields

In this chapter we will show how to work with ideals in quadratic number rings and how they can be applied to number theoretical problems.

6.1 Motivation

In Chap. 1, we have seen that

$$6 = 2 \cdot 3 = (1 + \sqrt{-5})(1 - \sqrt{-5})$$

is an example of nonunique factorization into irreducible elements in the domain $R = \mathbb{Z}[\sqrt{-5}]$. This factorization also provides a counterexample to other results that hold in unique factorization domains. We know, for example, that in unique factorization domains, it follows from $\gcd(a, c) = 1$ that $\gcd(a^2, c) = 1$. In our case, 2 and $1 + \sqrt{-5}$ are both irreducible; since they are not associate, they must be coprime. Yet $(1 + \sqrt{-5})^2 = -4 + 2\sqrt{-5}$ shows that $\gcd((1 + \sqrt{-5})^2, 2)$ is nontrivial since 2 is a common divisor.

If $\mathbb{Z}[\sqrt{-5}]$ were a principal ideal domain, we could write down such a factor immediately: $(2, 1 + \sqrt{-5}) = (\alpha)$ would imply $\alpha \sim \gcd(2, 1 + \sqrt{-5})$. But $\mathbb{Z}[\sqrt{-5}]$ is not a principal ideal domain, and the ideal $(2, 1 + \sqrt{-5})$ is not principal. Dedekind's idea was to regard the ideal $(2, 1 + \sqrt{-5})$ as the "correct" greatest common divisor of 2 and $(1 + \sqrt{-5})$. The introduction of such "ideal" factors then allows us to replace the non-existent unique factorization into elements by a unique factorization into prime ideals.

© The Author(s), under exclusive license to Springer Nature Switzerland AG 2021
F. Lemmermeyer, *Quadratic Number Fields*, Springer Undergraduate
Mathematics Series, https://doi.org/10.1007/978-3-030-78652-6_6

6.1.1 From Ideal Numbers to Ideals

Already in the first chapter, we have pointed out that Kummer's ideal numbers may be interpreted as ring homomorphisms $\mathcal{O}_k \longrightarrow \mathbb{F}_q$ from the ring of integers of a (quadratic) number field to finite fields and that Dedekind replaced these homomorphisms by their kernels. In this chapter we will have a closer look at the situation.

We may, for example, study the domain $\mathbb{Z}[i]$ by looking at ring homomorphisms from this ring to finite fields $\mathbb{F}_p = \mathbb{Z}/p\mathbb{Z}$. Recall that ring homomorphisms $f : R \longrightarrow S$ satisfy $f(r_1 + r_2) = f(r_1) + f(r_2)$ and $f(r_1 r_2) = f(r_1) f(r_2)$ and must have the property that the unit element of R is mapped to the unit element of S.

Each ring homomorphism $\kappa_2 : \mathbb{Z}[i] \longrightarrow \mathbb{Z}/2\mathbb{Z}$ satisfies, by definition, the equation $\kappa_2(1) = 1 + 2\mathbb{Z}$. Since $\kappa_2(i)^2 = \kappa_2(i^2) = \kappa_2(-1) = -1 + 2\mathbb{Z} = 1 + 2\mathbb{Z}$, we have $\kappa_2(i) = 1 + 2\mathbb{Z}$. Thus,

$$\kappa_2(a + bi) = \kappa_2(a) + \kappa_2(b)\kappa_2(i) = a + b + 2\mathbb{Z},$$

i.e., each Gaussian integer $a + bi$ is mapped to the residue class $a + b + 2\mathbb{Z}$. In particular, there exists a unique ring homomorphism from $\mathbb{Z}[i]$ to the field $\mathbb{F}_2$ with 2 elements.

Let us next ask whether there exist ring homomorphisms $\kappa_3 : \mathbb{Z}[i] \longrightarrow \mathbb{Z}/3\mathbb{Z}$. As above we find $\kappa_3(a) = a + 3\mathbb{Z}$ for all $a \in \mathbb{Z}$. Moreover $\kappa_3(i)^2 = \kappa_3(-1) = 2 + 3\mathbb{Z}$; but since $2 + 3\mathbb{Z}$ is not a square in $\mathbb{Z}/3\mathbb{Z}$, we arrive at a contradiction, and we conclude that there is no such ring homomorphism $\kappa_3 : \mathbb{Z}[i] \longrightarrow \mathbb{Z}/3\mathbb{Z}$. If we would like to construct a ring homomorphism κ_3 on $\mathbb{Z}[i]$ whose restriction to $\mathbb{Z}$ is reduction modulo 3, then its image must live in a field extension of $\mathbb{F}_3 = \mathbb{Z}/3\mathbb{Z}$ in which the residue class $2 + 3\mathbb{Z} = -1 + 3\mathbb{Z}$ is a square. The field $\mathbb{F}_9$ with 9 elements is such a field. We can think of $\mathbb{F}_9$ as the extension of $\mathbb{Z}/3\mathbb{Z}$ that is obtained by adjoining a square root i of -1 (modulo 3). The elements of this extension have the form $a + bi$ with $0 \leq a, b \leq 2$, and we do all calculations modulo 3. This construction provides us with a ring homomorphism $\kappa_3 : \mathbb{Z}[i] \longrightarrow \mathbb{F}_9$, which sends $a + bi$ to the residue class $a + bi$ modulo 3 in $\mathbb{F}_9$.

As our last example, we study the ring homomorphisms $\kappa_5 : \mathbb{Z}[i] \longrightarrow \mathbb{Z}/5\mathbb{Z}$. Here $\kappa_5(i)^2 = \kappa_5(-1) = -1 + 5\mathbb{Z} = 4 + 5\mathbb{Z} = (\pm 2 + 5\mathbb{Z})^2$, and hence we obtain two distinct ring homomorphisms, namely one with $\kappa_5(i) = 2 + 5\mathbb{Z}$ and another one defined by $\kappa_5'(i) = -2 + 5\mathbb{Z}$. Thus we have

$$\kappa_5(a + bi) = a + 2b + 5\mathbb{Z} \quad \text{and} \quad \kappa_5'(a + bi) = a - 2b + 5\mathbb{Z}.$$

The ideal numbers that Kummer introduced in cyclotomic number fields are essentially such ring homomorphisms; for Kummer, they were procedures for attaching residue classes to elements. Each such ring homomorphism possesses a kernel, and kernels of ring homomorphisms today are called ideals. If $f : R \longrightarrow S$ is a ring homomorphism, then its kernel I has the following properties:

- $I + I \subseteq I$: if $r_1, r_2 \in \ker f$, then $f(r_1 + r_2) = f(r_1) + f(r_2) = 0 + 0 = 0$, and so $r_1 + r_2$ is an element of the kernel: $r_1 + r_2 \in I$.
- $R \cdot I \subseteq I$: if $s \in \ker f$ and $r \in R$, then $f(r \cdot s) = f(r) \cdot f(s) = f(r) \cdot 0 = 0$, and hence $r \cdot s \in I$.

If, conversely, I is a subset of R with these properties, then we can define the quotient ring $S = R/I$ and obtain a natural map $f : R \longrightarrow R/I$ defined by $f(r) = r + I$; this map f is then, as can be easily checked, a ring homomorphism with kernel I.

For example, $\ker \kappa_2$ consists of all elements $a + bi \in \mathbb{Z}[i]$ with $a + b \equiv 0 \bmod 2$. These numbers are easily seen to be the multiples of $1 + i$. The ideal $\ker \kappa_3$ consists of the multiples of 3 and finally $\ker \kappa_5$ of the multiples of $1 + 2i$, whereas $\ker \kappa_5'$ consists of the multiples of $1 - 2i$.

We will denote the set of multiples of an element α in some domain R by aR or simply by (a) and call such ideals *principal*. All ideals in the rings $\mathbb{Z}$ and $\mathbb{Z}[i]$ are principal; domains with this property are called *principal ideal domains*. The domain $\mathbb{Z}[\sqrt{-5}\,]$ is not a principal ideal domain. In fact, the ring homomorphism κ_2 defined by $\kappa_2(a + b\sqrt{5}) = a + b + 2\mathbb{Z}$ has kernel $I = \ker \kappa_2 = \{a + b\sqrt{-5} : a \equiv b \bmod 2\}$, and this ideal is not principal. This can be seen as follows: If there exists an $\alpha \in \mathbb{Z}[\sqrt{-5}\,]$ with $\ker \kappa_2 = (\alpha)$, then $\kappa_2(2) = \kappa_2(1 + \sqrt{-5}) = 0$ implies that 2 and $1 + \sqrt{-5}$ are multiples of α, say $2 = \alpha\beta$ and $1 + \sqrt{-5} = \alpha\gamma$. Since we have already shown that 2 is irreducible in $\mathbb{Z}[\sqrt{-5}\,]$, we must have $\alpha = \pm 1$ or $\alpha = \pm 2$, but both alternatives are impossible. In fact, since $\kappa_2(\pm 1) = \pm 1 + 5\mathbb{Z}$, the elements ± 1 do not belong to the ideal $\ker \kappa_2$, and in the second case $1 + \sqrt{-5}$ would have to be a multiple of 2, which it is not.

6.1.2 Products of Ideals

For studying divisibility of elements in number rings without unique factorization, we must define products of ideals and investigate when an ideal is divisible by another.

For principal ideals, this is easy: The product of two principal ideals (α) and (β) is of course defined to be the principal ideal $(\alpha\beta)$. Using this definition we can characterize divisibility of elements and principal ideals as follows:

Proposition 6.1 *Let k be a quadratic number field with ring of integers $\mathcal{O}_k$. For $\alpha, \beta \in \mathcal{O}_k$, the following assertions are equivalent:*

1. $\alpha \mid \beta$;
2. $(\alpha) \supseteq (\beta)$.

This implies that the following assertions are also equivalent:

1. $(\alpha) = (\beta)$;
2. $\alpha \mid \beta$ and $\beta \mid \alpha$;
3. *There exists a unit* $\eta \in \mathcal{O}_k^\times$ *with* $\alpha = \beta\eta$.

The simple proof, which works in general domains and does not use any special properties of quadratic number rings, is left to the readers (see Exercise 4.32). We remark in passing that the transition from elements to principal ideals simplifies many questions concerning divisibility since the units of the domain do not play any role.

The product of two not necessarily principal ideals is defined in a similar manner: If A and B are ideals in some domain R, then the set,

$$AB = \{\alpha_1\beta_1 + \ldots + \alpha_m\beta_m : \alpha_j \in A, \beta_j \in B\},$$

of all *finite* sums of products $\alpha_j\beta_j$ is again an ideal in R. This is easily verified: Sums of finite linear combinations of products $\alpha_i\beta_j$ again have this form, so the set AB is an additive group. Moreover, AB is closed with respect to multiplication by ring elements; in fact, if $\alpha_1\beta_1 + \ldots + \alpha_m\beta_m \in AB$ and $r \in R$, then $r(\alpha_1\beta_1 + \ldots + \alpha_m\beta_m) = (r\alpha_1)\beta_1 + \ldots + (r\alpha_m)\beta_m \in AB$ since each $r\alpha_j \in A$.

An ideal A is finitely generated if there exist elements $\alpha_1, \ldots, \alpha_m \in A$ with $A = (\alpha_1, \ldots, \alpha_m)$. We will see that ideals in quadratic number rings have at most two generators; the ideal $(2, \sqrt{2}, \sqrt[4]{2}, \sqrt[8]{2}, \ldots)$ in $\mathbb{Z}[\sqrt{2}, \sqrt[4]{2}, \sqrt[8]{2}, \ldots]$, on the other hand, is not finitely generated. Products of finitely generated ideals can easily be written down.

Lemma 6.2 *If* $A = (\alpha_1, \ldots, \alpha_m)$ *and* $B = (\beta_1, \ldots, \beta_n)$ *are finitely generated ideals, then* $AB = (\alpha_1\beta_1, \alpha_1\beta_2, \ldots, \alpha_m\beta_n)$.

Proof Equality of ideals is usually proved by showing that each ideal is contained in the other. The proof therefore consists of two parts.

(1) $AB \subseteq (\alpha_1\beta_1, \ldots)$. Each element $\alpha \in A$ is an R-linear combination of the α_i and hence has the form $\alpha = \sum r_i\alpha_i$ with $r_i \in R$. Similarly, each $\beta \in B$ has the form $\beta = \sum s_i\beta_i$ with $s_i \in R$. Thus $\alpha\beta = \sum_{i,j} r_i s_j \alpha_i\beta_j$ is an R-linear combination of the $\alpha_i\beta_j$ and so is contained in the ideal $(\alpha_1\beta_1, \ldots)$.

(2) $AB \supseteq (\alpha_1\beta_1, \ldots)$. This is obvious since each generator $\alpha_i\beta_j$ in the ideal on the right is a product of elements of A and B.

$\square$

Without problems we verify the following properties:

Proposition 6.3 *Let* A, B, *and* C *be ideals in some domain* R. *Then* $AB = BA$, $(AB)C = A(BC)$, *and* $AR = A(1) = A$.

For every ideal $\mathfrak{a}$ in quadratic number rings, we can define the conjugate ideal $\mathfrak{a}^\sigma = \mathfrak{a}'$ (here σ is the nontrivial automorphism of $k/\mathbb{Q}$), which consists of all elements α' for which $\alpha \in \mathfrak{a}$. Again it is easy to check that conjugation commutes with multiplication, i.e., that $(ab)^\sigma = a^\sigma b^\sigma$.

It takes some time to get used to computing with ideals. Consider for example the ideals $\mathfrak{a} = (2, 1 + \sqrt{-5})$, $\mathfrak{b} = (3, 1 + \sqrt{-5})$, and $\mathfrak{c} = \mathfrak{b}^\sigma = (3, 1 - \sqrt{-5})$. Then

$$\mathfrak{a}^2 = (2 \cdot 2, 2(1 + \sqrt{-5}), 2(1 + \sqrt{-5}), -4 + 2\sqrt{-5})$$
$$= (4, 2(1 + \sqrt{-5}), -4 + 2\sqrt{-5})$$
$$= (2)(2, 1 + \sqrt{-5}, -2 + \sqrt{-5});$$

the last ideal contains $\sqrt{-5} = 2 + (-2 + \sqrt{-5})$ and thus also $1 = (1 + \sqrt{-5}) - \sqrt{5}$, and hence $\mathfrak{a}^2 = (2)(1) = (2)$.

In a similar way, we find

$$\mathfrak{b}\mathfrak{c} = (9, 3(1 + \sqrt{-5}), 3(1 - \sqrt{-5}), 6)$$
$$= (3)(3, 1 + \sqrt{-5}, 1 - \sqrt{-5}, 2) = (3)(1) = (3).$$

It is slightly more tricky to compute

$$\mathfrak{b}^2 = (9, 3(1 + \sqrt{-5}), (1 + \sqrt{-5})^2)$$
$$= (2 + \sqrt{-5})(2 - \sqrt{-5}, 1 - \sqrt{-5}, -2) = (2 + \sqrt{-5}).$$

The calculation of the products $\mathfrak{a}\mathfrak{b}$, $\mathfrak{a}\mathfrak{c}$, and $\mathfrak{c}^2$ is done in Exercise 6.3.

Let us have another look at the equation

$$2 \cdot 3 = (1 + \sqrt{-5})(1 - \sqrt{-5}).$$

If we form the ideals generated by the elements on the left and the right hand side, we obtain $(2)(3) = (2 \cdot 3) = (1 + \sqrt{-5})(1 - \sqrt{-5})$ (the fact that there is a product of ideals on the right hand side can only be seen by observing that there is a product of *ideals* on the left). If we plug in $(2) = \mathfrak{a}^2$ and $(3) = \mathfrak{b}\mathfrak{c}$, then we obtain the equation of ideals $(6) = \mathfrak{a}^2\mathfrak{b}\mathfrak{c}$; by writing the ideal factors in the form $\mathfrak{a}^2 \cdot (\mathfrak{b}\mathfrak{c})$, we obtain the decomposition $(6) = (2)(3)$ into two principal ideals, and $(\mathfrak{a}\mathfrak{b})(\mathfrak{a}\mathfrak{c}) = (1 + \sqrt{-5})(1 - \sqrt{-5})$ yields the second decomposition into principal ideals.

This shows that the two essentially different factorizations on the level of numbers correspond to different ways of taking products of the "prime" ideals $\mathfrak{a}$, $\mathfrak{b}$, and $\mathfrak{c}$. In the next section we will show that this holds in general quadratic number rings $\mathcal{O}_k$: even if the factorization into irreducible elements is not unique, the factorization into irreducible ideals is.

6.1.3 The Class Group at Work

If we look carefully at counterexamples to the Square Product Theorem in $\mathbb{Z}[\sqrt{-5}\,]$, then we will quickly see that there are infinitely many of them. For example, we have

$$
\begin{aligned}
2^2 + 5 \cdot 1^2 &= 3^2, \\
2^2 + 5 \cdot 3^2 &= 7^2, \\
22^2 + 5 \cdot 3^2 &= 23^2, \\
38^2 + 5 \cdot 9^2 &= 43^2, \\
2^2 + 5 \cdot 21^2 &= 47^2.
\end{aligned}
$$

In all these cases with $x^2 + 5y^2 = q^2$, the equations $x^2 + 5y^2 = q$ do not have an integral (not even a rational) solution, and it is not difficult to guess that the squares of all prime numbers $q \equiv 3, 7 \bmod 20$ are represented by the form $x^2 + 5y^2$. We will see in this chapter that this observation is a consequence of the fact that $\mathbb{Q}(\sqrt{-5}\,)$ has class number 2. The additional observation that the representation of primes by the form $x^2 + 5y^2$ only depends on their residue classes modulo 20 lies deeper and is a consequence of genus theory, which we will touch upon in Chap. 9.

Similar phenomena show up for other fields. The primes p with $(\frac{-23}{p}) = +1$, for example, can be written in the form $4p^3 = x^2 + 23y^2$ and some of them even as $4p = x^2 + 23y^2$. This is a consequence of the fact that $\mathbb{Q}(\sqrt{-23}\,)$ has class number 3. The class number is less visible in the observation that these primes are either represented by $4p = x^2 + 23y^2$ or by $8p = x^2 + 23y^2$: This is a consequence of the fact that the ideal class group is generated by the prime ideals above 2.

6.2 Unique Factorization into Prime Ideals

So far we have only discussed properties of ideals that hold in general domains. From now on we will exploit the fact that ideals in rings of integers of algebraic number fields have additional properties. For deriving them, it is useful to introduce the concept of $\mathbb{Z}$-modules. In arbitrary domains R, a $\mathbb{Z}$-module in R is just an additive subgroup of R.

6.2.1 Classification of Modules

Examples of $\mathbb{Z}$-modules in $\mathcal{O}_k$ are multiples of a single element (these are called $\mathbb{Z}$-modules of rank 1) such as $\mathbb{Z}$, the even integers $2\mathbb{Z}$, or $\mathbb{Z}\omega$, which consists of the multiples of ω. Similarly, the order $\mathbb{Z}[\sqrt{m}\,] = \mathbb{Z} \oplus \sqrt{m}\,\mathbb{Z}$ is a $\mathbb{Z}$-module of rank 2, as is the maximal order $\mathbb{Z} \oplus \omega\mathbb{Z}$ or $2\mathbb{Z} + 3\sqrt{m}\mathbb{Z}$.

Proposition 6.4 *The $\mathbb{Z}$-modules of $\mathbb{Z}$ are $m\mathbb{Z}$ for integers $m \geq 0$.*

Clearly $m\mathbb{Z}$ is a $\mathbb{Z}$-module for every integer m. If M is an arbitrary $\mathbb{Z}$-module, then either $M = 0 = 0\mathbb{Z}$ contains just the zero element or it contains an integer $n \neq 0$. If n is negative, then $-n \in M$ is positive. Let m be the smallest positive integer in M. We claim that $M = m\mathbb{Z}$. In fact, given any $a \in M$, we can write $a = mq + r$ with $0 \leq r < m$. If $r \neq 0$, then $r \in M$ is smaller than m contradicting our choice of m. Thus $r = 0$, which tells us that any $a \in M$ is a multiple of m, which is what we wanted to prove.

In the following, let $\mathcal{O}_k$ be the ring of integers in the quadratic number field $k = \mathbb{Q}(\sqrt{m})$. Given a $\mathbb{Z}$-module M in $\mathcal{O}_k$, we define two $\mathbb{Z}$-modules in $\mathbb{Z}$, namely $M \cap \mathbb{Z}$ and the coefficient module

$$\mathrm{coeff}(M) = \{s \in \mathbb{Z} : \text{ there is an } a \in \mathbb{Z} \text{ such that } a + s\omega \in M\},$$

where $\{1, \omega\}$ is an integral basis of $\mathcal{O}_k$. The $\mathbb{Z}$-module $\mathrm{coeff}(M)$ does not depend on the choice of ω: Replacing ω by $\omega_1' = \omega - r$ for an arbitrary integer r clearly leaves $\mathrm{coeff}(M)$ invariant.

The following table displays the $\mathbb{Z}$-modules $M \cap \mathbb{Z}$ and $\mathrm{coeff}(M)$ for a few choices of $\mathbb{Z}$-modules M in a quadratic number ring $\mathbb{Z}[\sqrt{m}]$:

M	$\mathbb{Z}$	$\mathbb{Z}[\sqrt{m}]$	$\mathcal{O}_k$	$5\mathbb{Z} \oplus (1 + 2i)\mathbb{Z}$
$M \cap \mathbb{Z}$	$\mathbb{Z}$	0	$\mathbb{Z}$	$5\mathbb{Z}$
$\mathrm{coeff}(M)$	0	$\mathbb{Z}$	$\mathbb{Z}$	$2\mathbb{Z}$

Actually we can define a group homomorphism $\mathrm{coeff} : M \longrightarrow \mathrm{coeff}(M)$ via $\mathrm{coeff}(a + b\omega) = b$. This map respects addition and is onto. Its kernel consists of all elements $a + b\omega \in M$ with $b = 0$, i.e., we have $\ker \mathrm{coeff} = M \cap \mathbb{Z}$. In the language of exact sequences that we will introduce in Chap. 9, this means that the sequence

$$0 \longrightarrow M \cap \mathbb{Z} \longrightarrow M \longrightarrow \mathrm{coeff}(M) \longrightarrow 0$$

is exact. This means little more than that there are maps $\iota : M \cap \mathbb{Z} \longrightarrow M$ (inclusion) and $\mathrm{coeff} : M \longrightarrow \mathrm{coeff}(M)$ (projection) with $\mathrm{im}\,\iota = \ker \mathrm{coeff}$.

Those familiar with the concept of direct sums in group theory will know that for writing M as a direct sum of two modules of rank 1, one needs a lift $\mathrm{coeff}(M) \longrightarrow M$. Clearly, given an element $m \in \mathrm{coeff}(M)$, there is an element $a + b\omega \in M$; this "lift" will occur in our proofs below, which do not use the exact sequence above or any other fancy tools from commutative algebra.

We now prove the following "basis theorem" for $\mathbb{Z}$-modules of $\mathcal{O}_k$:

Proposition 6.5 *Let M be a $\mathbb{Z}$-module in $\mathcal{O}_k$. Then there exist unique natural numbers $m, n \in \mathbb{N}_0$ and some $a \in \mathbb{Z}$ with $M = n\mathbb{Z} \oplus (a + m\omega)\mathbb{Z}$.*

This proposition claims that each such $\mathbb{Z}$-module possesses a $\mathbb{Z}$-basis; thus $\mathbb{Z}$-modules in $\mathcal{O}_k$ behave as subspaces of a vector space. The number of elements of a basis does not depend on the choice of the basis and is called the rank of the module. For example, the module $M = (0)$ has rank 0, the modules $\mathbb{Z}$ and $\sqrt{m} \cdot \mathbb{Z}$ have rank 1, and $\mathcal{O}_k$ has rank 2. Clearly $M = n\mathbb{Z} \oplus (a + m\omega)\mathbb{Z}$ has rank 2 if and only if $mn \neq 0$.

Proof of Proposition 6.5 Write $M \cap \mathbb{Z} = n\mathbb{Z}$ and coeff$(M) = m\mathbb{Z}$ for integers $m, n \in \mathbb{N}_0$. By construction there is an integer $a \in \mathbb{Z}$ with $a + m\omega \in M$; since we may change a by a multiply of n, the integer a is only determined modulo n.

Next we verify that these integers have the desired properties. We have to show that $M = n\mathbb{Z} \oplus (a + m\omega)\mathbb{Z}$. The fact that $M \supseteq n\mathbb{Z} \oplus (a + m\omega)\mathbb{Z}$ is clear. Assume therefore that $r + s\omega \in M$. Since $s \in$ coeff(M), we conclude that $s = um$ for some $u \in \mathbb{Z}$, and then $r - ua = r + s\omega - u(a + m\omega) \in M \cap \mathbb{Z}$, hence $r - ua = vn$. Now we obtain $r + s\omega = r - ua + u(a + m\omega) = vn + u(a + m\omega) \in n\mathbb{Z} \oplus (a + m\omega)\mathbb{Z}$. $\square$

Now let M be a $\mathbb{Z}$-module in $R = \mathcal{O}_k$. We consider the factor group R/M. This group consists of all expressions of the form $r + M$ with $r \in R$, where $r + M = s + M$ if and only if $r - s \in M$. This set becomes a group by setting $(r + M) + (s + M) = (r + s) + M$. The idea behind computing with factor groups such as this one is doing calculations in R and identifying elements that differ by an element in M.

The number of residue classes modulo M, i.e., the cardinality of the residue class group R/M, is called the norm of the module M, and we write $N(M) = (R : M)$. The norm of a module M need not be finite, as the example $R = \mathcal{O}_k$ and $M = \mathbb{Z}$ shows.

The importance of the numbers m and n in Proposition 6.5 is also emphasized by our next result.

Proposition 6.6 *Let $M = n\mathbb{Z} \oplus (a + m\omega)\mathbb{Z}$ be a $\mathbb{Z}$-module of rank 2 in $\mathcal{O}_k$. Then*

$$S = \{r + s\omega : r, s \in \mathbb{Z}; \ 0 \leq r < n, \ 0 \leq s < m\}$$

is a complete system of residues modulo M in R. In particular, the order of the residue class group R/M is $N(M) = mn$.

We have to show

(a) that each element of R is congruent modulo M to some element of S and
(b) that elements of S are congruent modulo M only if they are equal.

For proving the first claim, take an element $x + y\omega \in R$, and write $y = mq + s$ with $0 \leq s < m$ and $x - qa = np + r$ with $0 \leq r < n$. Then

$$x + y\omega - (np + q(a + m\omega)) = r + s\omega,$$

and since $np + q(a + m\omega) \in M$, the claim is true.

For proving the second claim, assume that $r + s\omega \equiv r' + s'\omega \mod M$ with $0 \le r, r' < n$ and $0 \le s, s' < m$. Then $r - r' + (s - s')\omega \in M$. Thus $r - r' \in m\mathbb{Z}$ and $s - s' \in n\mathbb{Z}$, and hence $r = r'$ and $s = s'$.

6.2.2   Ideals as Modules

Given a $\mathbb{Z}$-module M of rank 2 in $R = \mathcal{O}_k$, we have defined the (additive) residue class group R/M and determined its order. It is a natural question whether this group can actually be given a ring structure. Observe that we have added two residue classes $r + M$ and $s + M$ by adding the representatives, i.e., we have set

$$(r + M) + (s + M) = r + s + M.$$

What could prevent us from defining the product of these residue classes by

$$(r + M) \cdot (s + M) = rs + M?$$

All we have to do is verify that our product is well defined. To this end, we replace s by $s + m$ for some $m \in M$; then

$$(r + M) \cdot (s + m + M) = rs + rm + M,$$

and this is equal to $rs + M$ if and only if $rm \in M$ for every element $r \in R$. In other words, we can give the residue class ring R/M a ring structure only if M is closed with respect to multiplication by *arbitrary* elements of R. $\mathbb{Z}$-modules with this property are ideals by definition.

Our next result characterizes ideals in terms of their module basis.

Proposition 6.7 *The module* $\mathfrak{a} = n\mathbb{Z} + (a + m\omega)\mathbb{Z}$ *is an ideal if and only if* $m \mid n$, $m \mid a$ *(and thus* $a = mb$ *for some* $b \in \mathbb{Z}$*), and* $n \mid m \cdot N(b + \omega)$.

Proof We first show that $n\mathbb{Z} = \mathfrak{a} \cap \mathbb{Z} \subseteq \text{coeff}(\mathfrak{a}) = m\mathbb{Z}$, which then implies (to contain is to divide) that $m \mid n$. To this end assume that $c \in \mathfrak{a} \cap \mathbb{Z}$; then $c\omega \in \mathfrak{a}$, and by definition of the coefficient module, we conclude that $c \in \text{coeff}(\mathfrak{a})$.

For showing that $m \mid a$, we observe that $\omega^2 = x + y\omega$ for integers $x, y \in \mathbb{Z}$ since $\{1, \omega\}$ is an integral basis. Now $\mathfrak{a}$ is an ideal; thus if it contains $a + m\omega$, it will also contain $(a + m\omega)\omega = mx + (a + my)\omega$. Thus $a + my \in \text{coeff}(\mathfrak{a})$ is a multiple of m, and this implies immediately that $m \mid a$, and hence $a = mb$ for some $b \in \mathbb{Z}$.

For verifying the last divisibility relation we set $\alpha = a + m\omega = m(b + \omega)$. With $\alpha \in \mathfrak{a}$ clearly $\alpha(b + \omega')$ is contained in the ideal $\mathfrak{a}$. Since $\frac{1}{m}N\alpha = m(b+\omega)(b+\omega') \in \mathfrak{a} \cap \mathbb{Z}$ we find that $\frac{1}{m}N\alpha = m \cdot N(b + \omega)$ is a multiple of n. $\square$

Our next goal is the statement that the norm $\mathfrak{a}\mathfrak{a}'$ of an ideal $\mathfrak{a}$ is generated by an integer. For principal ideals this is clear since $(\alpha)(\alpha)' = (\alpha)(\alpha') = (\alpha\alpha') = (N\alpha)$.

A key step in proving unique factorization for ideals in general number fields is showing that for each integral ideal $\mathfrak{a} \neq (0)$ there is an ideal $\mathfrak{b} \neq (0)$ such that $\mathfrak{a}\mathfrak{b} = (\alpha)$ is principal.

Proposition 6.8 *Let* $\mathfrak{a} \neq (0)$ *be an ideal in* $\mathcal{O}_k$. *Then there is an* $a \in \mathbb{N}$ *with* $\mathfrak{a}\mathfrak{a}' = (a)$.

Remark Here the notation (a) is slightly ambiguous since it is not clear whether we are talking about the ideal $a\mathbb{Z}$ in $\mathbb{Z}$ or the ideal $a\mathcal{O}_k$ generated by a in $\mathcal{O}_k$. Since on the left side there is an ideal in $\mathcal{O}_k$, clearly (a) must be the ideal $(a) = a\mathcal{O}_k$.

For the proof of Proposition 6.8, we use the following lemma[1] due to A. Hurwitz:

Lemma 6.9 (Hurwitz's Lemma) *Assume that* $\alpha, \beta \in \mathcal{O}_k$ *and* $m \in \mathbb{N}$. *If* $N\alpha$, $N\beta$ *and* $\mathrm{Tr}\,\alpha\beta'$ *are divisible by* m, *then* $m \mid \alpha\beta'$ *and* $m \mid \alpha'\beta$.

Proof Let $\gamma = \alpha\beta'/m$; then $\gamma' = \alpha'\beta/m$, and we know that $\gamma + \gamma' = (\mathrm{Tr}\,\alpha\beta')/m$ and $\gamma\gamma' = \frac{N\alpha}{m}\frac{N\beta}{m}$ are integers. If the norm and the trace of an element of a quadratic number field are integers, the element must be an algebraic integer, hence $\gamma \in \mathcal{O}_k$, and the claim follows.   □

Proof of Proposition 6.8 We write $\mathfrak{a} = (\alpha, \beta)$ for $\alpha, \beta \in \mathcal{O}_k$ (we can do so by Proposition 6.5). Then $\mathfrak{a}' = (\alpha', \beta')$, and thus $\mathfrak{a}\mathfrak{a}' = (N\alpha, \alpha\beta', \alpha'\beta, N\beta)$. If we set $a = \gcd(N\alpha, N\beta, \mathrm{Tr}\,\alpha\beta')$ (in $\mathbb{Z}$), then by Hurwitz's Lemma 6.9, the two numbers $\frac{\alpha\beta'}{a}$ and $\frac{\alpha'\beta}{a}$ are algebraic integers; thus we obtain $\mathfrak{a}\mathfrak{a}' = (a)(\frac{N\alpha}{a}, \frac{N\beta}{a}, \frac{\alpha\beta'}{a}, \frac{\alpha'\beta}{a})$, where the last ideal lies in $\mathcal{O}_k$ by Hurwitz's Lemma. In order to show that $\mathfrak{a}\mathfrak{a}' = (a)$, it is enough to show that $1 \in (\frac{N\alpha}{a}, \frac{N\beta}{a}, \frac{\alpha\beta'}{a}, \frac{\alpha'\beta}{a})$. But 1 is a $\mathbb{Z}$-linear combination of $\frac{N\alpha}{a}$, $\frac{N\beta}{a}$ and $\frac{\mathrm{Tr}\,\alpha\beta'}{a}$, hence a $\mathcal{O}_k$-linear combination of $\frac{N\alpha}{a}$, $\frac{N\beta}{a}$ and $\frac{\alpha\beta'}{a} + \frac{\alpha'\beta}{a}$, and the claim follows.   □

The natural number a in Proposition 6.8 is called the norm of the ideal $\mathfrak{a}$; we thus have $\mathfrak{a}\mathfrak{a}' = (N\mathfrak{a})$. Since $(N\mathfrak{a}\mathfrak{b}) = (\mathfrak{a}\mathfrak{b})(\mathfrak{a}\mathfrak{b})' = (\mathfrak{a}\mathfrak{a}')(\mathfrak{b}\mathfrak{b}') = (N\mathfrak{a})(N\mathfrak{b})$, the ideal norm is multiplicative. Other important properties of the norm of ideals are the following:

- $N\mathfrak{a} = 1 \iff \mathfrak{a} = (1)$: In fact, $N\mathfrak{a} = 1$ implies $(1) = \mathfrak{a}\mathfrak{a}' \subseteq \mathfrak{a} \subseteq \mathcal{O}_k = (1)$, and the converse is clear.
- $N\mathfrak{a} = 0 \iff \mathfrak{a} = (0)$: It follows from $\mathfrak{a}\mathfrak{a}' = (0)$ that $N\alpha = \alpha\alpha' = 0$ for all $\alpha \in \mathfrak{a}$.

The following property shows that the norm of an ideal can easily be computed from its $\mathbb{Z}$-basis:

Proposition 6.10 *Let* $\mathfrak{a}$ *be an ideal in* $\mathcal{O}_k$. *Write* $\mathfrak{a} \cap \mathbb{Z} = n\mathbb{Z}$ *and* $\mathrm{coeff}(\mathfrak{a}) = m\mathbb{Z}$ *for positive integers* m *and* n. *Then* $N\mathfrak{a} = mn$.

---

[1]This lemma is related to Dedekind's "Prague Theorem"; see [80]. At this point we are using the fact that the ring $\mathcal{O}_k$ is integrally closed, i.e., is equal to the maximal order.

For proving this claim, we write $\mathfrak{a} = n\mathbb{Z} + (a + m\omega)\mathbb{Z}$ as in Prop. 6.7 and set $\alpha = m(b + \omega)$. Then $\mathfrak{a} = (n, \alpha)$, $\mathfrak{a}' = (n, \alpha')$ and

$$\mathfrak{aa}' = (n^2, mn(b + \omega'), mn(b + \omega), m^2 N(b + \omega))$$

$$= (mn)\left(\frac{n}{m}, b + \omega, b + \omega', \frac{1}{n}N(b + \omega)\right).$$

Proposition 6.7 implies that the last ideal is contained in $\mathcal{O}_k$, and hence $(N\mathfrak{a}) = \mathfrak{aa}' \subseteq (mn)\mathcal{O}_k = (mn)$ and thus $mn \mid N\mathfrak{a}$.

For proving the converse $N\mathfrak{a} \mid mn$, we proceed as follows: Let $A = N\mathfrak{a}$, i.e., $\mathfrak{aa}' = (A)$. Since $\alpha \in \mathfrak{a}$ and $n \in \mathfrak{a}'$, we have $n\alpha \in \mathfrak{aa}' = (A)$, and hence $A \mid n\alpha = na + nm\omega$. Since $\{1, \omega\}$ is an integral basis of $\mathcal{O}_k$, this implies $A \mid na$ and $A \mid nm$.

This shows that the norm of an ideal is equal to the ideal generated by the norm of an ideal interpreted as a module.

6.2.3 The Cancellation Law

Now we approach the theorem of unique factorization into prime ideals. The idea behind the proof is the same as for numbers. In that case we could immediately conclude from an equation $\alpha\beta = \alpha\gamma$ with $\alpha \neq 0$ that $\beta = \gamma$ (we just multiply by the inverse of α); in the case of ideals, this is not yet possible since we do not have an "inverse ideal" at our disposal. The fact that the "cancellation law" is nevertheless correct is the content of the next proposition.

Proposition 6.11 (The Cancellation Law) *If $\mathfrak{a}$, $\mathfrak{b}$, and $\mathfrak{c}$ are nonzero ideals in $\mathcal{O}_k$ with $\mathfrak{ab} = \mathfrak{ac}$, then $\mathfrak{b} = \mathfrak{c}$.*

Proof Assume first that $\mathfrak{a} = (\alpha)$ is principal; then $\mathfrak{ab} = \alpha\mathfrak{b}$, and hence $\mathfrak{b} = \alpha^{-1}\mathfrak{ab}$ and $\mathfrak{c} = \alpha^{-1}\mathfrak{ac} = \alpha^{-1}\mathfrak{ab} = \mathfrak{b}$.

If $\mathfrak{a}$ is an arbitrary ideal, then $\mathfrak{ab} = \mathfrak{ac}$ immediately implies that $(\mathfrak{aa}')\mathfrak{b} = (\mathfrak{aa}')\mathfrak{c}$. Since $\mathfrak{aa}' = (N\mathfrak{a})$ is principal, the claim now follows from the first part of the proof. $\qquad\square$

Thus the ideals in $\mathcal{O}_k$ form a monoid with cancellation. Such monoids can be completed to a group in a formal way by imitating the construction of $\mathbb{Q}$ from $\mathbb{Z}$, namely by considering expressions of the form $\mathfrak{a}/\mathfrak{b}$, which are multiplied via $\mathfrak{a}/\mathfrak{b} \cdot \mathfrak{c}/\mathfrak{d} = \mathfrak{ac}/\mathfrak{bd}$. It is possible to interpret an element $\mathfrak{ab}^{-1}$ of this group as a set by setting $\mathfrak{ab}^{-1} = \frac{1}{b}\mathfrak{ab}'$, where b is the norm of $\mathfrak{b}$, and defining $\frac{1}{m}\mathfrak{a} = \{\frac{\alpha}{m} : \alpha \in \mathfrak{a}\}$. Such sets are called *fractional ideals*.

6.2.4   Divisibility of Ideals

Now that we have defined products of ideals we can study divisibility questions. Of course we say that an ideal $\mathfrak{b}$ is divisible by an ideal $\mathfrak{a}$ if there exists an ideal $\mathfrak{c}$ such that $\mathfrak{b} = \mathfrak{a}\mathfrak{c}$. Since $\mathfrak{c} \subseteq \mathcal{O}_k$, it follows from $\mathfrak{a} \mid \mathfrak{b}$ that $\mathfrak{b} = \mathfrak{a}\mathfrak{c} \subseteq \mathfrak{a}(1) = \mathfrak{a}$: To divide is to contain. The converse also holds.

Proposition 6.12 *If $\mathfrak{a}$ and $\mathfrak{b}$ are ideals $\neq (0)$ with $\mathfrak{a} \supseteq \mathfrak{b}$, then $\mathfrak{a} \mid \mathfrak{b}$.*

Proof It follows from $\mathfrak{a} \supseteq \mathfrak{b}$ that $\mathfrak{b}\mathfrak{a}' \subseteq \mathfrak{a}\mathfrak{a}' = (a)$ with $a = N\mathfrak{a}$. Then $\mathfrak{c} = \frac{1}{a}\mathfrak{b}\mathfrak{a}'$ is an ideal since $\frac{1}{a}\mathfrak{a}'\mathfrak{b} \subseteq \mathcal{O}_k$ (the algebraic properties of ideals are easily verified). Now the claim follows from $\mathfrak{a}\mathfrak{c} = \frac{1}{a}\mathfrak{b}\mathfrak{a}\mathfrak{a}' = \mathfrak{b}$. $\qquad\square$

The notions of irreducible, maximal, and prime ideals are perhaps known from commutative algebra. We remind the readers that an ideal $\mathfrak{a} \neq (0), (1)$ is called

- irreducible if $\mathfrak{a} \neq \mathfrak{b}\mathfrak{c}$ for ideals $\mathfrak{b}, \mathfrak{c} \neq (1)$, i.e., if the ideal is not a product of nontrivial ideals;
- maximal if $\mathfrak{a} \subseteq \mathfrak{b} \subseteq (1)$ implies $\mathfrak{b} = \mathfrak{a}$ or $\mathfrak{b} = (1)$;
- prime if $\mathfrak{a} \mid \mathfrak{b}\mathfrak{c}$ implies $\mathfrak{a} \mid \mathfrak{b}$ or $\mathfrak{a} \mid \mathfrak{c}$.

The rings of integers in algebraic number fields have the pleasant property that these three notions coincide (in a domain, the zero ideal (0) is prime but not maximal).

- Irreducible ideals are maximal: If $\mathfrak{a}$ is not maximal, then there is an ideal $\mathfrak{b}$ with $\mathfrak{a} \subsetneq \mathfrak{b} \subsetneq (1)$, but then $\mathfrak{b} \mid \mathfrak{a}$ with $\mathfrak{b} \neq (1), \mathfrak{a}$.
- Maximal ideals are irreducible: It follows from $\mathfrak{a} = \mathfrak{b}\mathfrak{c}$ that $\mathfrak{a} \subsetneq \mathfrak{b} \subsetneq (1)$.

It also follows from the definition that maximal ideals are always prime:

- Irreducible (and thus maximal) ideals are prime. Assume that the ideal $\mathfrak{a}$ is irreducible and that $\mathfrak{a} \mid \mathfrak{b}\mathfrak{c}$, but $\mathfrak{a} \nmid \mathfrak{b}$; we have to show that $\mathfrak{a} \mid \mathfrak{c}$. To this end, we observe that the ideal $\mathfrak{a} + \mathfrak{b} = \{\alpha + \beta : \alpha \in \mathfrak{a}, \beta \in \mathfrak{b}\}$ (it is easily checked that this is an ideal; once we know that the factorization into prime ideals exists and is unique, we will see that $\mathfrak{a} + \mathfrak{b}$ is the greatest common divisor of $\mathfrak{a}$ and $\mathfrak{b}$) contains $\mathfrak{a}$ and hence divides it. On the other hand, $\mathfrak{a} + \mathfrak{b} \neq \mathfrak{a}$ since this would imply that $\mathfrak{a} = \mathfrak{a} + \mathfrak{b} \supseteq \mathfrak{b}$ and hence $\mathfrak{a} \mid \mathfrak{b}$, contradicting our assumptions. Since $\mathfrak{a}$ is irreducible, we must have $\mathfrak{a} + \mathfrak{b} = (1)$. This implies that there exist $\alpha \in \mathfrak{a}$ and $\beta \in \mathfrak{b}$ such that $1 = \alpha + \beta$. If $\gamma \in \mathfrak{c}$ is arbitrary, then $\gamma = \alpha\gamma + \beta\gamma$; but $\alpha\gamma \in \mathfrak{a}$ and $\beta\gamma \in \mathfrak{b}\mathfrak{c} \subseteq \mathfrak{a}$, and hence $\gamma \in \mathfrak{a}$. Thus we have shown that $\mathfrak{c} \subseteq \mathfrak{a}$, which implies $\mathfrak{a} \mid \mathfrak{c}$.

The proof that prime ideals in rings of integers of quadratic number fields are maximal uses Proposition 6.12, which does not hold in general domains (see Exercise 6.31):

- Prime ideals are irreducible and hence maximal. In fact, it follows from $\mathfrak{a} = \mathfrak{b}\mathfrak{c}$ and $\mathfrak{a} \nmid \mathfrak{b}$ that $\mathfrak{a} \mid \mathfrak{c}$, and since $\mathfrak{c} \mid \mathfrak{a}$ (to divide is to contain), we obtain $\mathfrak{a} = \mathfrak{c}$ and thus $\mathfrak{b} = (1)$.

It is not obvious how to conclude from $\mathfrak{a} \mid \mathfrak{c}$ and $\mathfrak{c} \mid \mathfrak{a}$ that $\mathfrak{a} = \mathfrak{c}$ without Proposition 6.12. From $\mathfrak{a} = \mathfrak{c}\mathfrak{d}$ and $\mathfrak{c} = \mathfrak{a}\mathfrak{e}$, we obtain $\mathfrak{a} = \mathfrak{d}\mathfrak{e}\mathfrak{a}$. But without the cancellation law, this does not allow us to conclude that $\mathfrak{d}\mathfrak{e} = (1)$.

Now we prove the main theorem of the theory of ideals in quadratic number fields.

Theorem 6.13 *Each ideal $\mathfrak{a} \neq (0)$ in the ring of integers $\mathcal{O}_k$ in some quadratic number field k can be written as a product of prime ideals, and this factorization is unique up to order.*

Proof We begin by proving the existence of a factorization into irreducible ideals. If $\mathfrak{a}$ is irreducible, we are done. If not, then $\mathfrak{a} = \mathfrak{b}\mathfrak{c}$; if $\mathfrak{b}$ and $\mathfrak{c}$ are irreducible, we are done. If not, we go on factoring the ideals. Since $N\mathfrak{a} = N\mathfrak{b}N\mathfrak{c}$ and $1 < N\mathfrak{b}, N\mathfrak{c} < N\mathfrak{a}$, etc., this procedure must terminate since norms are natural numbers and so cannot decrease indefinitely.

Now assume that we are given two factorizations $\mathfrak{a} = \mathfrak{p}_1 \cdots \mathfrak{p}_r = \mathfrak{q}_1 \cdots \mathfrak{q}_s$ of $\mathfrak{a}$ into prime ideals. Since $\mathfrak{p}_1$ is prime, it divides some $\mathfrak{q}_j$ on the right side. Rearranging the order of the factors, we may assume that $\mathfrak{p}_1 \mid \mathfrak{q}_1$. Since $\mathfrak{q}_1$ is irreducible, we must have $\mathfrak{p}_1 = \mathfrak{q}_1$, and the cancellation law yields $\mathfrak{p}_2 \cdots \mathfrak{p}_r = \mathfrak{q}_2 \cdots \mathfrak{q}_s$. The claim now follows by induction on the number of prime ideal factors of $\mathfrak{a}$.   □

Remark The assumption that $\mathcal{O}_k$ is the full ring of integers is important. The domain $R = \mathbb{Z}[\sqrt{-3}\,]$, for example, does not have unique factorization into irreducible ideals. In fact, we have $(2)(2) = (1 + \sqrt{-3})(1 - \sqrt{-3})$, and the ideal (2) is irreducible. We cannot have $(2) = (1 + \sqrt{-3})$ since then $\frac{1+\sqrt{-3}}{2} \in R$, which is not true.

Working in residue class rings modulo ideals I is easy. We write $a \equiv b \bmod I$ if $a - b \in I$. In order to reduce $17 + 19\sqrt{-5}$ modulo $I = (3, 1 + \sqrt{-5})$, we first reduce modulo $1 + \sqrt{-5}$; since $1 + \sqrt{-5} \in I$, we have $\sqrt{-5} \equiv -1 \bmod I$, and hence $17 + 19\sqrt{-5} \equiv 17 - 19 \equiv -3 \bmod I$. Reducing the result modulo 3 then shows that $17 + 19\sqrt{-5} \equiv -2 \equiv 1 \bmod I$.

Observe that this generalizes the usual notion of congruences: If $I = mR$ is a principal ideal, then $a - b \in mR$ is equivalent to $m \mid (a - b)$. The set of residue classes of a domain modulo an ideal I forms a ring which we denote by R/I.

6.2.5 Description of Prime Ideals

If $\mathfrak{p}$ is a prime ideal in $\mathcal{O}_k$, then there is a unique prime number $p > 0$ with $\mathfrak{p} \mid (p)$. In fact, $\mathfrak{p} \mid \mathfrak{p}\mathfrak{p}' = (N\mathfrak{p})$; factoring $N\mathfrak{p}$ in the integers and observing that $\mathfrak{p}$ is prime, we deduce that there is a prime number p such that $\mathfrak{p} \mid p$. The fact that $\mathfrak{p}$ cannot divide two distinct prime numbers should be clear: If $\mathfrak{p} \mid (p)$ and $\mathfrak{p} \mid (q)$, then $p, q \in \mathfrak{p}$, hence $1 \in \mathfrak{p}$, and this is a contradiction.

If p is the prime number that $\mathfrak{p}$ divides, then we say that $\mathfrak{p}$ lies above p. Since the ideal (p) in $\mathcal{O}_k$ has norm p^2, it follows that each prime ideal above p has norm p or p^2.

The determination of all prime ideals in $\mathcal{O}_k$ is not difficult (the case $p = 2$ is taken care of in Exercise 6.21).

Theorem 6.14 *Let p be an odd prime number, m a squarefree integer, and $k = \mathbb{Q}(\sqrt{m})$ a quadratic number field with discriminant Δ. Then we have the following:*

- *If $p \mid \Delta$, then $(p) = (p, \sqrt{m})^2$; we say that p is* ramified.
- *If $(\Delta/p) = +1$, then $(p) = \mathfrak{p}\mathfrak{p}'$ for prime ideals $\mathfrak{p} \neq \mathfrak{p}'$; we say that p splits.*
- *If $(\Delta/p) = -1$, then the ideal (p) is prime; we say that p is inert.*

Proof Assume first that $p \mid \Delta$; since p is odd, we have $p \mid m$. Now $(p, \sqrt{m})^2 = (p^2, p\sqrt{m}, m) = (p)(p, \sqrt{m}, \frac{m}{p}) = (p)$ since the ideal $(p, \sqrt{m}, \frac{m}{p})$ contains the coprime integers p and $\frac{m}{p}$ and thus is equal to the unit ideal (1).

Now assume that $(\Delta/p) = 1$; then Δ is a quadratic residue modulo p, and since $\Delta = m$ or $\Delta = 4m$, so is m. Thus there is an $x \in \mathbb{Z}$ with $x^2 \equiv m \bmod p$. We set $\mathfrak{p} = (p, x + \sqrt{m})$ and find

$$\mathfrak{p}\mathfrak{p}' = (p^2, p(x + \sqrt{m}), p(x - \sqrt{m}), x^2 - m)$$
$$= (p)(p, x + \sqrt{m}, x - \sqrt{m}, (x^2 - m)/p).$$

Clearly, $2\sqrt{m} = x + \sqrt{m} - (x - \sqrt{m})$, and thus $4m = (2\sqrt{m})^2$ is contained in the last ideal. Since p and $4m$ are coprime, this ideal is the unit ideal, and we have $\mathfrak{p}\mathfrak{p}' = (p)$. If we had $\mathfrak{p} = \mathfrak{p}'$, then it would follow as above that $4m \in \mathfrak{p}$ and $\mathfrak{p} = (1)$: a contradiction.

Finally, assume that $(\Delta/p) = -1$. If there were an ideal $\mathfrak{p}$ with norm p, then by Proposition 6.7 it would have the form $\mathfrak{p} = (p, b + \omega)$ with $p \mid N(b + \omega)$. If $\omega = \sqrt{m}$, this means $b^2 - m \equiv 0 \bmod p$, hence $(\Delta/p) = (4m/p) = (m/p) = +1$ contradicting our assumption. If $\omega = \frac{1}{2}(1 + \sqrt{m})$, then $(2b + 1)^2 \equiv m \bmod p$, and we get a contradiction as above.   □

We can combine the two cases $p = 2$ and $p \neq 2$ by using the *Kronecker symbol* (Δ/p). Recall that this symbol coincides with the Legendre symbol for odd values of p; for $p = 2$ and $\Delta \equiv 1 \bmod 4$, it is defined by $(\Delta/2) = (-1)^{(\Delta-1)/4}$, and for $\Delta \not\equiv 1 \bmod 4$, we set $(\Delta/2) = 0$. Using the Kronecker symbol, a prime number p splits, ramifies, or is inert according as $(\frac{\Delta}{p}) = +1, 0,$ or -1, respectively.

6.3 Ideal Class Groups

As we have seen, we may think of ideals as a substitute for greatest common divisors. The elements 2 and $1 + \sqrt{-5}$, for example, do not have a common divisor

$\neq \pm 1$, whereas the ideal $(2, 1 + \sqrt{-5})$ generated by them describes the "correct" greatest common divisor.

There are also pairs of elements with common divisors but without a greatest one. For example,

$$6 = 2 \cdot 3 = (1 + \sqrt{-5})(1 + \sqrt{-5}) \quad \text{and} \quad -4 + 2\sqrt{-5} = 2(-2 + \sqrt{-5}) = (1 + \sqrt{-5})^2$$

have common divisors 2 and $1 + \sqrt{-5}$, but there is no greatest common divisor.

On the other hand, there are elements in $\mathbb{Z}[\sqrt{-5}]$ that do have a greatest common divisor, for example, $2 + 2\sqrt{-5}$ and $3 + 3\sqrt{-5}$. Here we have

$$\gcd(2 + 2\sqrt{-5}, 3 + 3\sqrt{-5}) = (1 + \sqrt{-5})\gcd(2, 3) \sim 1 + \sqrt{-5}.$$

How can we decide whether such a greatest common divisor of two elements α and β exists? To answer this question, we consider the ideal (α, β) generated by them and check whether it is principal. If $(\alpha, \beta) = (\delta)$, then δ is an "honest" greatest common divisor of α and β. One goal of this (and the next) chapter is providing a method for testing whether an ideal is principal or not.

6.3.1 Equivalence of Ideals

We have seen that the set of integral ideals $\neq (0)$ in $\mathcal{O}_k$ forms a monoid with the cancellation law. Such monoids can be made into a group I_k in a rather formal way resembling the construction of the field $\mathbb{Q}$ of rational numbers from the multiplicative monoid $\mathbb{Z}$. Such quotients of ideals are called fractional ideals. Formally, two such ideals $\mathfrak{a}/\mathfrak{b}$ and $\mathfrak{c}/\mathfrak{d}$ are multiplied in the same way as fractions of numbers, and of course we may cancel common factors. Principal ideals of the form $(\alpha) = \alpha\mathcal{O}_k$, where $\alpha \in k^\times$ is not necessarily an algebraic integer, are called principal fractional ideals, and they form the subgroup P_k in I_k. The quotient group $\mathrm{Cl}(k) = I_k/P_k$ is called the *ideal class group* of k.

Those who do not like such a formal approach may describe fractional ideals as sets. In fact, write a fractional ideal $\mathfrak{a}\mathfrak{b}^{-1}$ as $\mathfrak{a}\mathfrak{b}'(\mathfrak{b}\mathfrak{b}')^{-1} = \frac{1}{b}\mathfrak{a}\mathfrak{b}$, where $b = N\mathfrak{b}$ denotes the norm of $\mathfrak{b}$. Then we define $\frac{1}{\alpha}\mathfrak{c} := \{\frac{\gamma}{\alpha} : \gamma \in \mathfrak{c}\}$. On the set of fractional ideals $\neq (0)$, we define products as for integral ideals; then we show that they form a group.

We will use a third approach that does not use any fractional ideals. In fact, the definition of the ideal class group above implies that two ideals $\mathfrak{a}$ and $\mathfrak{b}$ belong to the same class modulo the group P_k of principal ideals if $\mathfrak{a} = \xi\mathfrak{b}$ for some $\xi \in k^\times$. If we write $\xi = \beta/\alpha$ with $\alpha, \beta \in \mathcal{O}_k$, then this is equivalent with $\alpha\mathfrak{a} = \beta\mathfrak{b}$.

Such equations define an equivalence relation on the set of nonzero integral ideals: We will call ideals $\mathfrak{a}$ and $\mathfrak{b}$ equivalent and write $\mathfrak{a} \sim \mathfrak{b}$ if there exist elements

$\alpha, \beta \in \mathcal{O}_k$ such that $\alpha\mathfrak{a} = \beta\mathfrak{b}$. Of course we have to verify the usual axioms: symmetry, reflexivity, and transitivity (see Exercise 6.20).

With respect to this notion of equivalence, all principal ideals belong to the same equivalence class. In fact, if $\mathfrak{a} = (\alpha)$, then $1 \cdot \mathfrak{a} = \alpha \cdot (1)$, and this shows that $\mathfrak{a} \sim (1)$. Conversely, each ideal equivalent to the unit ideal is principal: $\alpha\mathfrak{a} = \beta(1)$ implies $\mathfrak{a} = (\frac{\beta}{\alpha})$, and since $\mathfrak{a}$ is an integral ideal, $\frac{\beta}{\alpha} = \gamma$ must be integral, as well. Thus all principal ideals are contained in the class of the unit ideal (1).

This implies that $\mathcal{O}_k$ has exactly one equivalence class of ideals if and only if each ideal is principal.

Proposition 6.15 *The ring $\mathcal{O}_k$ of integers in a quadratic number field k is a principal ideal domain if and only if k has class number 1.*

In particular, there are at least two ideal classes in $\mathbb{Z}[\sqrt{-5}]$: Since the prime ideal $\mathfrak{p} = (2, 1 + \sqrt{-5})$ is not principal, $\mathfrak{p}$ cannot belong to the equivalence class of principal ideals. Observe, however, that $\mathfrak{p}$ and $\mathfrak{q} = (3, 1 + \sqrt{-5})$ belong to the same ideal class: In order to satisfy the condition $\alpha\mathfrak{p} = \beta\mathfrak{q}$, we multiply this equation by $\mathfrak{p}$ and find, using $\mathfrak{p}^2 = (2)$ and $\mathfrak{p}\mathfrak{q} = (1 + \sqrt{-5})$, that $\alpha(2) = \beta(1 + \sqrt{-5})$. Thus it is sufficient to set $\alpha = 1 + \sqrt{-5}$ and $\beta = (2)$, and in fact the equation $(1 + \sqrt{-5})\mathfrak{p} = 2\mathfrak{q}$ is correct since both ideals have the same prime ideal factorization $\mathfrak{p}^2\mathfrak{q}$.

On the set of equivalence classes of ideals, we now introduce a group structure as follows. Given ideal classes c and d, we choose ideals $\mathfrak{a} \in c$ and $\mathfrak{b} \in d$ and call the class $cd = [\mathfrak{a}\mathfrak{b}]$ the product of c and d. We have to check that the product does not depend on the choice of the representatives (this is a simple exercise). Clearly, the class of principal ideals is the neutral element; associativity is inherited from the associativity of the multiplication of ideals, and the existence of the inverse follows from the fact that $\mathfrak{a}\mathfrak{a}' = (a)$ is principal; in other words, we have $[\mathfrak{a}]^{-1} = [\mathfrak{a}']$.

This shows that the ideal classes of a quadratic number field k form a group, which is called the *ideal class group* of k, and which is denoted by $\mathrm{Cl}(k)$. Together with the unit group, it is the most important invariant of a number field. The goal of this section is proving that the *class number* $h_k = \#\mathrm{Cl}(k)$ is finite. Our proof is constructive and will allow us to actually compute the ideal class group of a given quadratic number field.

6.3.2   Finiteness of the Class Number

We will now show that each ideal class in the ring of integers of a quadratic number field k contains an ideal whose norm is bounded by a constant depending only on k. This immediately implies that the class number is finite. Let us call an ideal *primitive* if it is not divisible by an ideal of the form $(m) \neq (1)$ with $m \in \mathbb{Z}$. Clearly, each ideal class is represented by a primitive ideal since dividing an ideal by the principal ideal (m) does not change its class.

By Proposition 6.7, each ideal $\mathfrak{a}$ possesses a $\mathbb{Z}$-basis of the form $\{n, m(b + \omega)\}$ with $m \mid n$; in particular, $\mathfrak{a}$ is primitive if and only if $m = 1$. In other words:

Proposition 6.16 *If the ideal $\mathfrak{a}$ is primitive, then there exist $n \in \mathbb{N}$ and $b \in \mathbb{Z}$ with $\mathfrak{a} = n\mathbb{Z} \oplus (b + \omega)\mathbb{Z}$, and we have $N\mathfrak{a} = n$. In particular, we have $\mathfrak{a} \cap \mathbb{Z} = (N\mathfrak{a})$ in this case.*

Now we claim the following:

Theorem 6.17 *Let $m \in \mathbb{Z}$ be squarefree and $k = \mathbb{Q}(\sqrt{m})$ a quadratic number field with ring of integers $\mathcal{O}_k = \mathbb{Z}[\omega]$ and discriminant Δ. Define the Gauss bound μ_k by*

$$\mu_k = \begin{cases} \sqrt{\Delta/5}, & \text{if } \Delta > 0, \\ \sqrt{-\Delta/3}, & \text{if } \Delta < 0. \end{cases}$$

Then each ideal class of k contains an integral ideal $\neq (0)$ with norm $\leq \mu_k$; in particular, the number h of ideal classes is finite.

Before we prove this result, we will present a few applications. Clearly, the bounds are best possible since for $\Delta = 5$ and $\Delta = -3$, they cannot be improved. If $\mu_k < 2$, then each ideal class contains an integral ideal $\neq (0)$ with norm < 2, hence with norm 1. The only such ideal is (1), and this implies that there is a single ideal class, namely the class of principal ideals. Thus in this case, $\mathcal{O}_k$ is a unique factorization domain. Theorem 6.17 tells us that this holds for all fields k with discriminant $-12 \leq \Delta \leq 20$, i.e., for $m \in \{-11, -7, -3, -2, -1, 2, 3, 5, 13, 17\}$.

Next consider the ring of integers $R = \mathbb{Z}[\sqrt{-5}]$ in the quadratic number field $k = \mathbb{Q}(\sqrt{-5})$ with $\Delta = -20$; according to Theorem 6.17, each ideal class contains an ideal with norm $< \sqrt{20/3}$ and so with norm ≤ 2. Since there are only two such ideals, namely the principal ideal (1) and the nonprincipal ideal $(2, 1 + \sqrt{-5})$, the field k has class number 2.

Proof of Theorem 6.17 Let $c = [\mathfrak{a}]$ be an ideal class represented by an ideal $\mathfrak{a}$. Without loss of generality, we may assume that $\mathfrak{a}$ is primitive. Thus $\mathfrak{a} = (a, \alpha)$ with $a = N\mathfrak{a}$ and $\alpha = b + \omega = s + \frac{1}{2}\sqrt{\Delta}$ for some $s \in \mathbb{Q}$ with $2s \in \mathbb{Z}$. If $a \leq \mu_k$, then we are done; otherwise, we apply the Euclidean algorithm to the pair (s, a) and find an integer $q \in \mathbb{Z}$ with $s - qa = r$ and

$$|r| \leq \frac{a}{2} \text{ if } \Delta < 0,$$

$$\frac{a}{2} \leq |r| \leq a \text{ if } \Delta > 0.$$

Setting $\alpha_1 = r + \frac{1}{2}\sqrt{\Delta}$, we will show below that

(1) $\alpha_1 \in \mathfrak{a}$,

(2) $|N\alpha_1| \leq \frac{a^2 - \Delta}{4} \leq a^2$, and

(3) $\mathfrak{a}_1 := \frac{1}{a}\alpha_1'\mathfrak{a} \sim \mathfrak{a}$ is an integral ideal with $[\mathfrak{a}_1] = [\mathfrak{a}]$ and $N\mathfrak{a}_1 < N\mathfrak{a}$.

We repeat this step until we have found an ideal with norm $\leq \mu_k$; since the norm decreases at each step by at least 1, this process must terminate after finitely many steps.

Now clearly $\alpha_1 = \alpha - q a \in \mathfrak{a}$, which proves (1). The proof of the inequality (2) is easy: If $\Delta < 0$, we have $|N\alpha_1| = |r^2 - \frac{\Delta}{4}| \leq \frac{a^2 + |\Delta|}{4} < 1$ since $a^2 > \mu_k^2 = \frac{|\Delta|}{3}$, and if $\Delta > 0$, we clearly have $-a^2 = \frac{a^2 - 5a^2}{4} < r^2 - \frac{\Delta}{4} < a^2$.

It remains to show that the ideal $\mathfrak{a}_1$ is integral; but since

$$\frac{1}{a}\alpha_1' \mathfrak{a} \subseteq \mathcal{O}_k \quad \Longleftrightarrow \quad \alpha'\mathfrak{a} \subseteq (a) = \mathfrak{a}\mathfrak{a}' \quad \Longleftrightarrow \quad (\alpha') \subseteq \mathfrak{a}',$$

this is clear.   □

The following observation, which generalizes our results on representations of prime numbers in the form $x^2 + y^2$ or $x^2 + 3y^2$, is an important consequence of Theorem 6.17.

Corollary 6.18 *Let m be a squarefree integer and $k = \mathbb{Q}(\sqrt{m})$ a quadratic number field with class number h, and assume that $p\mathcal{O}_k = \mathfrak{p}\mathfrak{p}'$ in $\mathcal{O}_k$ splits. Then there exist integers $x, y \in \mathbb{N}$ with $\pm 4p^h = x^2 - my^2$.*

Proof The h-th power of each ideal in $\mathcal{O}_k$ is principal. In particular, $\mathfrak{p}^h = (\frac{x + y\sqrt{m}}{2})$, and taking norms, we obtain $p^h = |\frac{x^2 - my^2}{4}|$.   □

6.3.3 Class Group Calculations

We now show how to compute class groups in a given quadratic number field.

$k = \mathbb{Q}(\sqrt{-21})$, $\Delta = -84$ The Gauss bound is $\mu_k = \sqrt{84/3}$, so we have to consider ideals with norm ≤ 5. Since $2 \mid \Delta$, the prime 2 is ramified: $(2) = \mathfrak{a}^2$ for $\mathfrak{a} = (2, 1 + \sqrt{-21})$. Similarly, $(3) = \mathfrak{b}^2$ with $\mathfrak{b} = (3, \sqrt{-21})$. Finally, $(-21/5) = 1$ implies that $(5) = \mathfrak{c}\mathfrak{c}'$ splits into prime ideals of norm 5, namely $\mathfrak{c} = (5, 2 + \sqrt{-21})$ and its conjugate $\mathfrak{c}'$.

The ideals with norm ≤ 5 thus are (1) $\mathfrak{a}$, $\mathfrak{b}$, $\mathfrak{a}^2 = (2)$ $\mathfrak{c}$ and $\mathfrak{c}'$ ($\mathfrak{a}\mathfrak{b}$ already has norm 6). Since $\mathfrak{a}^2 \sim (1)$, it remains to investigate $\mathfrak{a}$, $\mathfrak{b}$, $\mathfrak{c}$, and $\mathfrak{c}'$. Clearly, none of these ideals is principal since $\mathcal{O}_k$ does not contain elements with norms 2, 3, or 5. Similarly, $\mathfrak{a} \not\sim \mathfrak{b}$ since otherwise $(2) = \mathfrak{a}^2 \sim \mathfrak{a}\mathfrak{b}$, and there would exist an element of norm 6, which is not the case.

Now $\mathfrak{a}\mathfrak{b}\mathfrak{c}$ is an ideal with norm 30, and this is the norm of the elements $3 \pm \sqrt{-21}$. Since there is only one ideal above 2 and 3, respectively, we must have $(3 + \sqrt{-21}) = \mathfrak{a}\mathfrak{b}\mathfrak{c}$ or $(3 + \sqrt{-21}) = \mathfrak{a}\mathfrak{b}\mathfrak{c}'$. For finding the correct factorization, we compute $3 + \sqrt{-21}$ mod $\mathfrak{c}$. From $\mathfrak{c} = (5, 2 + \sqrt{-21})$, we deduce that $\sqrt{-21} \equiv -2$ mod $\mathfrak{c}$, and hence $3 + \sqrt{-21} \equiv 3 - 2 \equiv 1$ mod $\mathfrak{c}$, and so we must have $(3 + \sqrt{-21}) = \mathfrak{a}\mathfrak{b}\mathfrak{c}'$. This can be confirmed by verifying that $3 + \sqrt{-21} \equiv 0$ mod $\mathfrak{c}'$; in fact, we have $\sqrt{-21} \equiv 2 \bmod \mathfrak{c}'$, and hence $3 + \sqrt{-21} \equiv 3 + 2 = 5 \equiv 0 \bmod \mathfrak{c}'$. Thus $\mathfrak{abc}' \sim 1$, and now $\mathfrak{c}' \sim \mathfrak{c}^{-1}$ implies $\mathfrak{ab} \sim \mathfrak{c}$.

Finally, $\mathfrak{c}' \sim \mathfrak{c}^{-1} \sim \mathfrak{a}^{-1}\mathfrak{b}^{-1} \sim \mathfrak{ab}$ since $\mathfrak{a}^2 \sim \mathfrak{b}^2 \sim 1$. Thus there exist exactly 4 ideal classes: the class of principal ideals and three classes $[\mathfrak{a}]$, $[\mathfrak{b}]$, and $[\mathfrak{a}][\mathfrak{b}]$ with order 2 in the class group. Thus the class group is isomorphic to $\mathbb{Z}/2\mathbb{Z} \times \mathbb{Z}/2\mathbb{Z}$, which is called Klein's four group.

$k = \mathbb{Q}(\sqrt{-17})$, $\Delta = -68$ Here we have to consider all ideals with norm ≤ 4. We have $(2) = \mathfrak{a}^2$ with $\mathfrak{a} = (2, 1 + \sqrt{-17})$ and $(3) = \mathfrak{bb}'$ with $\mathfrak{b} = (3, 1 + \sqrt{-17})$. The ideals with norm ≤ 4 thus are (1) $\mathfrak{a}$, $\mathfrak{b}$, $\mathfrak{b}'$, and $(2) = \mathfrak{a}^2$.

Now $\mathfrak{b}^2$ cannot be principal: The only elements with norm 9 are ± 3, but $\mathfrak{b}^2 \neq (3) = \mathfrak{bb}'$ by unique factorization into prime ideals. On the other hand, $(1 + \sqrt{-17}) = \mathfrak{ab}^2$ shows that $\mathfrak{b}^2 \sim \mathfrak{a}^{-1} \sim \mathfrak{a}$. Finally, $\mathfrak{b}' \sim \mathfrak{b}^{-1}$, and we see that the class $[\mathfrak{b}]$ generates the whole class group: $\mathfrak{b}^2 \sim \mathfrak{a}$, $\mathfrak{b}^4 \sim \mathfrak{a}^2 \sim 1$ and thus $\mathfrak{b}^3 \sim \mathfrak{b}^{-1} \sim \mathfrak{b}'$. The ideal class group of k therefore is cyclic of order 4, i.e., $\mathrm{Cl}(k) \simeq \mathbb{Z}/4\mathbb{Z}$.

$k = \mathbb{Q}(\sqrt{79})$, $\Delta = 316$ Here the class group is generated by ideals with norm ≤ 7. The ideal $\mathfrak{a} = (2, 1 + \sqrt{79}) = (9 + \sqrt{79})$ is principal. The ideals above the odd prime numbers ≤ 7 are, up to conjugates, $\mathfrak{b} = (3, 1 + \sqrt{79})$, $\mathfrak{c} = (5, 2 + \sqrt{79})$, and $\mathfrak{d} = (7, 3 + \sqrt{79})$. The computation of ideals with small norm and their prime ideal factorizations yields the following table:

α	Prime ideal factorization
$7 + \sqrt{79}$	$\mathfrak{abc}$
$8 + \sqrt{79}$	$\mathfrak{b}'\mathfrak{c}'$
$10 + \sqrt{79}$	$\mathfrak{bd}$
$11 + \sqrt{79}$	$\mathfrak{ab}'\mathfrak{d}'$
$17 + 2\sqrt{79}$	$\mathfrak{b}^3$

Since $[\mathfrak{a}] = [(1)]$, it follows from the first relation that $[\mathfrak{b}] = [\mathfrak{c}'] = [\mathfrak{c}]^{-1}$. The third relation shows that $[\mathfrak{b}] = [\mathfrak{d}'] = [\mathfrak{d}]^{-1}$. Thus the ideal class group is generated by the ideal $\mathfrak{b}$ whose third power is principal. It remains to check whether $\mathfrak{b}$ is principal. As we will see in the next chapter in connection with the bounds (7.7), this is not the case. Thus $\mathrm{Cl}(k) \simeq \mathbb{Z}/3\mathbb{Z}$.

Here is a small table with nontrivial class numbers for practicing class number calculations:

Δ	-52	-23	-20	-15	40	60
h	2	3	2	2	2	2

For another beautiful application of class groups, consider the field $k = \mathbb{Q}(\sqrt{-5})$. We have seen above that its ideal class group consists of the classes of (1) and $\mathfrak{a} = (2, 1 + \sqrt{-5})$. Let $\mathfrak{p}$ be a prime ideal with norm p (thus $(-5/p) = +1$ and

$p\mathcal{O}_k = \mathfrak{p}\mathfrak{p}'$). Then either $\mathfrak{p} = (a + b\sqrt{-5})$ is principal and thus $p = a^2 + 5b^2$, or $\mathfrak{p} \sim \mathfrak{a}$, and then $\mathfrak{a}\mathfrak{p} = (C + d\sqrt{-5})$ is principal. In this case we find $2p = C^2 + 5d^2$; but since C and d are odd, we can write $C = 2c + d$ for some $c \in \mathbb{Z}$, and then we find $2p = (2c + d)^2 + 5d^2 = 4c^2 + 4cd + 6d^2$, and hence $p = 2c^2 + 2cd + 3d^2$. In other words, if $(-5/p) = +1$, then p can be written in the form $p = a^2 + 5b^2$ or $p = 2c^2 + 2cd + 3d^2$.

Polynomials $Ax^2 + Bxy + Cy^2 \in \mathbb{Z}[x, y]$ are called binary quadratic forms. Their discriminant is defined by $\Delta = B^2 - 4AC$. In particular, the binary quadratic forms $x^2 + 5y^2$ and $2x^2 + 2xy + 3y^2$ both have discriminant $\Delta = -20$. This is not a coincidence: Gauss defined an equivalence relation on the set of binary quadratic forms with the same discriminant, and Dirichlet and Dedekind have shown that these equivalence classes correspond, in the case of fundamental discriminants, to ideal classes in quadratic number fields (with a technical complication in case of positive discriminants). For $\Delta = -20$, there exist two different classes, namely those represented by $x^2 + 5y^2$ and $2x^2 + 2xy + 3y^2$.

By the Modularity Theorem, we have

$$\left(\frac{-5}{p}\right) = +1 \quad \Longleftrightarrow \quad p \equiv 1, 3, 7, 9 \bmod 20.$$

If we investigate which prime numbers are represented by which of the forms above, then we find

$$p = \begin{cases} x^2 + 5y^2 & \text{if} \quad p \equiv 1, 9 \bmod 20, \\ 2x^2 + 2xy + 3y^2 & \text{if} \quad p \equiv 3, 7 \bmod 20. \end{cases}$$

Examples. $29 = 3^2 + 5 \cdot 2^2$, $41 = 6^2 + 5 \cdot 1^2$, $3 = 2 \cdot 1^2 + 2 \cdot 1 \cdot (-1) + 3 \cdot (-1)^2$, $7 = 2 \cdot 1^2 + 2 \cdot 1 \cdot 1 + 3 \cdot 1^2$, etc.

This observation can be proved easily. In fact, we have $p = x^2 + 5y^2 \equiv x^2 + y^2 \equiv 0, 1 \bmod 4$. If p is prime, then $p \equiv 1 \bmod 4$, and since we also have $p \equiv \pm 1 \bmod 5$ because of $(\frac{5}{p}) = +1$, this happens only for $p \equiv 1, 9 \bmod 20$. If $p = 2x^2 + 2xy + 3y^2$, on the other hand, then y is odd, and hence $p \equiv 2x^2 + 2x + 3 = 2x(x+1) + 3 \equiv 3 \bmod 4$ (since $x(x + 1)$ is always even).

Proposition 6.19 *The quadratic form* $Q_0(x, y) = x^2 + 5y^2$ *represents all prime numbers* p *with* $(\frac{-1}{p}) = (\frac{5}{p}) = +1$, *the form* $Q_1(x, y) = 2x^2 + 2xy + 3y^2$ *all with* $(\frac{-1}{p}) = (\frac{5}{p}) = -1$.

This pretty observation is a special case of genus theory, which we will investigate in Chap. 9. This observation also reflects the fact that the genus class field of $\mathbb{Q}(\sqrt{-5})$ is the biquadratic extension $\mathbb{Q}(\sqrt{-1}, \sqrt{5})$; for more in this direction, see the beautiful book [25] by D. Cox.

6.4   The Diophantine Equation $y^2 = x^3 - d$

Let us now investigate what we can say about the solutions of the Bachet–Mordell equation $y^2 = x^3 - d$ for integers $d > 0$, where we will make suitable assumptions on d in the course of our calculations.

We begin by factoring the right side and write

$$x^3 = y^2 + d = (y + \sqrt{-d})(y - \sqrt{-d}).$$

We would like the ideals $\mathfrak{a} = (y + \sqrt{-d})$ and $\mathfrak{a}'$ to be coprime. Clearly, each common prime ideal factor $\mathfrak{p}$ (with $\mathfrak{p} \mid p$) also divides the difference $2\sqrt{-d}$; since $\mathfrak{p} \mid \sqrt{-d}$ (and $p \neq 2$) immediately yields $p \mid d$, $p \mid y$, $p \mid x$ and finally $p^2 \mid d$, we may exclude this case by assuming that $\boxed{d \text{ is squarefree}}$. Thus only the possibility $\mathfrak{p} \mid 2$ remains; we now distinguish the following cases:

- $d \equiv 2 \bmod 4$: Then $\mathfrak{p} \mid (\sqrt{-d})$ (since $\mathfrak{p} = (2, \sqrt{-d})$), hence $\mathfrak{p} \mid y$, $p \mid y$, and finally $x^3 = y^2 + d \equiv 2 \bmod 4$: this is a contradiction since a cube cannot be divisible exactly by 2.
- $d \equiv 1 \bmod 4$: Then $\mathfrak{p} = (2, 1 + \sqrt{-d})$, and hence $\mathfrak{p} \mid (y + \sqrt{-d})$ if and only if y is odd. This implies $x^3 = y^2 + d \equiv 1 + 1 \equiv 2 \bmod 4$, and this is again a contradiction.
- $d \equiv 3 \bmod 4$: Here $y + \sqrt{-d}$ is divisible by $\mathfrak{p}$ (even by 2) if and only if y is odd. It follows from $d = x^3 - y^2$ that x must be even, and hence $d \equiv -y^2 \equiv -1 \bmod 8$. Thus if we assume that $\boxed{d \not\equiv 7 \bmod 8}$, then $\mathfrak{p} \mid 2$ cannot be a common divisor of $\mathfrak{a}$ and $\mathfrak{a}'$ also in this case.

Thus $\mathfrak{a}$ and $\mathfrak{a}'$ are in fact coprime. Since their product is a cube, there is an ideal $\mathfrak{b}$ such that $\mathfrak{a} = \mathfrak{b}^3$, which implies that we also have $\mathfrak{a}'^3 = \mathfrak{b}'^3$. Now we need the next assumption: If h denotes the class number of $\mathbb{Q}(\sqrt{-d})$, then we demand that $\boxed{3 \nmid h}$. Then $\mathfrak{b}^3$ and $\mathfrak{b}^h$ are principal ideals, hence so is $\mathfrak{b}^{3a+hb}$ for all $a, b \in \mathbb{Z}$, and since 3 and h are coprime we find, using Bézout, that $\mathfrak{b}$ itself must be principal, say $\mathfrak{b} = (\frac{r + s\sqrt{-d}}{2})$ for integers r, s with $r \equiv s \bmod 2$.

If we assume that $\boxed{d > 0, d \neq 1, 3}$, then ± 1 are the only units, and from the equation of ideals above, we obtain the equation of elements

$$y + \sqrt{-d} = \left(\frac{r + s\sqrt{-d}}{2}\right)^3,$$

where we have subsumed the sign into the cube. Comparing coefficients now yields $1 = \frac{1}{8}(3r^2 s - ds^3)$, and therefore $8 = 3r^2 s - ds^3 = s(3r^2 - ds^2)$. Clearly, we must have $s \mid 8$, and hence $s = \pm 1$ or $r \equiv s \equiv 0 \bmod 2$. In the first case we find $\pm 8 = 3r^2 - d$, hence $d = 3r^2 \mp 8$, and in the second case, we set $r = 2t$, $s = 2u$ and find $1 = u(3t^2 - du^2)$, that is, $u = \pm 1$ and $d = 3t^2 \mp 1$.

Thus we have shown: If d satisfies our assumptions and does not have the form $d = 3t^2 \pm 1$ or $d = 3t^2 \pm 8$, then the Diophantine equation $y^2 = x^3 - d$ does not have an integral solution.

What happens if d has one of these forms? Assume for example that $d = 3r^2 - 8$; then comparing coefficients immediately yields (observe that $s = 1$)

$$8y = r^3 - 3dr = r^3 - 9r^3 + 24r = 24r - 8r^3,$$

hence $y = (3 - r^2)r$, and

$$y^2 + d = r^6 - 6r^4 + 12r^2 - 8 = (r^2 - 2)^3,$$

hence $x = r^2 - 2$. Thus any representation $d = 3r^2 - 8$ corresponds to an integral solution $(r^2 - 2, \pm(3 - r^2)r)$ of our Diophantine equation. Similarly, the other representations $d = 3r^2 + 8, 3t^2 + 1$, and $3t^2 - 1$ correspond to the pairs of solutions $(r^2 + 2, \pm r(r^2 + 3))$, $(4t^2 + 1, \pm t(8t^2 + 3))$, and $(4t^2 - 1, \pm t(8t^2 - 3))$.

Of course we now can ask whether an integer d can have more than one such representation. The answer is surprisingly simple: Only $d = 11$ possesses two such representations, all other d have at most one. The proof is not difficult: Equations such as $3r^2 - 8 = 3t^2 - 1$ are impossible modulo 3; the remaining equations are $3r^2 - 8 = 3t^2 + 1$ (this yields $3(r^2 - t^2) = 9$, hence $r^2 - t^2 = (r - t)(r + t) = 3$, with the only solution $r = \pm 2, t = \pm 1$, which leads to $d = 4$, which is not squarefree), and $3r^2 + 8 = 3t^2 - 1$ (this yields similarly $3 = t^2 - r^2$, hence $t = \pm 2, r = \pm 1$, and thus $d = 3 + 8 = 3 \cdot 2^2 - 1 = 11$).

Thus we have proved the following result[2]:

Theorem 6.20 *Let $d \neq 1, 3$ be a squarefree natural number, and $d \not\equiv 7$ mod 8. If the class number of $\mathbb{Q}(\sqrt{-d}\,)$ is not divisible by 3, then the Diophantine equation $y^2 = x^3 - d$*

1. *has exactly two pairs of integral solutions $(3, \pm 4)$ and $(15, \pm 58)$ for $d = 11$;*
2. *has exactly one pair of integral solutions if $d \neq 11$ has the form $d = 3t^2 \pm 1$ or $d = 3t^2 \pm 8$:*

d	(x, y)
$3t^2 - 1$	$(4t^2 - 1, \pm t(8t^2 - 3))$
$3t^2 + 1$	$(4t^2 + 1, \pm t(8t^2 + 3))$
$3t^2 - 8$	$(t^2 - 2, \pm t(3 - t^2))$
$3t^2 + 8$	$(t^2 + 2, \pm t(3 + t^2))$

3. *does not have any integral solutions otherwise.*

---

[2]In a similar way it can be shown that the Fermat equation $x^p + y^p = z^p$ for prime exponents p has only trivial solutions with $xyz = 0$ if p does not divide the class number of the field $\mathbb{Q}(\zeta_p)$ of p-th roots of unity—this is essentially Kummer's approach to Fermat's Last Theorem.

Observe that Theorem 6.20 contains several results on this equation that we have obtained before: Since $2 = 3 \cdot 1^2 - 1$, the equation $y^2 = x^3 - 2$ has the only integral solution $(3, \pm 5)$.

If we look carefully at the case $d = 26 = 3 \cdot 3^2 - 1$, we see that $y^2 = x^3 - 26$ has the solution $(35, \pm 207)$ given by our theorem as well as the additional solutions $(3, \pm 1)$. This is not a contradiction: The theorem now implies that the class number of $\mathbb{Q}(\sqrt{-26})$ must be divisible by 3. In fact, the class number is equal to 6. This example can be generalized (see Exercise 6.27).

It is natural to ask whether the solutions found for $d = 26$ are the only ones. We cannot answer this question here, but we would like to show how to begin such an investigation.

As above we find $(y + \sqrt{-26})(y - \sqrt{-26}) = x^3$, and since the two factors on the left hand side are coprime, we must have $(y + \sqrt{-26}) = \mathfrak{a}^3$ for some ideal $\mathfrak{a}$. If $\mathfrak{a} = (\alpha)$ is principal, then the only solution is, as we have already seen, $(x, y) = (35, \pm 207)$. If $\mathfrak{a}$ is not principal, then it lies in some ideal class of order 3. One such class is generated by the ideal $\mathfrak{p} = (3, 1 + \sqrt{-26})$ the other one by its conjugate; in fact, $\mathfrak{p}^3 = (1 + \sqrt{-26})$, and clearly $\mathfrak{p}$ is not principal. Thus either $\mathfrak{p}\mathfrak{a} = (\alpha)$ or $\mathfrak{p}'\mathfrak{a} = (\alpha)$ is principal. In the first case, multiplying $(y + \sqrt{-26}) = \mathfrak{a}^3$ by $(1 + \sqrt{-26})$, we obtain the equation

$$(1 + \sqrt{-26})(y + \sqrt{-26}) = (\mathfrak{p}\mathfrak{a})^3 = (\alpha)^3,$$

and hence

$$y - 26 - (y + 1)\sqrt{-26} = (a - b\sqrt{-26})^3.$$

Comparing the coefficients of the real and imaginary parts, we obtain the equations

$$y - 26 = a(a^2 - 78b^2),$$
$$y + 1 = b(3a^2 - 26b^2).$$

Eliminating y, we then obtain

$$27 = -a^3 + 3a^2b + 78ab^2 - 26b^3.$$

Using $a = -3A + b$, we find

$$A^3 - 9Ab^2 + 2b^3 = 1. \tag{6.1}$$

At this point we invoke algebraic number theory. We consider the cubic number field $\mathbb{Q}(\omega)$, where ω is a root of the polynomial $f(x) = x^3 - 9x + 2 = 0$, and the domain $\mathbb{Z}[\omega]$. The norm of an element $\alpha = A - b\omega$ can be determined, as in the quadratic case, by computing the determinant of the linear map given by multiplication by α. We find

$$\alpha\omega = A\omega - b\omega^2,$$

$$\alpha\omega^2 = A\omega^2 - b\omega^3 = A\omega^2 - b(9\omega - 2)$$

because $\omega^3 = 9\omega - 2$, and hence

$$N\alpha = \begin{vmatrix} A & 0 & 2b \\ -b & A & -9b \\ 0 & -b & A \end{vmatrix} = A^3 - 9Ab^2 + 2b^3.$$

Thus (6.1) boils down to the question whether there is a unit of the form $A - b\omega$ in $\mathbb{Z}[\omega]$. Since f has three real roots, Dirichlet's unit theorem tells us that there exist two independent units. Using pari, we find the units $\varepsilon_1 = 3\omega^2 + 9\omega - 1$ and $\varepsilon_2 = 2\omega^2 + 4\omega - 1$. Thus the question (which is anything but easy to answer) is whether there exist exponents m and n with $\varepsilon_1^m \varepsilon_2^n = A - b\omega$.

As we have seen, it is a highly nontrivial problem to determine which integers are represented by a binary cubic form such as (6.1). Thue [122] showed in 1909 that an equation $F(U, V) = m$, where $F(U, V) = AU^3 + BU^2V + CUV^2 + DV^3$ is an irreducible cubic form, has only finitely many solutions in integers.

Let me say a few words about a connection between class numbers and elliptic curves of the form $y^2 = x^3 - m$. If we write this equation in the form $y^2 + m = x^3$, we see that for each integral point (x, y) on this elliptic curve, the principal ideal $(y + \sqrt{-m}\,)$ is, except for factors coming from common divisors with its conjugate ideal $(y - \sqrt{-m}\,)$, is a cube of an ideal. Ideals whose third powers are principal are sources for ideal classes of order 3 in the class group of $\mathbb{Q}(\sqrt{-m}\,)$.

In general, the equation $y^2 = x^3 - mz^2$ will lead to ideals $(y + z\sqrt{-m}\,)$ that often are cubes of ideals. For more on the connection between this equation and the 3-class group[3] of quadratic number fields, see [51] and the literature cited there.

If we factor the equation $y^2 = x^3 - m$ on the right side, then we have to study the 2-class group of pure cubic number fields $\mathbb{Q}(\sqrt[3]{m}\,)$ or their normal closure $\mathbb{Q}(\sqrt{-3}, \sqrt[3]{m}\,)$. This will require familiarity with the basic concepts of algebraic number theory and the arithmetic theory of elliptic curves.

---

[3]This is the 3-Sylow subgroup of the ideal class group, which consists of all ideal classes whose order is a power of 3.

6.4.1 Summary

This chapter contains the essential results of the basic arithmetic of quadratic number fields $k = \mathbb{Q}(\sqrt{m}\,)$:

- Ideals in the rings of integers $\mathcal{O}_k$ form a monoid with cancellation rule.
- The prime ideal factorization in $\mathcal{O}_k$ is unique.
- Rational prime numbers ramify, split, or are inert in $\mathcal{O}_k$ according as $(\Delta/p) = 0$, $+1$, or -1, respectively.
- Ideals $\mathfrak{a}$ and $\mathfrak{b}$ are called equivalent if there exist $\alpha, \beta \in \mathcal{O}_k \setminus \{0\}$ with $\alpha\mathfrak{a} = \beta\mathfrak{b}$. The equivalence classes of ideals form a group called the ideal class group.
- The class group of k is a finite group.

6.5 Exercises

6.1. Show that the elements $a + bi$ with $a + b \equiv 0 \bmod 2$ are exactly the multiples of $1 + i$.

6.2. Let R be a principal ideal domain and $(d) = (a, b)$ for $a, b, d \in R$. Show that d is a greatest common divisor of a and b.

6.3. Let $\mathfrak{a} = (1 + \sqrt{-5}, 2)$, $\mathfrak{b} = (1 + \sqrt{-5}, 3)$, and $\mathfrak{c} = (1 - \sqrt{-5}, 3)$. Verify $\mathfrak{a}\mathfrak{b} = (1 + \sqrt{-5})$, $\mathfrak{a}\mathfrak{c} = (1 - \sqrt{-5})$, and $\mathfrak{c}^2 = (2 - \sqrt{-5})$.

6.4. Show that the integer a in the basis of Proposition 6.5 is only determined modulo n.

6.5. Show that the ideal $I = (1 + 2i)$ properly contains the $\mathbb{Z}$-module $M = 5\mathbb{Z} + (1 + 2i)\mathbb{Z}$ by showing that $-2 + i \notin M$. Determine the $\mathbb{Z}$-basis of the ideal I.

6.6. Let $R = \mathcal{O}_k$ for $k = \mathbb{Q}(\sqrt{m}\,)$ and a squarefree integer m, and $M = \mathbb{Z}$. Show that the residue classes $b\sqrt{m}$ ($b \in \mathbb{Z}$) in R/M are all distinct and that $N(M) = \infty$.

6.7. Show that $(7, 1 + \sqrt{-5}) = (1)$. Show more generally that $(a, \alpha) = (1)$ for $a \in \mathbb{Z}$ and $\alpha \in \mathcal{O}_k$ if $\gcd(a, N\alpha) = 1$.

6.8. Determine the prime ideal factorization of $(4 + \sqrt{-5}\,)$.

6.9. Compute a greatest common divisor of $8 + \sqrt{-14}$ and $4 - \sqrt{-14}$ [107, S. 313].

6.10. The ideal $(21, 10 + \sqrt{-5}\,)$ has norm 21 and is divisible by prime ideals above 3 and 7 and hence by $(3, 1 \pm \sqrt{-5}\,)$ and $(7, 3 \pm \sqrt{-5}\,)$. Determine the exact prime ideal factorization [107, S. 350].

6.11. Let $m = a^2 + b^2$ be a sum of two squares, and assume that a is odd. Show that $(a, b + \sqrt{m}\,)^2 = (b + \sqrt{m}\,)$.

6.12. Explain $2 \cdot 3 = -\sqrt{-6}^2$ by factoring the elements into prime ideals.

6.13. Let $k = \mathbb{Q}(\sqrt{-23}\,)$; show that $(2) = \mathfrak{a}\mathfrak{a}'$ for $\mathfrak{a} = (2, \frac{1+\sqrt{-23}}{2})$ and $\mathfrak{a}^3 = (\frac{3-\sqrt{-23}}{2})$. Why is the ideal $\mathfrak{a}^2$ not principal?

6.14. Show that there exist homomorphisms $\phi_2 : R \longrightarrow \mathbb{F}_2$, $\phi_3 : R \longrightarrow \mathbb{F}_3$ and $\phi_3' : R \longrightarrow \mathbb{F}_3$ in $R = \mathbb{Z}[\sqrt{-5}]$ whose kernels are exactly the prime ideals $\mathfrak{a}$, $\mathfrak{b}$, and $\mathfrak{c}$ considered above.

6.15. Show that the equation of ideals

$$(2, 1 + \sqrt{-3})(2, 1 + \sqrt{-3}) = (2)(2, 1 + \sqrt{-3})$$

holds in $R = \mathbb{Z}[\sqrt{-3}]$, but that the cancellation law does not because $(2, 1 + \sqrt{-3}) \neq (2)$.

6.16. Generalize the last exercise to all orders $\mathbb{Z}[\sqrt{m}]$, where $m \equiv 1 \bmod 4$ is squarefree.

6.17. Consider the domain $\mathbb{Z}[3i] = \{a + 3bi : a, b \in \mathbb{Z}\}$. Show that $(2) = \mathfrak{p}^2$, that ideals (q) for primes $q \equiv 3 \bmod 4$ are inert, and that the ideals (p) for primes $p \equiv 1 \bmod 4$ split.

Show that $(5, 1 - 3i) = (5, 3 + 6i)$.

Show that $(3) \supset (3 + 6i)$, but that there does not exist an ideal A with $(3)A = (3 + 6i)$. In particular, the rule "to contain is to divide" does not hold in $\mathbb{Z}[3i]$.

Show that we do not have unique factorization into irreducible ideals in $\mathbb{Z}[3i]$.

6.18. Let R be a domain in which unique factorization into prime ideals holds (such domains are called *Dedekind domains*). Show that if $\mathfrak{A}$ and $\mathfrak{B}$ are coprime ideals with $\mathfrak{A}\mathfrak{B} = \mathfrak{c}^n$, then $\mathfrak{A} = \mathfrak{a}^n$ and $\mathfrak{B} = \mathfrak{b}^n$.

6.19. Let $\mathfrak{a}$ and $\mathfrak{b}$ be ideals in $\mathcal{O}_k$. Show that $\mathfrak{a} \cap \mathfrak{b} \supseteq \mathfrak{a}\mathfrak{b}$, and prove that we even have equality if $\mathfrak{a}$ and $\mathfrak{b}$ are coprime.

6.20. Recall that two ideals $\mathfrak{a}$ and $\mathfrak{b}$ are equivalent ($\mathfrak{a} \sim \mathfrak{b}$) if and only if there exist elements $\alpha, \beta \in \mathcal{O}_k$ with $\alpha\mathfrak{a} = \beta\mathfrak{b}$. Show that this is an equivalence relation on the nonzero ideals in $\mathcal{O}_k$, i.e., that the following assertions are true:

- symmetry: $\mathfrak{a} \sim \mathfrak{a}$;
- reflexivity: $\mathfrak{a} \sim \mathfrak{b}$ implies $\mathfrak{b} \sim \mathfrak{a}$;
- transitivity: $\mathfrak{a} \sim \mathfrak{b}$ and $\mathfrak{b} \sim \mathfrak{c}$ imply $\mathfrak{a} \sim \mathfrak{c}$.

6.21. Let $k = \mathbb{Q}(\sqrt{m})$ be a quadratic number field, and assume that m is squarefree.

- If $m \equiv 2 \bmod 4$, then $(2) = (2, \sqrt{m})^2$ is ramified.
- If $m \equiv 3 \bmod 4$, then $(2) = (2, 1 + \sqrt{m})^2$ is ramified.
- If $m \equiv 1 \bmod 8$, then $(2) = \mathfrak{a}\mathfrak{a}'$ with $\mathfrak{a} = (2, \frac{1 + \sqrt{m}}{2})$ and $\mathfrak{a} \neq \mathfrak{a}'$ (the ideal (2) splits).
- If $m \equiv 5 \bmod 8$, then (2) is prime (the ideal (2) is inert).

The first three claims can be verified by a simple calculation. For the last, you have to show that a prime ideal with norm 2 for $m \equiv 1 \bmod 4$ necessarily has the form $(2, a + \frac{1 + \sqrt{m}}{2})$. This implies $m \equiv (2a + 1)^2 \bmod 8$, and hence $m \equiv 1 \bmod 8$. For guidance, keep the proof of Theorem 6.14 in mind.

6.22. If $\Delta \equiv 5 \bmod 8$, then (2) remains prime, and there are no ideals of norm 2 in $\mathcal{O}_k$. Show that this implies that the fields with $\Delta = -19, 21, 29, 37$ have class number 1. Which fields do we obtain by demanding in addition that $\Delta \equiv 2 \bmod 3$?

6.23. Show that the class number of $\mathbb{Q}(\sqrt{-m})$ is even for each $m \equiv 1 \bmod 4$ with $m > 1$. To this end, show that $(2, 1 + \sqrt{-m})^2$ is principal but that $(2, 1 + \sqrt{-m})$ is not.

6.24. Let k be a complex quadratic number field with discriminant $\Delta < 0$. For some small values of Δ, compute the sum

$$ h = \frac{w}{2|\Delta|} \sum_{r=1}^{|\Delta|/2} \left(\frac{\Delta}{r}\right) r, $$

where w denotes the number of roots of unity contained in k (which is simply the order of the unit group in this case) and (Δ/r) is the Kronecker symbol. Compare h with the class number of k. We will return to this observation in our last chapter.

6.25. Show that the complex quadratic number fields $\mathbb{Q}(\sqrt{m})$ for $m = -1, -2, -3, -7, -11, -19, -43, -67$, and -163 have class number 1.

   The conjecture that there are no other complex quadratic fields with class number 1 essentially goes back to Gauss. The proof was obtained independently from each other by Kurt Heegner, Harold Stark, and Alan Baker.

6.26. Show that the prime ideals above (2) in $\mathbb{Q}(\sqrt{-m})$ for squarefree integers $m \equiv 7 \bmod 8$ are principal if and only if $m = 7$.

6.27. Show that the equation $y^2 = x^3 - d$ has an integral solution not listed in Theorem 6.20 for all $d = 3t^2 - 1$ with $t = 3c^3$. Deduce that the class number of $\mathbb{Q}(\sqrt{d})$ is divisible by 3 for all $d = 27c^6 - 1$ for which d or $4d$ is the discriminant of a quadratic number field (for example, if it is squarefree).

   How often a polynomial such as $f(x) = 27x^6 - 1$ attains squarefree values is an open problem, even in the case of polynomials of degree 4.

6.28. The class group of $k = \mathbb{Q}(\sqrt{-5})$ consists of the principal class and the class of order 2 generated by the ideal $\mathfrak{a} = (2, 1 + \sqrt{-5})$. Show that the following assertions hold:

   a. For each prime ideal $\mathfrak{p}$ of norm $p \neq 5$, either $\mathfrak{p}$ or $\mathfrak{p}\mathfrak{a}$ is principal.
   b. If p is a prime with $(-5/p) = +1$, then either $p = x^2 + 5y^2$ or $2p = x^2 + 5y^2$.
   c. If $p = x^2 + 5y^2$, then $p \equiv 1 \bmod 4$ and thus $p \equiv 1, 9 \bmod 20$; if $2p = x^2 + 5y^2$, on the other hand, then we must have $p \equiv 3, 7 \bmod 20$.
   d. Deduce that primes $p \equiv 1, 9 \bmod 20$ can be represented by the form $p = x^2 + 5y^2$ and primes $p \equiv 3, 7 \bmod 20$ by $2p = x^2 + 5y^2$.
   e. Verify that $(a^2 + 5b^2)(c^2 + 5d^2) = (ac - 5bd)^2 + 5(ad + bc)^2$.
   f. If $2p = a^2 + 5b^2$, show that p^2 can be written in the form $p^2 = x^2 + 5y^2$.
   g. If $2p = a^2 + 5b^2$ and $2q = c^2 + 5d^2$, show that $pq = x^2 + 5y^2$.

6.29. Solve the preceding exercise for the fields $\mathbb{Q}(\sqrt{-6})$ and $\mathbb{Q}(\sqrt{-10})$.

6.30. Verify the following assertions for small prime numbers p:

Let p be an odd prime number with $(\frac{-23}{p}) = +1$. Then the two prime ideals above p in $\mathbb{Q}(\sqrt{-23})$ are principal if and only if the polynomial $f(x) = x^3 - x + 1$ splits into three linear factors modulo p.

Observe that f has discriminant -23. This result is a consequence of class field theory.

6.31. Show that $(2, x) \supseteq (2)$ in $R = \mathbb{Z}[x]$, but that there does not exist an ideal I in R with $(2, x)I = (2)$.

6.32. Consider the set $\mathcal{S}$ of all sequences of rational numbers. This set is a ring with respect to addition and multiplication defined as follows:

$$(a_1, a_2, a_3, \ldots) + (b_1, b_2, b_3, \ldots) = (a_1 + b_1, a_2 + b_2, a_3 + b_3, \ldots),$$

$$(a_1, a_2, a_3, \ldots) \cdot (b_1, b_2, b_3, \ldots) = (a_1b_1, a_2b_2, a_3b_3, \ldots).$$

Show that the following subsets of $\mathcal{S}$ are subrings:

a. the set $\mathcal{N}$ of all null sequence (sequences that converge to 0);
b. the set $\mathcal{D}$ of all sequences converging in $\mathbb{Q}$;
c. the set $\mathcal{C}$ of all Cauchy sequences;
d. the set $\mathcal{B}$ of all bounded sequences.

Observe that $\mathcal{N} \subset \mathcal{D} \subset \mathcal{C} \subset \mathcal{B} \subset \mathcal{S}$. Which of these subrings are actually ideals in $\mathcal{B}$ (or in $\mathcal{C}$ or $\mathcal{D}$, respectively)?

Show that all these rings contain zero divisors and that $\mathcal{N}$ is a maximal ideal in $\mathcal{C}$ (actually, we have $\mathcal{C}/\mathcal{N} \simeq \mathbb{R}$).

6.33. Consider the equation $y^2 = x^3 - d$ for $d = 4f$ and squarefree $f \equiv 3 \bmod 8$ with $f \geq 11$. Assume that the class number h of $\mathbb{Q}(\sqrt{-f})$ is not divisible by 3.

a. Show that x and y are odd.
b. Show that we must have $(y + 2\sqrt{-f}) = \mathfrak{a}^3$. Conclude that $\mathfrak{a} = (\alpha)$ is principal.
c. Write $\alpha = \frac{r+s\sqrt{-f}}{2}$ and deduce that $s(3r^2 - fs^2) = 16$.
d. Show that the only integral solutions are given by

   • $x = r^2 - 4$, $y = \pm(r^3 - 6r)$ if $f = 3r^2 - 16 \geq 11$;
   • $x = r^2 + 4$, $y = \pm(r^3 + 6r)$ if $f = 3r^2 + 16$.

6.34. Let $m = 85 = 9^2 + 4 \cdot 1^2 = 7^2 + 4 \cdot 3^2$. Consider the two ideals $\mathfrak{a}_1 = (2 + \sqrt{85}, 9)$ and $\mathfrak{a}_2 = (6 + \sqrt{85}, 7)$. Show that $\mathfrak{a}_1$ is principal and $\mathfrak{a}_2$ is not.

6.35. Let $m = 5 \cdot 13 \cdot 17$. Write m in the form $m = a_j^2 + 4b_j^2$ in all possible ways, and investigate which of the ideals $\mathfrak{a}_j = (2b_j + \sqrt{m}, a_j)$ are principal and which are not.

6.36. Show that the Diophantine equation $x^3 + 4 = py^2$ for primes $p \equiv 5 \bmod 8$ has the only solution $x = 1$, $y = \pm 1$ and $p = 5$ if the class number of $\mathbb{Q}(\sqrt{p})$ is not divisible by 3.

6.37. Let $k \equiv 7 \bmod 8$ be a squarefree negative integer, and assume that the class number of $K = \mathbb{Q}(\sqrt{k})$ is divisible by 3 (Hall [47]). Show that the equation $y^2 = x^3 - k$ does not have an integral solution with y odd if the ideal class generated by the prime ideal $\mathfrak{p} = (2, \frac{1+\sqrt{k}}{2})$ is not a cube of another ideal class.

6.38. Show that the Diophantine equation $y^2 = x^3 - 31$ does not have an integral solution (Hall [47]).

Chapter 7
The Pell Equation

Complex quadratic number rings have finitely many units; in the real quadratic case, the rings of integers of $\mathbb{Q}(\sqrt{m})$ seem to contain a unit ε of infinite order:

m	2	3	5	6	7
ε	$1 + \sqrt{2}$	$2 + \sqrt{3}$	$\frac{1}{2}(1 + \sqrt{5})$	$5 + 2\sqrt{6}$	$8 + 3\sqrt{7}$

The existence of nontrivial units $\varepsilon = \frac{x+y\sqrt{m}}{2}$ in real quadratic number fields $\mathbb{Q}(\sqrt{m})$ is equivalent to the solvability of the *Pell equation* $x^2 - my^2 = \pm 4$ in nonzero integers for all squarefree values of $m > 0$. In this chapter we will prove that the equation $x^2 - my^2 = 1$ has a nontrivial[1] solution in integers whenever $m > 0$ is not a square, and we will provide methods for computing units in real quadratic number fields.

Before we prove the solvability of the Pell equation, we make a few remarks on the connection between the equations $x^2 - my^2 = \pm 4$ and the equation $x^2 - my^2 = 1$ and on how to compute the fundamental unit of $\mathbb{Q}(\sqrt{m})$ from the minimal nontrivial solution of $x^2 - my^2 = 1$ and vice versa.

Consider for example the case $m = 13$. Here the fundamental unit of the ring of integers $\mathcal{O}_k$ of the quadratic number field $k = \mathbb{Q}(\sqrt{13})$ is $\varepsilon = \frac{3+\sqrt{13}}{2}$, which corresponds to the solution $(3, 1)$ of the Pell equation $x^2 - 13y^2 = -4$. The unit $\varepsilon^3 = 18 + 5\sqrt{13}$ is a unit in the order $\mathbb{Z}[\sqrt{13}]$ and corresponds to the fundamental solution $(18, 5)$ of the Pell equation $x^2 - 13y^2 = -1$. Finally the unit $\varepsilon^6 = 649 + 180\sqrt{13}$ corresponds to the minimal nontrivial solution $(649, 180)$ of the Pell equation $x^2 - 13y^2 = 1$.

---

[1] By this we mean solutions with $y \neq 0$.

© The Author(s), under exclusive license to Springer Nature Switzerland AG 2021
F. Lemmermeyer, *Quadratic Number Fields*, Springer Undergraduate
Mathematics Series, https://doi.org/10.1007/978-3-030-78652-6_7

Proposition 7.1 *If* $k = \mathbb{Q}(\sqrt{m})$ *for some squarefree integer* $m \equiv 1 \bmod 4$, *and if* ε *is a unit in* $\mathcal{O}_k$, *then* ε^3 *is a unit in the order* $\mathbb{Z}[\sqrt{m}]$.

In other words, if $t^2 - mu^2 = \pm 4$, *then* $T^2 - mU^2 = \pm 1$, *where*

$$T + U\sqrt{m} = \left(\frac{t + u\sqrt{m}}{2}\right)^3.$$

Proof If $t^2 - my^2 = \pm 4$ for odd integers t and u, then reducing this equation modulo 8 shows that $m \equiv 5 \bmod 8$. In this case, the prime ideal (2) is inert in $\mathbb{Q}(\sqrt{m})$; hence, the group of coprime residue classes modulo 2 has order $N(2) - 1 = 3$, and this in turn implies that $\varepsilon^3 \equiv 1 \bmod 2$. □

Clearly, if ε is a unit with norm -1, then ε^2 is a unit with norm $+1$. Thus if we want to compute a fundamental unit from the unit $x + y\sqrt{m}$ corresponding to the smallest positive solution of the Pell equation $x^2 - my^2 = 1$, then we have to check whether ε is a square, a cube, or a sixth power of a unit in $\mathbb{Q}(\sqrt{m})$.

We will explain how to do this in the case at hand. Assume we have the minimal positive solution $(649, 180)$ of the Pell equation $x^2 - 13y^2 = 1$; then $\eta = 649 + 180\sqrt{13}$ is a unit with norm 1 (and the smallest positive unit with norm 1 in the domain $\mathbb{Z}[\sqrt{13}]$).

For checking whether $\sqrt{\eta} \in \mathbb{Q}(\sqrt{13})$, we use the real approximations

$$\eta = 649 + 180\sqrt{13} \approx 1297.9992295\ldots,$$
$$\eta' = 649 - 180\sqrt{13} \approx \quad\ 0.0007704\ldots.$$

Clearly, the trace $\eta + \eta' = 2 \cdot 649$ is an integer. If η is a square, then the trace of $\sqrt{\eta}$ must also be an integer. We find

$$\sqrt{\eta} \approx 36.0277563773\ldots,$$
$$\sqrt{\eta'} \approx \ \ 0.0277563773\ldots.$$

This shows[2] that $\frac{1}{2}(\sqrt{\eta} - \sqrt{\eta'}) \approx 18$ and $\frac{\sqrt{\eta}+\sqrt{\eta'}}{2\sqrt{13}} \approx 5$ are very close to integers, which in turn suggests that $\sqrt{\eta} = 18 + 10\sqrt{13}$. Now we can readily verify that $(18 + 5\sqrt{13})^2 = \eta$. Observe that $18 - 5\sqrt{13} < 0$, which is why $18 - 5\sqrt{3} \approx -\sqrt{\eta'}$.

In a similar way we can check that η is a cube and in fact a sixth power:

$$\sqrt[6]{\eta} \approx 3.30277563773\ldots,$$
$$\sqrt[6]{\eta'} \approx 0.30277563773\ldots,$$

---

[2]See Exercise 7.1.

and this time

$$\frac{\sqrt[6]{\eta} - \sqrt[6]{\eta'}}{2} \approx 1.5 \quad \text{and} \quad \frac{\sqrt[6]{\eta} + \sqrt[6]{\eta'}}{2\sqrt{13}} \approx 0.5$$

suggests that $\sqrt[6]{\eta} = \frac{3+\sqrt{13}}{2}$, which can then be verified.

7.1 The Solvability of the Pell Equation

The history of the Pell equation in Europe[3] begins with Fermat's challenge in 1657. In that year, Fermat posed the following problem (among others) and asked his contemporaries, in particular the English mathematicians John Wallis and William Brouncker, for a solution:

> Given an arbitrary natural number, which is assumed to be not a square, there are infinitely many square numbers with the property that after adding 1 to the product of one of these square numbers with the given number, another square is produced [...]. We ask e.g. for a square that produces another square after adding 1 to the product with 149 or 109 or 433 etc.

Thus Fermat asked for solutions of equations such as $Na^2 + 1 = b^2$ for positive nonsquare integers N, in particular for the values $N = 149, 109,$ and 433. Brouncker and Wallis solved these equations in rational numbers,[4] and Fermat remarked that he hardly would have posed a problem that any "three-day-arithmetician" could have solved. Brouncker then succeeded in solving Fermat's equation for any given value, but Fermat complained that Brouncker had not shown that his method would always work' and claimed that such a proof is possible using his method of infinite descent. Whether Fermat himself had such a proof is an open question.

Long before Fermat, Indian mathematicians such as Brahmagupta (ca. 598–670) and Bhaskara II (1114–1185) had developed a method for solving the Diophantine equation $Nx^2 + 1 = y^2$; this became known in Europe only rather late and did not have, as far as we know, any influence on the development of number theory in the West.[5]

Leonhard Euler later studied the equation $Na^2 + 1 = b^2$ in several articles, and Joseph-Louis Lagrange succeeded in proving the solvability in integers. His first proof, which already used what later became known as Dirichlet's pigeonhole

---

[3] Strictly speaking, the investigation of Platon's side and diagonal numbers by Theon may be seen as the only serious investigation of a Pell equation in ancient Greece. Equations of Pell type also figure prominently in the Cattle Problem of Archimedes; it is not known, however, whether there were any attempts at solving this problem before it was discovered by Lessing in 1773.

[4] We have derived the rational parametrization of Pell conics in Theorem 3.1.

[5] An excellent account of Indian mathematics was given by Kim Plofker [104]. For an investigation of the Indian method of solving the equation $Nx^2 + 1 = y^2$, see [114].

principle, was streamlined and generalized by Dirichlet: In his unit theorem, he proved the existence of nontrivial units in all number fields except $\mathbb{Q}$ and complex quadratic number fields.

Below we will prove the solvability of the Pell equation $x^2 - my^2 = 1$ for all natural numbers m that are not squares. The essential idea behind this proof is due to Lagrange, who derived the necessary lemmas from the theory of continued fractions. Dirichlet later replaced Lagrange's arguments by repeated applications of his pigeonhole principle, which simplified the proof considerably. The proof is more or less by descent: From the solvability of an equation $x^2 - my^2 = c$ for some value of $c > 1$, we will deduce the solvability of $x^2 - my^2 = c'$ for some $c' < c$. In order to make this argument work, we will have to exploit the fact that these equations $x^2 - my^2 = c$ have infinitely many solutions.

Dirichlet's pigeonhole principle[6] may be stated as follows:

> If $N + 1$ pearls are put into N pigeonholes, then there must be a pigeonhole containing at least two pearls.

For finding solutions of equations such as $x^2 - my^2 = c$, we observe that if m is large and c is small, then $x^2 \approx my^2$ implies that $\frac{x}{y}$ is an approximation of $\sqrt{m}$. In order to find such approximations, one may use, as Lagrange did, the theory of continued fractions; if we are content with proving the existence of solutions, we may use Dirichlet's pigeonhole principle.

Theorem 7.2 *The equation $x^2 - my^2 = 1$ is solvable in nonzero integers x, y whenever $m > 0$ is not a square.*

We begin with the following lemma.

Lemma 7.3 *If ξ_1 and ξ_2 are two nonzero real numbers such that ξ_1/ξ_2 is irrational, then for any $N \in \mathbb{N}$ there exist integers $A, B \in \mathbb{Z}$, which are not both 0 and satisfy the following inequalities:*

$$|A\xi_1 + B\xi_2| \le \frac{1}{N}(|\xi_1| + |\xi_2|), \quad |A| \le N, \ |B| \le N. \tag{7.1}$$

Proof We assume that ξ_1 and ξ_2 are both positive (otherwise we just have to change the signs of a and b in the proof below). The irrationality of ξ_1/ξ_2 implies that the function

$$f : \mathbb{Z} \times \mathbb{Z} \longrightarrow \mathbb{R} : (a, b) \longmapsto a\xi_1 + b\xi_2 \tag{7.2}$$

is injective (see Exercise 7.7.17).

There are $(N + 1)^2$ pairs of integers $(a, b) \in [0, N] \times [0, N]$, and for these we have $0 \le f(a, b) \le N(|\xi_1| + |\xi_2|)$. If we divide the interval $[0, N(|\xi_1| + |\xi_2|)]$

---

[6]It seems that this principle was given a name rather late (in the twentieth century?); a pigeonhole is a drawer, so the last thing you would like to put there are pigeons.

into N^2 subintervals of equal length $\frac{1}{N}(|\xi_1| + |\xi_2|)$, then since $(N + 1)^2 > N^2$ there must exist, according to Dirichlet's pigeonhole principle, at least two pairs $(a, b) \neq (a', b')$ with $|f(a, b) - f(a', b')| \leq \frac{1}{N}(|\xi_1| + |\xi_2|)$. Now we set $A = a - a'$ and $B = b - b'$; these integers have the desired properties. $\qquad\square$

Corollary 7.4 *Assume that $m \in \mathbb{N}$ is not a square. Then there exists an integer c such that the equation $A^2 - mB^2 = c$ has infinitely many solutions $(A, B) \in \mathbb{Z} \times \mathbb{Z}$.*

Proof By the preceding lemma, there exist numbers $A, B \in \mathbb{Z}$, not both 0, that satisfy the inequalities

$$|A - B\sqrt{m}| \leq \frac{1}{N}(1 + \sqrt{m}), \quad |A| \leq N, \quad |B| \leq N. \tag{7.3}$$

The triangle inequality shows that

$$|A + B\sqrt{m}| \leq |A| + |B\sqrt{m}| \leq (1 + \sqrt{m}) \cdot N, \tag{7.4}$$

and multiplying (7.3) and (7.4) yields

$$|A^2 - mB^2| \leq (1 + \sqrt{m})^2. \tag{7.5}$$

Now let $N \to \infty$; then infinitely many distinct pairs (A, B) must occur, since if we had only finitely many, then the set $\{|A - B\sqrt{m}| : A, B \in \mathbb{Z}\}$ would possess a minimum, which is impossible because of (7.3).

Since $|A^2 - mB^2|$ is bounded from above by (7.3), there must exist an integer c with $|c| \leq (1 + \sqrt{m})^2$ for which $A^2 - mB^2 = c$ has infinitely many solutions in integers. $\qquad\square$

Now we can prove Theorem 7.2. According to Corollary 7.4, there exists an integer $c \neq 0$ such that there are infinitely many pairs (A, B) with $A^2 - mB^2 = c$; here we may clearly assume that $A > 0$. Among these infinitely many solutions, we choose $(c + 1)^2$ solutions and consider the residue classes of A and B modulo c; by Dirichlet's pigeonhole principle, there must exist pairs $(A_1, B_1) \neq (A_2, B_2)$ with $A_1 \equiv A_2 \bmod c$ and $B_1 \equiv B_2 \bmod c$. The elements $\eta_j = A_j + B_j\sqrt{m}$ then have the same norms $N\eta_1 = N\eta_2 = c$ and satisfy the congruence $\eta_1 \equiv \eta_2 \bmod c$. It follows from $N(\eta_1/\eta_2) = 1$ that η_1/η_2 is a unit if we can show that this is an algebraic integer. To this end, observe that $\eta_1/\eta_2 = 1 + (\eta_1 - \eta_2)/\eta_2 = 1 + (\eta_1 - \eta_2)\eta_2'/c$. Since the difference $\eta_1 - \eta_2$ is divisible by c by construction, η_1/η_2 is indeed an algebraic integer and thus a unit.

It remains to show that $\eta_1/\eta_2 \neq \pm 1$ is a nontrivial unit. But $\eta_1/\eta_2 \neq 1$ follows from $\eta_1 \neq \eta_2$, and $\eta_1/\eta_2 \neq -1$ follows from the fact that A_1 and A_2 are both positive. This concludes the proof of Theorem 7.2.

We now know that there exist nontrivial units in each real quadratic number field. In fact, it is possible to determine the abstract structure of the unit group: For real quadratic number fields k, we have $\mathcal{O}_k^{\times} \simeq (\mathbb{Z}/2\mathbb{Z}) \times \mathbb{Z}$. As we will show in a moment, each unit $\eta \in \mathcal{O}_k^\times$ can be written in the form $\eta = (-1)^s \varepsilon^t$ for some "fundamental unit" ε, and then the map $\lambda : E_k \longrightarrow (\mathbb{Z}/2\mathbb{Z}) \times \mathbb{Z}$ defined by $\lambda(\eta) = (s, t)$ provides us with an isomorphism of abelian groups. This is the content of our next theorem.

Theorem 7.5 *If k is a real quadratic number field, then there is a unit $\varepsilon \in \mathcal{O}_k^\times$ with the property that every unit $\eta \in \mathcal{O}_k^\times$ can be written uniquely in the form $\eta = \pm \varepsilon^t$ for some $t \in \mathbb{Z}$. In particular,*

$$\mathcal{O}_k^\times \simeq \mathbb{Z}/2\mathbb{Z} \oplus \mathbb{Z}.$$

We immediately see that if ε has the property in Theorem 7.5, then so do the units $\pm \varepsilon^{\pm 1}$. Among these four units, there are two that are positive, and among these exactly one is > 1. This unit $\varepsilon > 1$ will be called the fundamental unit of k.

Proof We identify the numbers $a + b\sqrt{m}$ with those real numbers that correspond to the positive square root of m. The only units $\eta \in \mathcal{O}_k^\times$ with $|\eta| = 1$ then are $\eta = \pm 1$, which follows from irrationality of $\sqrt{m}$.

We claim that among the units with $|\eta| > 1$, there is one with minimal absolute value. Otherwise there would exist a unit (in fact, infinitely many) with $1 < |\eta| < \frac{5}{4}$ (just pick two units that are sufficiently close to the infimum of the absolute values and consider their quotient). Since $|\eta \eta'| = 1$, this implies $\frac{4}{5} < |\eta'| < 1$. If we write $\eta = a + b\sqrt{m}$ (where $2a, 2b \in \mathbb{Z}$), then $2|a| = |\eta + \eta'| \leq |\eta| + |\eta'| < \frac{9}{4}$, and hence $|a| \leq 1$. Since $a = 0$ is not possible, we must have $a = \pm 1$. Then it follows immediately from $1 < |\eta| < \frac{5}{4}$ that $b = 0$, and hence $\eta = 1$ in contradiction to our assumption.

Let ε be a unit with minimal absolute value > 1. We claim that ε has the properties listed in Theorem 7.5. Otherwise there would exist a unit η with $\varepsilon^n < |\eta| < \varepsilon^{n+1}$ for some $n \in \mathbb{N}$ (the proof is similar to that of Theorem 2.6). But then $\eta \varepsilon^{-n}$ is a unit whose absolute value lies strictly between 1 and $|\varepsilon|$, and this contradicts the choice of ε.

Uniqueness is clear: $\pm \varepsilon^t = \pm \varepsilon^u$ implies $|\varepsilon^{t-u}| = 1$, which in turn implies $t = u$ since ε is irrational. But then the signs must also coincide. □

Remark The proof of the solvability of the Pell equation $t^2 - mu^2 = 1$ given here does not provide us with a method of computing the fundamental unit, except for very small values of m. For example, $\varepsilon = 48842 + 5967\sqrt{67}$ is the fundamental unit of $\mathbb{Q}(\sqrt{67})$, and this solution is hard to find by solving the Pell equation by brute force (i.e., looking for an integer $m = 1, 2, 3, \ldots$ such that $mu^2 + 1 = t^2$ is a square). A much better way of computing the fundamental unit of quadratic number fields with modest discriminants is based on the theory of continued fractions. For number fields of higher degree, the computation of the unit group becomes time consuming with growing degree and discriminant even when using the best algorithms that are known today.[7]

---

[7]Good sources for the state of the art are [20, 91], and, in particular, [66].

7.1.1 The Negative Pell Equation

The equation $t^2 - mu^2 = -1$ is called the *negative Pell equation* or sometimes the anti-Pell equation. In this section we will show how to derive solvability conditions for the negative Pell equation from the solvability of the usual Pell equation.

We begin by considering the equation $t^2 - pu^2 = 1$ for prime values of p. We can write this equation in the form

$$pu^2 = t^2 - 1 = (t - 1)(t + 1).$$

The greatest common divisor of $t + 1$ and $t - 1$ divides their difference 2, and hence one of the following four possibilities must occur:

$$t + 1 = a^2, \qquad\qquad t - 1 = pb^2,$$
$$t + 1 = pb^2, \qquad\qquad t - 1 = a^2,$$
$$t + 1 = 2a^2, \qquad\qquad t - 1 = 2pb^2,$$
$$t + 1 = 2pb^2, \qquad\qquad t - 1 = 2a^2.$$

We choose the integers a and b positive. Subtracting the right equation from the left, we find that at least one of the following four equations must be solvable in integers:

$$a^2 - pb^2 = 2; \quad a^2 - pb^2 = -2; \quad a^2 - pb^2 = 1; \quad a^2 - pb^2 = -1.$$

If we assume that (t, u) is the smallest positive solution of the Pell equation, then we can exclude the equation $a^2 - pb^2 = 1$ since $t + 1 = 2a^2$ implies that $a < t$.

A necessary condition for the equation $a^2 - pb^2 = 2$ to be solvable is that $p \equiv \pm 1 \bmod 8$. Similar considerations yield the following table:

	$p \equiv 3 \bmod 8$	$p \equiv 5 \bmod 8$	$p \equiv 7 \bmod 8$
$a^2 - pb^2 = +2$	x	x	
$a^2 - pb^2 = -2$		x	x
$a^2 - pb^2 = -1$	x		x

Here "x" represents the unsolvability of the corresponding equation.

If $p \equiv 1 \bmod 8$, it follows from $t^2 - pu^2 = 1$ that t must be odd and thus $\gcd(t - 1, t + 1) = 2$. Therefore the first three cases are impossible, and we end up with the equation $a^2 - pb^2 = -1$.

Since, as we have seen, one of the three equations must be solvable, we obtain the following:

Proposition 7.6 *The solvability of $t^2 - pu^2 = 1$ for odd prime numbers p implies the solvability of*

$$a^2 - pb^2 = -1 \qquad\qquad for\ p \equiv 1 \bmod 4,$$
$$a^2 - pb^2 = -2 \qquad\qquad for\ p \equiv 3 \bmod 8,$$
$$a^2 - pb^2 = +2 \qquad\qquad for\ p \equiv 7 \bmod 8.$$

Even if the equation $a^2 - pb^2 = \pm 2$ for some prime $p \equiv 1 \bmod 8$ is not solvable in integers, the ring of integers in $k = \mathbb{Q}(\sqrt{p})$ might contain elements of norm ± 2, as the example $N(\frac{5+\sqrt{17}}{2}) = 2$ shows.

7.2   Which Numbers Are Norms?

The only method we know so far for showing the unsolvability of the norm equation $x^2 - my^2 = c$ for given values of $m \in \mathbb{N}$ and $c \in \mathbb{Z}$ is reducing the equation modulo n for some choice of n, where n in general is a divisor of m or c, and showing that the congruence does not have a solution. For example, $x^2 - 10y^2 = \pm 2$ is not solvable in integers since the congruence $x^2 \equiv \pm 2 \bmod 5$ does not have solutions. This method does not work in the case of the equation $x^2 - 79y^2 = \pm 3$, the reason being that $x^2 - 79y^2 = -3$ has the rational solution $x = \frac{2}{5}$, $y = \frac{1}{5}$; in particular, this equation is solvable module each modulus coprime to 5. Similarly, the solutions $x = \frac{13}{7}$ and $y = \frac{2}{7}$ show that the congruence is solvable for each modulus coprime to 7. This implies by the Chinese remainder theorem that $x^2 - 79y^2 \equiv -3 \bmod m$ is solvable for each nonzero modulus $m \in \mathbb{Z}$.

Remark Hasse's Local–Global Principle for quadratic forms implies that equations such as $x^2 - my^2 = c$ have rational solutions if and only if the congruence $x^2 - my^2 \equiv c \bmod N$ is solvable for each modulus N. The example above implies that there is no similar Local–Global Principle for integral solutions of such equations. For cubic equations such as $y^2 = x^3 - m$ even the Local–Global Principle for rational solutions does not hold. In the case of integral solutions of $x^2 - my^2 = c$, the class group $\mathrm{Cl}(k)$ of $k = \mathbb{Q}(\sqrt{m})$ is a measure for the obstruction to the Local–Global Principle in the sense that if the class group of k is trivial, the equation $x^2 - my^2 = c$ is solvable in integers if and only if the congruence $x^2 - my^2 \equiv c \bmod N$ is solvable for each modulus N. In the case of elliptic curves, there is a similar group called the Tate–Shafarevich group. Understanding the failure of Local–Global principles is a central area of research in modern number theory.

In order to show that $x^2 - 79y^2 = -3$ does not have integral solutions, we have to employ a different technique. Let us consider an arbitrary real quadratic number field $k = \mathbb{Q}(\sqrt{m})$, and let $\varepsilon = t + u\sqrt{m} > 1$ be the fundamental unit (we allow t and u to be half-integers). Assume moreover that $\alpha \in \mathcal{O}_k$ is a solution of the equation

$|N\alpha| = c$. The basic idea is to choose the exponent n in $\beta = \alpha\varepsilon^n$ in such a way that the coefficients of β with respect to the basis $\{1, \sqrt{m}\}$ become as small as possible. It is clear from geometric considerations that there exists an exponent $n \in \mathbb{Z}$ such that

$$1 \leq |\varepsilon^n \alpha| < \varepsilon.$$

If we set $\beta = \varepsilon^n \alpha$ and write $\beta = a + b\sqrt{m}$ (again a and b are allowed to be half-integers), then

$$|\beta'| = \frac{|\beta\beta'|}{|\beta|} = \frac{c}{|\beta|},$$

and we obtain the bounds

$$\frac{c}{\varepsilon} < |\beta'| \leq c.$$

The triangle inequality now yields

$$|2a| = |\beta + \beta'| \leq |\beta| + |\beta'| < \varepsilon + c,$$
$$|2b|\sqrt{m} = |\beta - \beta'| \leq |\beta| + |\beta'| < \varepsilon + c. \tag{7.6}$$

This immediately yields bounds for a and b, and now the problem can be solved in finitely many steps by simply checking the possible values of a and b one by one. Before we do this in our example, we will improve the bounds on a and b.

To this end, we set $\beta = \varepsilon^n \alpha$ and choose the exponent $n \in \mathbb{Z}$ in such a way that

$$\frac{\sqrt{c}}{\sqrt{\varepsilon}} \leq |\beta| < \sqrt{c\varepsilon}.$$

As above, this implies the bounds

$$|\beta| < \sqrt{c\varepsilon} \quad \text{and} \quad |\beta'| \leq \sqrt{c\varepsilon},$$

and now we obtain $|2a| < 2\sqrt{c\varepsilon}$, which is a lot better than $|2a| < \varepsilon + c$.

As a matter of fact these bounds may be improved again by using the following lemma:

Lemma 7.7 *If $x, y \in \mathbb{R}$ satisfy the inequalities $0 < x \leq r$, $0 < y \leq r$, and $0 < xy \leq s$, then $x + y \leq r + \frac{s}{r}$.*

This claim follows from the observation $0 < (r-x)(r-y) = r^2 - r(x+y) + xy \leq r^2 + s - r(x+y)$.

In our case, we have $r = \sqrt{c\varepsilon}$ and $s = c$; thus $|\beta + \beta'| \leq \sqrt{c}\left(\sqrt{\varepsilon} + \frac{1}{\sqrt{\varepsilon}}\right)$. Since $\frac{1}{\sqrt{\varepsilon}} < 1$, this bound improves the previous one by a factor of about 2. We have proved the following:

Theorem 7.8 *Let k be a quadratic number field with a unit $\varepsilon > 1$; then for each $\alpha \in \mathcal{O}_k$ with norm $|N\alpha| = c$, there exists an associate $\beta = a + b\sqrt{m}$ (with integers or half-integers a, b) such that*

$$|a| \leq \frac{1}{2}\sqrt{c}\left(\sqrt{\varepsilon} + \frac{1}{\sqrt{\varepsilon}}\right) \quad and \quad |b| \leq \frac{1}{2\sqrt{m}}\sqrt{c}\left(\sqrt{\varepsilon} + \frac{1}{\sqrt{\varepsilon}}\right). \qquad (7.7)$$

If there is an element $\alpha \in \mathbb{Z}[\sqrt{79}]$ with norm ± 3, then (set $m = 79$, $\varepsilon = 80 + 9\sqrt{79}$, and $c = 3$) there is an element $a + b\sqrt{79}$ with norm ± 3 such that $|b| < 1.25$. Thus it is sufficient to consider $b = 1$, but the equation $a^2 - 79 \cdot 1^2 = \pm 3$ is not solvable in integers since 79 ± 3 is not a square. Thus $\mathbb{Z}[\sqrt{79}]$ does not contain an element with norm ± 3, and hence 3 is irreducible, but not prime since $3 \mid (2 - \sqrt{79})(2 + \sqrt{79})$.

Remark Theorem 7.8 goes back to Pafnuty Chebyshev [18]; the corresponding result in general number fields but with weaker bounds had been obtained before by Dirichlet [30]. Chebyshev is best known for his contributions to the proof of the prime number theorem. This theorem states that the number $\pi(x)$ of all prime numbers $\leq x$ is asymptotically equal to $\pi(x) \sim \frac{x}{\log x}$ in the sense that the quotient of these functions has limit 1 as $x \to \infty$; here $\log x$ denotes the natural logarithm. Chebyshev proved that if the limit of $\frac{\pi(x)}{x/\log x}$ for $x \to \infty$ exists, then it must be equal to 1. The existence of this limit and thus the prime number theorem was established independently in 1896 by Jacques Hadamard and Charles-Jean de la Vallée-Poussin.

7.2.1 Davenport's Lemma

Using Theorem 7.8, it is easy to prove a result going back to Harold Davenport:

Proposition 7.9 *Let m, n, and t be natural numbers with $m = t^2 + 1$; if the Diophantine equation $x^2 - my^2 = \pm n$ has integral solutions with $n < 2t$, then $x + y\sqrt{m}$ is associated with a rational integer a and $n = a^2$ is a perfect square.*

This result tells us that the only norms less than $2t$ in absolute value are the obvious ones, namely elements associated with rational integers a, which have norms $\pm a^2$. The norms of all other elements have absolute value $\geq 2t$.

For a proof, set $\xi = x + y\sqrt{m}$; we will show that if $|N\xi| = n$ is not a square, then $n \geq 2t$. Assume therefore that $n < 2t$; since $\varepsilon = t + \sqrt{m} > 1$ is a unit in $\mathbb{Z}[\sqrt{m}\,]$, we can find a power η of ε for which $\xi\eta = a + b\sqrt{m}$ has coefficients a and b that satisfy the bounds from Theorem 7.8. Because of $2t < \varepsilon < 2\sqrt{m}$, we find

$$|b| \leq \frac{\sqrt{n}}{2\sqrt{m}}\left(\sqrt{\varepsilon} + \frac{1}{\sqrt{\varepsilon}}\right) < 1 + \frac{1}{t}.$$

Since the claim is trivial for $t = 1$, we may assume that $t \geq 2$, and then the last inequality gives $|b| \leq 1$. If $b = 0$, then $(x + y\sqrt{m})\eta = a$ is associated with a rational integer, and $|N\xi| = a^2$ is a square. If $b = \pm 1$, then $\alpha = \xi\eta = a \pm \sqrt{m}$. Now $|N\xi| = |N\alpha| = |a^2 - m|$ is minimal for values of a close to $\sqrt{m}$, and we find

$$n = |a^2 - m| = \begin{cases} 2t & \text{if } a = t \pm 1; \\ 1 & \text{if } a = t. \end{cases}$$

Thus either $n = 1$ (which we have excluded) or $n \geq 2t$. This proves our claims.

Proposition 7.9 was used by Ankeny, Chowla, and Hasse [2] for constructing quadratic number fields with nontrivial class groups.

Proposition 7.10 *The quadratic number field $k = \mathbb{Q}(\sqrt{m})$ with $m = t^2 + 1$ and $t = 2lq$, where q is prime and $l > 1$, has class number > 1.*

Since $m \equiv 1 \bmod q$, the prime q splits in k, and we have $(q) = \mathfrak{q}\mathfrak{q}'$. If $\mathfrak{q}$ is principal, then the equation $x^2 - my^2 = \pm 4q$ has integral solutions. But since $4q < 2t = 4lq$ is not a square, this contradicts Proposition 7.9.

Examples In the following examples, $m = t^2 + 1$ is prime. The ambiguous class number formula (see Chap. 9) will explain why the class number h is odd in this case.

q	l	$t^2 + 1$	h	q	l	$t^2 + 1$	h
3	4	577	7	5	2	401	5
	6	1297	11		4	1601	7
	9	2917	3		9	8101	13
	11	4357	5		11	12101	5
	14	7057	21		12	14401	43

The following result[8] shows that even a simple result such as Proposition 7.9 allows us to deduce astonishingly simple lower bounds for class numbers of fields of Richaud–Degert type.

---

[8]This theorem is due to Halter-Koch [48] and the proof presented here to Mollin [95].

Theorem 7.11 *Let t be an odd integer with prime factorization $t = p_1^{e_1} \cdots p_s^{e_s}$ and set $m = t^2 + 1$, and assume that $m = t^2 + 1$ is squarefree. Then the class number h of $\mathbb{Z}[\sqrt{m}\,]$ satisfies $h \geq S = 2\tau(n) - 2$, where*

$$\tau(n) = (e_1 + 1)(e_2 + 1) \cdots (e_s + 1)$$

denotes the sum of all divisors of n.

If $t = q^e$ is a prime power, then the class number is divisible by S.

Proof For each prime $p_1 \mid m$, we have $(m/p_j) = +1$, and hence p_j splits in $\mathbb{Z}[\sqrt{m}\,]$ as $(p_j) = \mathfrak{p}_j \mathfrak{p}_j'$, where $\mathfrak{p}_j = (p_j, 1 + \sqrt{m})$. Let $\mathfrak{p}_0 = (2, \sqrt{m})$ denote the prime ideal above 2 and set $e_1 = 1$. We now consider the ideals $\mathfrak{a} = \mathfrak{p}_0^{a_0} \mathfrak{p}_1^{a_1} \cdots \mathfrak{p}_s^{a_s}$ with $0 \leq a_j \leq e_j$; clearly there are $(e_0 + 1)(e_1 + 1) \cdots (e_s + 1)$ such ideals; since $(e_0 + 1)(e_1 + 1) \cdots (e_s + 1) = \tau(n)$ is the number of divisors of n and since $e_0 + 1 = 2$, the number of ideals is $2\tau(n)$.

The class number bound will follow if we can show that only two among these ideals $\mathfrak{a}$ can be principal:

- the unit ideal $\mathfrak{a} = \mathfrak{p}_0^0 \mathfrak{p}_1^0 \cdots \mathfrak{p}_s^0 = (1)$;
- the ideal $\mathfrak{a} = \mathfrak{p}_0 \mathfrak{p}_1^{e_1} \cdots \mathfrak{p}_s^{e_s}$ with norm $2t$.

In fact, if $\mathfrak{a} = \mathfrak{p}_0^{a_0} \mathfrak{p}_1^{a_1} \cdots \mathfrak{p}_s^{a_s} = (x + y\sqrt{m})$ is principal, then $|x^2 - my^2| = N\mathfrak{a} \leq N(\mathfrak{p}_0 \mathfrak{p}_1^{e_1} \cdots \mathfrak{p}_s^{e_s}) = 2t$. If $\mathfrak{a}$ is not one of the two ideals with norm 1 or $2t$, then $1 < |x^2 - my^2| < 2t$; but then $y = 0$ and $\mathfrak{a} = (x)$ for some rational integer x. Now $\mathfrak{a} = \mathfrak{a}^\sigma$ for the the nontrivial automorphism of $k/\mathbb{Q}$ implies $a_1 = \ldots = a_s = 0$, and hence $\mathfrak{a} = \mathfrak{p}_0$; but the prime ideal above 2 is not principal.

For proving the last claim, assume that $t = q^e$ for an odd prime number q. The relation $\mathfrak{p}_0 q^e = (t + 1 + \sqrt{m})$, together with the fact that no ideal of the form $2q^j$ with $0 \leq j < e$ is principal, implies that the ideal class of $\mathfrak{q}$ has order $2e$; thus we obtain the lower bound $2e \mid h$, where $2e = 2\tau(t) - 2$.   □

The following table compares the lower bound S in Theorem 7.11 with the class number h for a few small values of m:

t	m	S	h	t	m	S	h	t	m	S	h
3	10	2	2	21	442	6	8	37	1370	2	4
5	26	2	2	23	530	2	4	39	1522	6	12
9	82	4	4	25	626	4	4	45	2026	10	14
11	122	2	2	27	730	6	12	47	2210	2	8
13	170	2	4	29	842	2	6	49	2402	4	8
15	226	6	8	31	962	2	4	51	2602	6	10
17	290	2	4	33	1090	6	12	53	2810	2	8
19	362	2	2	35	1226	6	10	55	3026	6	16

7.3   Computing the Solution of the Pell Equation

The computation of the units in a real quadratic number field $\mathbb{Q}(\sqrt{m})$, that is, solving the corresponding Pell equation $x^2 - my^2 = \pm 1$, is usually a quite difficult problem.

The basic idea behind the computation of the fundamental unit of a quadratic number field that we will present here is, as in the proof of the solvability of the Pell equation, the construction of sufficiently many elements of small norm. If we have many such elements, then we will look for elements α and β that not only have the same norm but also generate the same principal ideal. In this case, the quotient $\frac{\alpha}{\beta}$ will be a (possibly trivial) unit.

In order to convey the main idea, we consider the ring $\mathbb{Z}[\sqrt{11}]$. We look for solutions of the equation $x^2 - 11y^2 = n$ for small values of n. If we pick $y = 1$, the expression $x^2 - 11y^2$ will be small if $x \approx \sqrt{11}$, that is, for $x = 3$ and $x = 4$. Thus

$$3^2 - 11 = -2,$$

$$4^2 - 11 = +5.$$

For $y = 2$, we choose $x \approx 2\sqrt{11}$, and we find

$$6^2 - 11 \cdot 2^2 = -8,$$

$$7^2 - 11 \cdot 2^2 = +5.$$

Thus we already have found elements $4 \pm \sqrt{11}$ and $7 \pm 2\sqrt{11}$ with the same norm 5. Which of these generate the same ideal? One possibility of finding the right choice of signs is simply computing the quotients:

$$\frac{7 + 2\sqrt{11}}{4 + \sqrt{11}} = \frac{(7 + 2\sqrt{11})(4 - \sqrt{11})}{(4 + \sqrt{11})(4 - \sqrt{11})} = \frac{6 + \sqrt{11}}{5},$$

which is not an algebraic integer; thus $7 + 2\sqrt{11}$ and $4 + \sqrt{11}$ generate distinct prime ideals above 5. On the other hand,

$$\frac{7 + 2\sqrt{11}}{4 - \sqrt{11}} = \frac{(7 + 2\sqrt{11})(4 + \sqrt{11})}{(4 + \sqrt{11})(4 - \sqrt{11})} = \frac{50 + 15\sqrt{11}}{5} = 10 + 3\sqrt{11},$$

and we have found the nontrivial unit $\varepsilon = 10 + 3\sqrt{11}$.

Here is a more elegant way of verifying that $7 + 2\sqrt{11}$ and $4 - \sqrt{11}$ generate the same ideal: We know that these elements have norm 5, and hence they generate prime ideals above 5. There are only two such ideals, namely $5_1 = (5, 1 + \sqrt{11})$

and $52 = (5, 1 - \sqrt{11})$. Thus $\sqrt{11} \equiv -1 \bmod 5_1$ and $\sqrt{11} \equiv +1 \bmod 5_2$, hence

$$7 + 2\sqrt{11} \equiv 0 \bmod 5_1, \qquad\qquad 7 + 2\sqrt{11} \equiv 4 \bmod 5_2,$$

$$4 + \sqrt{11} \equiv 3 \bmod 5_1, \qquad\qquad 4 + \sqrt{11} \equiv 0 \bmod 5_2,$$

and this shows that $(7 + 2\sqrt{11}) = (4 - \sqrt{11}) = 5_1$.

Another possibility of finding a nontrivial unit is based on the observation that $(2) = 2^2$ is ramified in K. Since $3 + \sqrt{11}$ has norm -2, we must have $2 = (3 + \sqrt{11})$, and then $(2) = 2^2 = (3 + \sqrt{11})^2 = (20 + 6\sqrt{11})$ shows that $\frac{20 + 6\sqrt{11}}{2} = 10 + 3\sqrt{11}$ is a unit.

Now let us see how this method works for larger values of m, say for $m = 3431$. Again we begin by collecting elements with small norms:

α	$N\alpha$	α	$N\alpha$
$55 + \sqrt{m}$	$-2 \cdot 7 \cdot 29$	$60 + \sqrt{m}$	13^2
$56 + \sqrt{m}$	$-5 \cdot 59$	$61 + \sqrt{m}$	$2 \cdot 5 \cdot 29$
$57 + \sqrt{m}$	$-2 \cdot 7 \cdot 13$	$62 + \sqrt{m}$	$7 \cdot 59$
$58 + \sqrt{m}$	-67	$63 + \sqrt{m}$	$2 \cdot 269$
$59 + \sqrt{m}$	$2 \cdot 5^2$	$64 + \sqrt{m}$	$5 \cdot 7 \cdot 19$

We remark in passing that $60^2 - m = 13^2$ is a square; this implies that $m = 60^2 - 13^2 = (60 - 13)(60 + 13) = 47 \cdot 73$. Fermat's method of factorization is based on this idea.

The fact that 3 does not occur among these prime factors is explained by the observation that there is not even an ideal with norm 3 in $\mathbb{Q}(\sqrt{m})$ since $(\frac{m}{3}) = -1$. For the same reason, the primes 11 and 17 do not show up as factors. Instead of waiting until elements with the same norm occur, we will use an idea that was already used by Fermat and his contemporaries in their search for numbers whose sums of divisors are squares or cubes. We factor the elements with small norm into primes. It is easy to write down a list of prime ideals with small norms; in our case, these are $2 = (2, 1 + \sqrt{m})$, $5_1 = (5, 1 + \sqrt{m})$, $5_2 = (5, 1 - \sqrt{m})$, $7 = (7, 1 + \sqrt{m})$, and $7_2 = (7, 1 - \sqrt{m})$. Now we factor all elements with small norm that are only divisible by 2, 5, and 7:

α	2	5_1	5_2	7_1	7_2
$1 + \sqrt{m}$	1	1	0	3	0
$1 - \sqrt{m}$	1	0	1	0	3
$41 + \sqrt{m}$	1	3	0	0	1
$41 - \sqrt{m}$	1	0	3	1	0
$59 + \sqrt{m}$	1	0	2	0	0
$59 - \sqrt{m}$	1	2	0	0	0

The first line in this table records the prime ideal decomposition

$$(1 + \sqrt{m}) = 2^1 \cdot 5_1^1 \cdot 7_1^3.$$

If we look carefully at this table, then we can see that

$$(1 + \sqrt{m})(41 + \sqrt{m})^3 = 2^4 5_1^{10} 7_1^3 7_2^3.$$

Since $2^2 = (2)$ und $7_1 7_2 = (7)$, the element

$$\frac{(1 + \sqrt{m})(41 + \sqrt{m})^3}{2^2 \cdot 7^3} = 21549 + 364\sqrt{m}$$

has the prime ideal factorization 5_1^{10}. But then $(59 - \sqrt{m})^5 = 2^5 5^{10}$ shows that

$$\alpha = \frac{(59 - \sqrt{m})^5}{21549 + 364\sqrt{m}} = 49316884 - 841948\sqrt{m}$$

is an algebraic integer with the factorization 2^5. Since the ideal 2 is ramified, the element $\varepsilon = 2^5/\alpha^2$ must be a unit, and we have

$$\varepsilon = 152009690466840 + 2595140740627\sqrt{m}.$$

Observe that this method gives us not only a nontrivial unit but also something called a "compact representation" of this unit:

$$\varepsilon = \frac{2(1 + \sqrt{m})^2(41 + \sqrt{m})^6}{7^6(59 - \sqrt{m})^{10}}.$$

Also observe that the prime ideal factorization in quadratic number fields is an essential component of this method of solving the completely elementary equation $x^2 - my^2 = 1$.

After having found a nontrivial unit ε, the question remains how we can check that this unit is fundamental. So far we only know that $\varepsilon = \pm\eta^n$ for some integer n, where η is the fundamental unit. Since $\varepsilon > 1$, the positive sign must hold, and we have $n \geq 1$. Clearly, ε is not a square as we can read off from its compact representation. Thus we only have to check whether ε is an n-th power for the values $n = 3, 5, 7, \ldots$, and the first problem is bounding this exponent.

The following bound is simple and best possible:

Lemma 7.12 *Let $\eta > 1$ be the fundamental unit of a real quadratic number field with discriminant d. Then*

$$\log \eta > \begin{cases} \log \sqrt{d}, & \text{if } N\varepsilon = +1, \\ \log(\sqrt{d} - 1), & \text{if } N\varepsilon = -1. \end{cases}$$

Proof Let $K = \mathbb{Q}(\sqrt{m})$ with $m \equiv 2, 3 \bmod 4$ and $N\varepsilon = +1$. Then the smallest possible value of ε is $a + \sqrt{m}$ with $a \approx \sqrt{m}$. If $N\varepsilon = +1$, then $a > \sqrt{m}$, and this implies $\varepsilon > 2\sqrt{m} = \sqrt{d}$.

Observe that the family of quadratic number rings $\mathbb{Z}[\sqrt{m}]$ with $m = t^2 - 1$ has units $\varepsilon_m = t + \sqrt{m}$ for which $\varepsilon_m = \sqrt{m+1} + \sqrt{m} > 2\sqrt{m} = \sqrt{d}$ is best possible since $\varepsilon_m - \sqrt{d} = \sqrt{m+1} - \sqrt{m} = \frac{1}{\sqrt{m+1}+\sqrt{m}} < \frac{1}{2\sqrt{m}}$.

The other cases are discussed similarly.   □

In our case we have $m = 3431 = 47 \cdot 73$; since m is divisible by the prime number $47 \equiv 3 \bmod 4$, we must have $N\varepsilon = +1$ and thus $\log \varepsilon \geq 4.763\ldots$, and hence $n = \log \varepsilon / \log \eta \leq 33.3/4.763 = 6.991\ldots$. Therefore $n \leq 6$, and since we already know that ε is not a square, we must have $m \leq 5$.

Thus it remains to show that ε is not a cube or a fifth power. Perhaps the easiest way of doing this is finding a prime ideal $\mathfrak{p}$ modulo which ε is not a cube or a fifth power.

- Since $\varepsilon \equiv 0 - 3\sqrt{m} \equiv 3 \bmod 51$, the unit ε is not a square; in fact, $(\frac{3}{5}) = -1$ implies that 3 is not a square modulo 51.
- For showing that ε is not a cube, we need to find a prime ideal $\mathfrak{p}$ with norm $N\mathfrak{p} \equiv 1 \bmod 3$. Now $\varepsilon \equiv 3 + \sqrt{m} \equiv 2 \bmod 71$, and since 2 is not a cubic residue modulo 7 and $\mathcal{O}_K/71 \simeq \mathbb{Z}/7\mathbb{Z}$, the unit ε is not a cubic residue modulo 71. In particular, ε is not a cube.
- Let $\mathfrak{q} = (61, 25+\sqrt{m})$; then $\mathfrak{p}$ has norm $N\mathfrak{p} = 61$, and we have $\varepsilon \equiv 40-3\sqrt{m} \equiv 54 \bmod \mathfrak{q}$. Since 54 is not a fifth power modulo 61, the unit ε cannot be a fifth power.

Instead of working with residue classes, we can compute with real numbers. To this end, we determine the real approximations

$$\varepsilon \approx 304\,019\,380\,933\,679.999\,999\,999\,999\,996\,711$$
$$1/\varepsilon \approx \qquad\qquad\qquad 0.000\,000\,000\,000\,003\,289.$$

Now $\varepsilon + 1/\varepsilon$ is an integer; in fact, if we write $\varepsilon = a + b\sqrt{m}$, then $1/\varepsilon = a - b\sqrt{m} = \varepsilon'$ and thus $\varepsilon + \frac{1}{\varepsilon} = 2a$. If $\varepsilon = \eta^5$ were a fifth power, then $\eta + 1/\eta = \eta + \eta'$ would be an integer. Since $\eta \approx 788.098052\ldots$ and $1/\eta \approx 0.001268877\ldots$, we have $\eta + 1/\eta \approx 788.0993\ldots$. Thus ε is not a fifth power, and the cases $n = 2$ and $n = 3$ can be treated similarly.

7.4 Parametrized Units

It is rather easy to construct families of quadratic number fields in which the fundamental unit can be written down explicitly. The simplest way of finding such a family is looking at the unit equation $t^2 - mu^2 = \pm 1$ and setting $u = 1$; then $m = t^2 \mp 1$ (with $t \geq 2$), and in fact $\varepsilon_t = t + \sqrt{t^2 \mp 1}$ is a unit in $\mathbb{Z}[\sqrt{m}\,]$.

Proposition 7.13 *If $m = t^2 - 1$ is squarefree for $t \geq 2$, then $\varepsilon_m = t + \sqrt{m}$ is the fundamental unit of $\mathbb{Z}[\sqrt{m}\,]$.*

This result also holds if m is not squarefree, but then $\mathbb{Z}[\sqrt{m}\,]$ is not the ring of integers of $\mathbb{Q}(\sqrt{m})$.

Since we have already shown that ε_m is a unit, it only remains to show that ε_m is fundamental. But since $\varepsilon_m > 1$, this unit can only be not fundamental if $\varepsilon_m = (r + s\sqrt{m}\,)^k$ for some exponent $k \geq 2$, and in that case the coefficient of $\sqrt{m}$ in ε would have to be strictly greater than 1; for example, we have $(r + s\sqrt{m}\,)^2 = r^2 + ms^2 + 2rs\sqrt{m}$.

The case $m = t^2 + 1$ is slightly more complicated.

Proposition 7.14 *Assume that $m = t^2 + 1$ is squarefree for $t \geq 1$. If t is odd, then $\varepsilon_m = t + \sqrt{m}$ is the fundamental unit of $\mathbb{Z}[\sqrt{m}\,]$. If t is even, then ε_m is the fundamental unit of $\mathbb{Z}[\frac{1+\sqrt{m}}{2}\,]$ except for $t = 2$, when $\varepsilon_5 = 2 + \sqrt{5} = (\frac{1+\sqrt{5}}{2})^3$.*

The proof that ε_m is fundamental if t is odd (and thus if $m \equiv 2 \bmod 4$) is exactly as above. Assume therefore that t is even and $m = t^2 + 1 \equiv 1 \bmod 4$. If ε_m is not fundamental, then $\varepsilon_m = \left(\frac{r+s\sqrt{m}}{2}\right)^k$ for some exponent $k \geq 2$. If r and s are even, the proof above works. Assume therefore that r and s are odd. Then the smallest power of $\frac{r+s\sqrt{m}}{2}$ that lies in $\mathbb{Z}[\sqrt{m}\,]$ is 3, and in fact k must be a multiple of 3. The case $k \geq 6$ cannot occur (the same proof as above), and if $k = 3$, then

$$\left(\frac{r + s\sqrt{m}}{2}\right)^3 = \frac{r^3 + 3rms^2 + (3r^2s + ms^3)\sqrt{m}}{8}$$

shows that we must have $3r^2s + ms^3 = 8$. Since s is odd, this implies $s = 1$, and then $3r^2 + m = 8$ yields $m = 5$ and $r = 1$ as the only integral solution.

We obtain a slightly less trivial family by writing the Pell equation $t^2 - mu^2 = 1$ in the form $mu^2 = t^2 - 1 = (t - 1)(t + 1)$. Setting $t - 1 = u^2$ and $t + 1 = m$, we find $m = u^2 + 2$ and $t = u^2 + 1$. In this way we obtain the following proposition.

Proposition 7.15 *Assume that $m = t^2 + 2$ is squarefree. Then $\varepsilon_m = t^2 + 1 + t\sqrt{m}$ is the fundamental unit of $\mathbb{Z}[\sqrt{m}\,]$.*

Here $\varepsilon_m = (r + s\sqrt{m}\,)^k$ is impossible for $k \geq 2$ because already the coefficient of $\sqrt{m}$ in $(r + s\sqrt{m}\,)^2 = r^2 + ms^2 + 2rs\sqrt{m}$ is too large: We must have $r^2 + ms^2 > t^2 + 1$ since otherwise $rs = 0$, which is impossible.

In the examples above, the units are rather small. For finding fields with larger fundamental units,[9] we construct elements α and β with $N\alpha = \pm a^n$ and $N\beta = a$; using some additional conditions, we can make sure that the quotient $\varepsilon = \alpha/\beta^n$ is integral and therefore a unit.

For finding fields $K = \mathbb{Q}(\sqrt{m})$ containing elements with norm $\pm a^n$, we can simply write $m = r^2 + a^n$; then $\alpha = r + \sqrt{m}$ has norm $N\alpha = -a^n$. For finding elements with norm $\pm a$ in $\mathbb{Z}[\sqrt{m}]$, we observe that if $Q(x, y) = Ax^2 + Bxy + Cy^2$ (below, we will often abbreviate this form by $Q = (A, B, C)$) is a binary quadratic form with discriminant $\Delta = B^2 - 4AC = 4m$, and if $Q(s, t) = 1$, then

$$As^2 + Bst + Ct^2 = 1 \quad \text{implies} \quad 4A = (2As + Bt)^2 - \Delta t^2,$$

and hence $As + \frac{1}{2}Bt + t\sqrt{m}$ has norm A. Clearly, the quadratic form $Q = (a, 2r, a^{n-1})$ has discriminant $4m$, so all we need is a solution of the equation $1 = Q(s, t) = as^2 + 2rst + a^{n-1}t^2$ in integers. Before we construct such solutions, we prove that we do in fact obtain units in this way:

Proposition 7.16 *Assume that* $m = r^2 + a^n \equiv 2, 3 \bmod 4$ *for coprime integers* $a > 1$ *and* $n \geq 2$. *Assume moreover that* $Q(s, t) = 1$, *where* s *and* t *are nonzero integers and where* $Q = (-a, 2r, a^{n-1})$. *Set* $\alpha = r + \sqrt{m}$ *and* $\gamma = as - rt - t\sqrt{m}$; *then* $N\alpha = -a^n$, $N\gamma = -a$, *and the element*

$$\varepsilon = \frac{\gamma^n}{\alpha}$$

is a nontrivial unit in $\mathbb{Z}[\sqrt{m}]$.

There is, of course, a similar result for $m \equiv 1 \bmod 4$. If r is odd, then one has to consider elements of the form $\alpha = \frac{r+\sqrt{m}}{2}$.

Proof Set $\mathfrak{a} = (a, r + \sqrt{m})$. Clearly,

$$\gamma = as - rt - t\sqrt{m} = as - t(r + \sqrt{m}) \in \mathfrak{a},$$

hence $\mathfrak{a} \mid (\gamma)$. Moreover,

$$\mathfrak{a}^n = (a^n, a^{n-1}\alpha, \ldots, a\alpha^{n-1}, \alpha^n) = (\alpha)(r - \sqrt{m}, a^{n-1}, \ldots, \alpha^{n-1}) = (\alpha)$$

since $(\alpha', \alpha^{n-1}) \supseteq (\alpha', \alpha) = (1)$. In fact, if $\mathfrak{p}$ is a prime ideal dividing (α', α), then either $\mathfrak{p}$ is ramified, or $m \equiv 1 \bmod 4$ and $\mathfrak{p} \mid (2)$. Since we have excluded the last

---

[9]The class number formula roughly implies that fields with large fundamental units tend to have small class numbers; constructing families of fields with large fundamental units is therefore important with respect to Gauss's conjecture that there are infinitely many real quadratic number fields with class number 1.

case, $\mathfrak{p}$ is ramified. Since $\mathfrak{p} \mid N\alpha = a^n$, this implies $\mathfrak{p} \mid (a)$. But then $\mathfrak{p} \mid (m)$ implies $\mathfrak{p} \mid (r)$ contradicting the assumption that r and a are coprime.

Now $\gamma \in \mathfrak{a}$ and $|N\gamma| = N\mathfrak{a}$ implies $\mathfrak{a} = (\gamma)$. This shows that $(\alpha) = (\gamma)^n$, and hence ε is a unit as claimed.

It remains to show that the unit ε is nontrivial, i.e., that $\varepsilon \neq \pm 1$. But $\varepsilon = \pm 1$ is equivalent to $\pm \gamma^n = \alpha = r + \sqrt{m}$, and this is impossible for $n \geq 2$ as soon as $a > 1$. Clearly, $\pm(t + u\sqrt{m})^2 = r + \sqrt{m}$ is impossible; similarly, $\pm(\frac{t+u\sqrt{m}}{2})^3 = r + \sqrt{m}$ implies $t = u = 1$ and $m = 5$, which in turn is only possible if $r = 2$ and $a = 1$. $\quad\square$

Now let $m = r^2 + a^3$ and $Q = (-a, 2r, a^2)$. Setting $Q(x, 1) = 1$ and solving for r, we obtain

$$r = \frac{1 - a^2 + ax^2}{2x}.$$

This value of r is an integer, e.g., when $x = a - 1$ is even. Then $r = \frac{a^2 - 2a - 1}{2}$, and we have $Q(a - 1, 1) = 1$ for the quadratic form $Q = (-a, a^2 - 2a - 1, a^2)$, and hence

$$\gamma = a(a - 1) - \frac{a^2 - 2a - 1}{2} - \sqrt{m} = \frac{a^2 + 1 - 2\sqrt{m}}{2}$$

has norm $-a$.

An explicit calculation yields the unit

$$\varepsilon = \frac{a^5 - a^4 + 3a^3 + a^2 + 2}{2} + (a^3 - a^2 + 2a)\sqrt{m}.$$

The first few examples are given in the following table:

a	r	m	γ	ε
5	7	174	$13 - \sqrt{m}$	$1451 + 110\sqrt{174}$
9	31	1690	$41 - \sqrt{m}$	$27379 + 666\sqrt{1690}$
13	71	7238	$85 - \sqrt{m}$	$174747 + 2054\sqrt{7238}$
17	127	21042	$145 - \sqrt{m}$	$675683 + 4658\sqrt{21042}$

There are many other choices of r, each of which yields a similar family of units. Now let $\Delta = (2a + 1)^2 + 4 \cdot 2^n$ for some integer a. Then

$$(2a + 1)^2 - \Delta = -4 \cdot 2^n, \quad \text{or} \quad a^2 + a + \frac{1 - \Delta}{4} = -2^n.$$

This shows that $(a, 1)$ is an integral point on the conic $x^2 + xy + \frac{1-\Delta}{4} = -2^n$; equivalently, the element $\alpha = \frac{2a+1+\sqrt{\Delta}}{2}$ has norm $N\alpha = -2^n$.

Next we look at conics of the form $Q(x, y) = 1$ with $Q = (2, b, c)$ and

$$\text{disc } Q = b^2 - 8c = (2a + 1)^2 + 4 \cdot 2^n$$

that have an integral point. The simplest possible form is $Q = (2, -2a - 1, -2^{n-1})$, and the simplest possible integral points are those with $y = \pm 1$. A necessary condition for the existence of an integral solution of $Q(x, \pm 1) = 1$, that is, of

$$2x^2 \pm (2a + 1)x - 2^{n-1} = 1,$$

is that the discriminant of the quadratic equation in x is a square:

$$(2a + 1)^2 + 8 \cdot (2^{n-1} - 1) = 4a^2 + 4a + 4 \cdot 2^n + 9 = \square.$$

Setting this expression equal to $(2a + 3)^2$ quickly yields $a = 2^{n-1}$. In this case, the quadratic equation

$$2x^2 - (2^n + 1)x - (2^{n-1} - 1) = 0$$

has the solutions

$$x_{1,2} = \frac{2^n + 1 \pm (2^n + 3)}{4}, \quad i.e., \quad x_1 = -\frac{1}{2}, \quad x_2 = 2^{n-1} + 1.$$

Thus we now have $\Delta = (2^n + 3)^2 - 8 = (2^n + 1)^2 + 4 \cdot 2^n$, and the conic $Q(x, y) = 1$ with $Q = (2, -2^n - 1, -2^{n-1})$ has the integral point $(2^{n-1} + 1, 1)$. Since

$$8Q(x, y) = (4x - (2^n + 1)y)^2 - \Delta y^2,$$

this provides us with the element

$$\gamma = \frac{2^n + 3 + \sqrt{\Delta}}{2}$$

with norm $N\gamma = -2$. Since $\alpha = \frac{1}{2}[(2^n + 1) - \sqrt{\Delta}]$ has norm -2^n, the solution of the Pell equation is given by

$$\varepsilon = -\frac{\gamma^n}{\alpha},$$

which is a unit with norm -1. This family is due to Michael Nyberg [100] and (independently) to Daniel Shanks [115].

n	Δ	ε
1	17	$4 + \sqrt{\Delta}$
2	41	$32 + 5\sqrt{\Delta}$
3	113	$776 + 73\sqrt{\Delta}$
4	353	$71264 + 3793\sqrt{\Delta}$
5	1217	$27628256 + 791969\sqrt{\Delta}$
6	4481	$46496952832 + 694603585\sqrt{\Delta}$

In this case, the discriminant Δ is asymptotically equal to $\Delta \sim 2^{2n}$, and we have $\gamma \sim 2^n$. Moreover, $\frac{1}{\alpha} = \frac{\alpha'}{N\alpha} \sim -\frac{2^n}{2^n} = -1$ is bounded, and hence $\varepsilon \sim \gamma^n \sim (2^n)^n \sim \Delta^{n/2}$. Thus $\log \varepsilon / \log \Delta$ is not bounded.

7.5 Factorization Algorithms

The same idea that we have used for computing the fundamental unit of a real quadratic number field can be applied directly for factoring large integers. As a modest example, we choose $N = 4469$ and begin by factoring the integers $a^2 - N$ for $a \approx \sqrt{4469} \approx 67$. We keep only those factorizations that involve sufficiently small prime numbers:

a	-1	2	5
62	1	0	4
63	1	2	3
67	0	2	1

The first line in this table encodes the factorization $62^2 - N = -5^4$.

Already the Indian mathematician Narayana Pandit (ca. 1340–1400) and later Pierre Fermat had used a similar method for factoring integers that do not have small factors. They checked whether any of the numbers $a^2 - N$ for $N = 1, 2, 3, \ldots$ is a square number: If $a^2 - N = b^2$, then we obtain the factorization $N = a^2 - b^2 = (a - b)(a + b)$.

The essential idea behind the modern factorization methods based on this idea (see, e.g., [130]) is the observation that we do not need a solution of the equation $a^2 - N = b^2$ but only a solution of the congruence $a^2 \equiv b^2 \bmod N$. Once we have found such a pair of integers a and b, the numbers $\gcd(a+b, N)$ and $\gcd(a-b, N)$ are (possibly trivial) factors of N. Now observe that

$$(63^2 - N)(67^2 - N) = -2^4 \cdot 5^4$$

implies the congruence

$$63^2 \cdot 67^2 \equiv -2^4 \cdot 5^4 \bmod N.$$

Moreover we have $62^2 \equiv -5^4 \bmod N$, and hence $63^2 \cdot 67^2 \equiv 4^2 \cdot 62^2 \bmod N$, and we find only the trivial factor $\gcd(63 \cdot 67 - 4 \cdot 62, N) = 1$.

By enlarging our factor base, we obtain

a	-1	2	5	11	13
62	1	0	4	0	0
63	1	2	3	0	0
67	0	2	1	0	0
71	0	2	0	1	1
72	0	0	1	1	1
83	0	2	1	2	0

Now we see $67^2 \cdot 72^2 \equiv 71^2 \cdot 5^2 \bmod N$, but this solution gives us once again just the trivial factorization. We are more lucky with $67^2 \cdot 11^2 \equiv 83^2 \bmod N$ since now $\gcd(67 \cdot 11 - 83, N) = 109$, and in fact we have $N = 41 \cdot 109$.

Finding such relations is essentially linear algebra: We interpret the exponents in the factorizations as elements of an $\mathbb{F}_2$-vector space, and then finding squares boils down to finding linear dependent vectors. The factorization method based on this idea is called the quadratic sieve.

Factoring Integers with the Pell Equation The computation of the fundamental unit is, for many values of m, about as difficult as factoring m. Indeed it follows from $x^2 - my^2 = 1$ that $my^2 = x^2 - 1 = (x-1)(x+1)$, and $\gcd(m, x-1)$ is a (possibly trivial) factor of m. For $m = 91$, for example, the fundamental unit is $\varepsilon = 1574 + 165\sqrt{91}$, and we have $\gcd(91, 1573) = 13$. The Bohemian mathematician Franz von Schafgotsch [128] factored $a = 909\,191$ by solving the Pell equation for $m = 5a = 4\,545\,955$; he obtained

$$7904827417056517386293496562684929005511866785872458337976 0874^2 =$$

$$m \cdot 3707488617933672582804872308816078480451363428966076349 8655^2 + 1,$$

and used the Euclidean algorithm to find the greatest common divisor of

$$79048274170565173862934965626849290055118667858724583379760874 + 1$$

and $m = 1315$, which gave him the factorization

$$909\,191 = 263 \cdot 3457.$$

7.6 Diophantine Equations

There is a large class of Diophantine equation whose solution depends crucially on the structure of the unit group of quadratic number fields. To give the readers an idea of a few elementary techniques in this area, we will prove a result due to J.H.E. Cohn (see, e.g., [21]).

Let $m \equiv 5 \bmod 8$ be a squarefree natural number, and assume that the fundamental unit $\varepsilon = \frac{a+b\sqrt{m}}{2}$ has norm -1 and satisfies $a \equiv b \equiv 1 \bmod 2$. We will consider the sequence of integers

$$V_n = \varepsilon^n + \varepsilon'^n,$$

where $\varepsilon' = \frac{a-b\sqrt{m}}{2} = -\frac{1}{\varepsilon}$. The first three elements of the sequence (V_n) are $V_0 = 2$, $V_1 = a$, and $V_2 = a^2 + 2$.

If $m = 5$ (and therefore $a = 1$), these are called *Lucas numbers* after Édouard Lucas (1842–1891):

n	0	1	2	3	4	5	6	7
V_n	2	1	3	4	7	11	18	29

One consequence of the theorem we are about to prove is that the only squares in this sequence are $V_1 = 1$ and $V_3 = 4$.

We will need the following observations:

Proposition 7.17 *For all $k, n \in \mathbb{Z}$, we have*

$$V_{n+2} = aV_{n+1} + V_n, \tag{7.8}$$

$$V_{2n} = V_n^2 - 2(-1)^n, \tag{7.9}$$

$$V_{n+2k} \equiv (-1)^{k+1} V_n \bmod V_k. \tag{7.10}$$

Equation (7.8) follows from

$$V_{n+2} = \varepsilon^{n+2} + \varepsilon'^{n+2} = (\varepsilon + \varepsilon')(\varepsilon^{n+1} + \varepsilon'^{n+1}) - \varepsilon\varepsilon'^{n+1} - \varepsilon'\varepsilon^{n+1}$$
$$= V_1 V_{n+1} - \varepsilon\varepsilon' V_n = V_1 V_{n+1} + V_n$$

since $V_1 = a$ and $\varepsilon\varepsilon' = -1$. Next

$$V_{2n} = \varepsilon^{2n} + \varepsilon'^{2n} = (\varepsilon^n + \varepsilon'^n)^2 - 2(\varepsilon\varepsilon')^n = V_n^2 - 2(-1)^n.$$

Equation (7.9) immediately implies that the numbers V_{2n} cannot be squares. Finally,

$$V_{n+2k} = \varepsilon^{n+2k} + \varepsilon'^{n+2k} = (\varepsilon^k + \varepsilon'^k)(\varepsilon^{n+k} + \varepsilon'^{n+k}) - \varepsilon^k \varepsilon'^{n+k} - \varepsilon'^k \varepsilon^{n+k}$$
$$= V_k V_{n+k} - (-1)^k V_n \equiv (-1)^{k+1} V_n \bmod V_k.$$

Next we observe that V_n is even if and only if n is divisible by 3. This follows from the recursion

$$V_{n+2} = a V_{n+1} + V_n$$

by induction.

In the following, k will always denote an integer not divisible by 3. Thus

$$V_{2k} = V_k^2 - 2(-1)^k \equiv \begin{cases} 3 \bmod 8 & \text{if } k \text{ is odd,} \\ 7 \bmod 8 & \text{if } k \text{ is even.} \end{cases}$$

Theorem 7.18 *Let $m \equiv 5 \bmod 8$ be squarefree, and let $\varepsilon = \frac{a+b\sqrt{m}}{2}$ denote the fundamental unit of $\mathbb{Q}(\sqrt{m})$, where we assume that a and b are odd. The number $V_n = \varepsilon^n + \varepsilon'^n$ is a square only in the following cases:*

1. $n = 1$ and a is a square;
2. $n = 3$ and $a(a^2 + 3)$ is a square.

Using sage, it is possible to show that the elliptic curve $y^2 = a(a^2+3)$ has exactly four integral points, namely $(0, 0)$, $(1, 0)$, $(3, 6)$, and $(12, 42)$. A proof by hand leads to the Diophantine equation $x^4 - 3y^4 = -2$, which seems to be difficult to solve with the methods presented here.

For the proof of Theorem 7.18, we will distinguish several cases.

1. $a \equiv 5, 7 \bmod 8$ and $n \equiv 3 \bmod 4$ We write $n = 2 \cdot 3^r k - 1$ for an even integer k not divisible by 3. Then $V_k \equiv 3 \bmod 4$ and

$$V_n = V_{2 \cdot 3^r k - 1} \equiv -V_{-1} \equiv a \bmod V_k$$

by (7.10); hence,

$$\left(\frac{V_n}{V_k}\right) = \left(\frac{a}{V_k}\right) \qquad \text{since } V_n \equiv a \bmod V_k,$$

$$= \left(\frac{-1}{a}\right)\left(\frac{V_k}{a}\right) \qquad \text{since } \left(\frac{a}{V_k}\right) = \left(\frac{-V_k}{a}\right)$$

$$= \left(\frac{-2}{a}\right) \qquad \text{since } V_k \equiv 2 \bmod a \text{ if } k \text{ is even.}$$

$$= -1 \qquad \text{since } a \equiv 5, 7 \bmod 8.$$

But this implies our claim that V_n is not a square in this case.

2. $a \equiv 5, 7 \bmod 8$ **and** $n \equiv 1 \bmod 4$ Here we write $n = -3 + 2 \cdot 3^r k$ for some even integer n not divisible by 3 and find $V_n \equiv -V_{-3} = V_3 \bmod V_k$. Now $V_3 = a(a^2 + 3) = a \cdot 4b$ for some odd integer b; hence,

$$\left(\frac{V_3}{V_k}\right) = \left(\frac{a}{V_k}\right)\left(\frac{b}{V_k}\right) \qquad \text{since } \left(\frac{4}{V_k}\right) = 1$$

$$= \left(\frac{a}{V_k}\right)\left(\frac{-V_k}{b}\right) \qquad \text{since } V_k \equiv 3 \bmod 4$$

$$= \left(\frac{a}{V_k}\right) \qquad \text{since } V_k \equiv V_2 = a^2 + 2 \equiv -1 \bmod b.$$

$$= -1 \qquad \text{since } \left(\frac{a}{V_k}\right) = \left(\frac{-V_k}{a}\right) = \left(\frac{-2}{a}\right) = -1 \text{ as above.}$$

3. $a \equiv 1, 3 \bmod 8$ **and** $n \equiv 1 \bmod 4$ We write $n = 2 \cdot 3^r k + 1$ for an even integer k not divisible by 3; since $V_n \equiv -V_1 \equiv -a \bmod V_k$ if $n > 1$, we have

$$\left(\frac{V_n}{V_k}\right) = \left(\frac{-a}{V_k}\right) = -\left(\frac{a}{V_k}\right) = -\left(\frac{-2}{a}\right) = -1.$$

4. $a \equiv 1, 3 \bmod 8$ **and** $n \equiv 3 \bmod 4$ If $n \neq 3$, we can write $n = 3 + 2 \cdot 3^r k$, where k is even and not divisible by 3. Then $V_n \equiv -V_3 \equiv -a \bmod V_k$, and again we find

$$\left(\frac{V_n}{V_k}\right) = \left(\frac{-V_3}{V_k}\right) = -\left(\frac{V_3}{V_k}\right) = -\left(\frac{-2}{a}\right) = -1.$$

If $n = 3$, then $V_3 = a^3 + 3a$ must be a square.

7.6.1   Summary

In this chapter we have shown

- that the Pell equation $x^2 - my^2 = 1$ has nontrivial solutions in integers for each nonsquare natural number m, that the unit group E_K of a real quadratic number field K is isomorphic to $E_K \simeq \mathbb{Z}/2\mathbb{Z} \times \mathbb{Z}$, and
- how to find the solutions of the Pell equation for modest values of m by studying elements of K with small norms.

For learning more about the Pell equation, see [8, 66, 91], and [46], as well as the series of articles [78].

7.7   Exercises

7.1. Let k be a real quadratic number field; assume that $\eta = \alpha^2$ and $N\alpha < 0$ for elements $\eta, \alpha \in k^\times$. Show that, as real numbers, $\mathrm{Tr}\,\alpha = \sqrt{\eta} - \sqrt{\eta'}$.

7.2. Show that if $m = n^2$ is a square, then the equation $x^2 - my^2 = 1$ has only the trivial solutions $x = \pm 1$ in integers.

7.3. Show using Dirichlet's pigeonhole principle that for each real number x, there exist infinitely many pairs $(p, q) \in \mathbb{Z} \times \mathbb{Z}$ such that $|x - \frac{p}{q}| < \frac{1}{q^2}$.

Hint: Consider the remainders modulo 1 of the numbers $0, x, 2x, \ldots, nx$; these $n + 1$ remainders lie in the n intervals $[0, \frac{1}{n}), [\frac{1}{n}, \frac{2}{n}), \ldots, [\frac{n-1}{n}, 1)$.

7.4. Find elements with small nontrivial norm in the family of quadratic number fields $\mathbb{Q}(\sqrt{m})$ with $m = t^2 - 1$ and $m = t^2 \pm 4$.

Use this result for finding examples of real quadratic number fields with large class number.

7.5. Prove the following lemma (Hasse [58]): If $m > 0$ is not a square and $\varepsilon = \frac{t + u\sqrt{m}}{2}$ the fundamental unit of $\mathbb{Q}(\sqrt{m})$, and if n is the smallest positive nonsquare for which $x^2 - my^2 = \pm 4n$ is solvable in nonzero integers, then

$$ n \geq \begin{cases} \frac{t}{u^2}, & \text{if } N\varepsilon = -1, \\ \frac{t-2}{u^2}, & \text{if } N\varepsilon = +1. \end{cases} $$

7.6. Show: If $m = 2p$ for primes $p \equiv 5 \bmod 8$, then $N\varepsilon_m = -1$ for the fundamental unit ε_m of $\mathbb{Q}(\sqrt{m})$.

7.7. Show: If $m = 2p$ for primes $p \equiv 3 \bmod 4$, then either $x^2 - my^2 = 2$ or $x^2 - my^2 = -2$ is solvable in nonzero integers. Also show that, in this case, $2\varepsilon_m$ is a square in $K = \mathbb{Q}(\sqrt{m})$, where ε_m denotes the fundamental unit in K.

7.8. Compute the fundamental units of $\mathbb{Q}(\sqrt{m})$ for $m = 3, 19, 43, 67, 131, 159, 199$.

7.9. Show: If $\varepsilon = \frac{t+u\sqrt{m}}{2}$ is the fundamental unit of $\mathbb{Q}(\sqrt{m})$ for $m \equiv 1 \bmod 8$, then t and u are even.

7.10. Let $m = n^2 - 1$ for some natural number $n \geq 2$. Show that $\varepsilon = n + \sqrt{m}$ is a unit in $\mathbb{Z}[\sqrt{m}]$ and that it is the fundamental unit of $\mathbb{Q}(\sqrt{m})$ if m is squarefree.

More generally, find units for $m = n^2 \pm 1$ and $m = n^2 \pm 4$.

7.11. Compute the class number and the fundamental unit of $K = \mathbb{Q}(\sqrt{478})$.

Hint: Consider the prime ideal above (2) and the prime ideals above 3 and 7. Determine the prime ideal factorizations of $(a + \sqrt{478})$ for $a = 10, 17, 22, 24,$ and 25, and conclude that K has class number 1.

7.12. The solvability of the Pell equation $x^2 - my^2 = 1$ for positive nonsquares m may be formulated as follows: The part of the Euclidean plane defined by the hyperbolas $x^2 - my^2 = 1$ and $x^2 - my^2 = -1$ that contains their asymptotes contains infinitely many lattice points. In this formulation, the claim even holds when m is a square; in this case, all integral points lie on the asymptotes.

Show that the region between the two hyperbolas $2x^2 - 5y^2 = 1$ and $2x^2 - 5y^2 = -1$ does not contain any lattice point except $(0, 0)$.

7.13. Show that the continued fraction expansion of $\sqrt{m}$ for $m = t^2 - 1$ is given by

$$\sqrt{m} = [t - 1; 1, 2t - 2, 1, 2t - 2, 1, 2t - 2, \ldots] = [t - 1; \overline{1, 2t - 2}].$$

For example,

$$\sqrt{3} = 1 + \cfrac{1}{1 + \cfrac{1}{2 + \cfrac{1}{1 + \cfrac{1}{2 + \cfrac{1}{1 + \cfrac{1}{2 + \ldots}}}}}}.$$

7.14. Show that the continued fraction expansion of $\sqrt{m}$ for $m = t^2 + 2$ is given by

$$\sqrt{m} = [t; \overline{t, 2t}].$$

7.15. Show that $x^2 = a^3 + 3a$ for odd integers a is equivalent to

$$\left(\frac{a-1}{2}\right)^3 + \left(\frac{a+1}{2}\right)^3 = y^2,$$

where $y = \frac{x}{2}$.

Show moreover that the equation $y^2 = x(x^2 + 3)$ has the only integral points $(0, 0)$, $(1, \pm 2)$, $(3, \pm 6)$, and $(12, \pm 42)$ assuming that the equation $r^3 - 3s^4 = -2$ has the only integral solution $r = s = s$.

7.16. Let $p \equiv 3 \bmod 4$ be a prime number, and let $\varepsilon = t + u\sqrt{p}$ denote the fundamental unit of $\mathbb{Q}(\sqrt{p})$. Show that t is even and that $t \equiv 1 - (\frac{2}{p}) \bmod 4$.

7.17. Show that the function f defined in (7.2) is injective.

7.18. Let $\varepsilon = 2 + \sqrt{3}$ be the fundamental unit of $\mathbb{Z}[\sqrt{3}]$. Define the numbers $V_n = \varepsilon^n + \varepsilon'^n$ for $n \geq 0$. Show that these numbers satisfy the recurrence relation

$$V_{n+1} = 4V_n - V_{n-1}.$$

Also show that $V_{2n} = V_n^2 - 2$ and that the subsequence V_{2^n} consists of the numbers occurring in the Lucas–Lehmer test.

Chapter 8
Catalan's Equation

In this chapter we will show how to apply the arithmetic of quadratic number fields to special cases of Catalan's conjecture.

In 1844, Catalan conjectured that the only powers of (positive) natural numbers that differ by 1 are $2^3 = 8$ and $3^2 = 9$; in other words, the Diophantine equation

$$x^p - y^q = 1$$

has $3^2 - 2^3 = 1$ as its only nontrivial solution. This conjecture was proved by Preda Mihailescu [94] in 2004. His proof uses the work of many other mathematicians and in particular results about the arithmetic of cyclotomic number fields that are beyond the scope of the present book.

We will, however, be able to cover the equations $x^p - y^q = 1$ with $p = 2$ or $q = 2$ because these cases can be attacked using the arithmetic of quadratic number fields.

8.1 Lebesgue's Theorem

Already in 1850, Victor-Amédée Lebesgue [74] published a proof of the following special case of Catalan's conjecture:

Theorem 8.1 *The equation $y^2 + 1 = x^m$ does not have any solutions in natural numbers x, y with $m \geq 2$.*

The following proof based on the arithmetic of the ring $\mathbb{Z}[i]$ of Gaussian integers is essentially the one given by Lebesgue. Clearly, y cannot be odd since otherwise $y^2 + 1 \equiv 2 \bmod 4$ cannot be a nontrivial power. Thus y is even.

The exponent m must be odd; in fact, if m is even, then y^2 and x^m are consecutive squares, which implies $x = 1$ and $y = 0$, a case we have excluded.

© The Author(s), under exclusive license to Springer Nature Switzerland AG 2021
F. Lemmermeyer, *Quadratic Number Fields*, Springer Undergraduate
Mathematics Series, https://doi.org/10.1007/978-3-030-78652-6_8

Now we factor our equation:

$$x^m = y^2 + 1 = (y + i)(y - i).$$

Since the two factors are coprime (recall that y is even), we deduce that there exists a Gaussian integer $a + bi$ and a unit i^k with

$$y + i = i^k(a + bi)^m \quad \text{and} \quad y - i = i^{-k}(a - bi)^m.$$

Observe that $2 + i$ has odd norm, and hence a and b must have different parity. These equations imply

$$2i = i^k(a + bi)^m - i^{-k}(a - bi)^m = i^k[(a + bi)^m - (-1)^k(a - bi)^m]. \quad (8.1)$$

Now there are two cases:

- $k = 2r$ is even; then comparing the coefficients of i in (8.1) shows that

$$1 = (-1)^r \left[ma^{m-1}b - \binom{m}{3} a^{m-2}b^3 + \ldots \pm b^m \right]. \quad (8.2)$$

  Since the expression in the bracket is divisible by b, we must have $b = \pm 1$, and a must be even.
- $k = 2r + 1$ is odd; then we find

$$1 = (-1)^r \left[a^m - \binom{m}{2} a^{m-2}b^2 + \ldots \pm mab^{m-1} \right]. \quad (8.3)$$

In this case we find $a = \pm 1$, and b must be even.

Both cases lead to an equation of the form

$$1 - \binom{m}{2} c^2 + \binom{m}{4} c^4 - \ldots \pm mc^{m-1} = \pm 1,$$

where c is an even integer. If the right side of this equation is -1, then c^2 would divide 2, which is nonsense. Thus we can subtract 1 from both sides and divide through by $-c^2$; in this way, we obtain

$$\binom{m}{2} - \binom{m}{4} c^2 + \ldots \pm mc^{m-3} = 0.$$

If $m = 3$, this equation says $\binom{3}{2} = 0$, which is nonsense. Thus $m \geq 5$, and then $\binom{m}{2}$ must be divisible by 4 since c is even. But m is odd, and hence $\binom{m}{2} = \frac{m(m-1)}{2}$ is divisible by 4 if and only if $m \equiv 1 \bmod 8$.

We will now finish the proof by showing that $\binom{m}{2}$ is divisible by a smaller power of 2 than all the other terms. This will then imply that the sum cannot vanish.

Observe that the factor $\frac{(m-2)(m-3)}{3\cdot 4}c^2$ of

$$\binom{m}{4}c^2 = \binom{m}{2}\cdot\frac{(m-2)(m-3)}{3\cdot 4}c^2$$

is even: In fact, the factor 4 in the denominator cancels against c^2, and the numerator $(m-2)(m-3)$ is even; thus $\binom{m}{4}c^2$ is divisible by a higher power of 2 than $\binom{m}{2}$. In general we have

$$\binom{m}{2r}c^{2r-2} = \binom{m}{2}\binom{m-2}{2r-2}\cdot\frac{2}{2r(2r-1)}c^{2r-2} = \binom{m}{2}\binom{m-2}{2r-2}\cdot\frac{1}{r(2r-1)}c^{2r-2}.$$

Now c^{2r-2} is divisible by 2^{2r-2}; for $r \geq 3$, the factor r is not divisible by 2^{2r-2} since $r < 2^{2r-2}$. This implies our claims.

8.2 Euler's Theorem

Euler solved the equation $y^2 = x^3 + 1$ in rational numbers; using the basic machinery from the theory of elliptic curves,[1] such proof can be given in a couple of lines. Direct proofs of this result, on the other hand, tend to be rather technical.

Theorem 8.2 *The only rational solutions of the Diophantine equation $x^2 = y^3 + 1$ are $(x, y) = (0, -1), (\pm 1, 0), \text{ and } (\pm 3, 2)$.*

Euler's proof is based on the key fact that the equation

$$z^2 = p^4 + 9p^2q^2 + 27q^4$$

does not have solutions in nonzero integers. The proof is quite involved but only uses the arithmetic of the ordinary integers. Wakulicz [131] was apparently unaware of the fact that his proof is essentially that of Euler.

8.2.1 Monsky's Proof

We will now present Paul Monsky's beautiful solution of the Diophantine equation $y^2 = x^3 + 1$. Euler first showed that the $(2, 3)$ is the only solution in positive rational numbers.

---

[1] Everything you need is contained in [117].

The proof of Theorem 8.2 is based on a reformulation of the theorem similar to the one we have used in our proof of Fermat's Last Theorem for cubes:

Theorem 8.3 *Let $\alpha, \beta, \gamma \in \mathbb{Q}(\rho)^{\times}$. If $\alpha + \beta + \gamma = 0$ and $\alpha\beta\gamma = 2\mu^3$, then after a suitable permutation of the three numbers $\alpha = 0$ or $\beta = \gamma$.*

Observe that the equations $\alpha + \beta + \gamma = 0$ and $\alpha\beta\gamma = 2\mu^3$ are homogeneous, and hence it will be sufficient to prove the result for algebraic integers $\alpha, \beta, \gamma \in \mathbb{Z}[\rho]$.

For proving that Theorem 8.3 implies Theorem 8.2, assume that $x^2 = y^3 + 1$ for rational numbers x, y. Setting $\alpha = 1 - x$, $\beta = 1 + x$, and $\gamma = -2$, we find $\alpha + \beta + \gamma = 0$ and $\alpha\beta\gamma = 2(x^2 - 1) = 2y^3$, and hence $\alpha = 0$ (and $x = 1$), $\beta = 0$ (and $x = -1$), $\alpha = \beta$ (and $x = 0$), $\alpha = \gamma$ (and $x = 3$), or $\beta = \gamma$ (and $x = -3$). Thus we obtain Theorem 8.2, and in fact we have also proved that the only rational points on the curve $x^2 = y^3 + 1$ are the integral points given in Theorem 8.2.

Proof Let (α, β, γ) be a counterexample. Then α, β, and γ are pairwise coprime in $\mathbb{Z}[\rho]$, and so (after a suitable permutation of the numbers) we find that there exist $A_1, B_1, C_1 \in \mathbb{Z}[\rho]$ with

$$\alpha = 2\rho^a A_1^3, \quad \beta = \rho^b B_1^3, \quad \gamma = \rho^c C_1^3$$

because 2 is prime in $\mathbb{Z}[\rho]$. Among all such counterexamples, we choose one in which $N\alpha$ is minimal. Dividing all three elements by ρ^a, we obtain $a = 0$.

Now we observe that all cubes in $\mathbb{Z}[\rho]$ are either $\equiv 0$ or $\equiv 1 \bmod 2$ (Exercise 5.14). This implies that $0 \equiv \alpha \equiv \beta + \gamma \equiv \rho^b + \rho^c \bmod 2$, which is only possible if $b = c$. Since $\alpha\beta\gamma$ is two times a cube, we must even have $a = b = c$ and hence $a = b = c = 0$.

Therefore

$$\alpha = 2A_1^3, \quad \beta = B_1^3, \quad \gamma = C_1^3.$$

Since $B_1^3 \equiv C_1^3 \equiv 1 \bmod 2$, we may assume according to Proposition 5.10 that $B_1 \equiv C_1 \equiv 1 \bmod 2$.

Now we set $\alpha_1 = B_1 + C_1$, $\beta_1 = \rho B_1 + \rho^2 C_1$, and $\gamma_1 = \rho^2 B_1 + \rho C_1$. Then

- $\alpha_1 + \beta_1 + \gamma_1 = B_1(1 + \rho + \rho^2) + C_1(1 + \rho + \rho^2) = 0$.
- $\alpha_1\beta_1\gamma_1 = B_1^3 + C_1^3 = \beta + \gamma = -\alpha = 2(-A_1)^3$.
- $\beta_1 + \gamma_1 = (B_1 + C_1)(\rho + \rho^2) = -(B_1 + C_1) \neq 0$ since $\beta + \gamma = -\alpha \neq 0$.
- $N(\alpha_1\beta_1\gamma_1) = N\alpha \mid N(\alpha\beta\gamma)$; if we had equality, it would follow that $N(\beta) = N(\gamma) = 1$ and thus $\beta, \gamma = \pm 1$. But this yields $\beta = 1$, $\gamma = -1$, and $\alpha = 0$ contradicting our assumption.

Thus $(\alpha_1, \beta_1, \gamma_1)$ is a solution with $N(\alpha_1\beta_1\gamma_1) < N(\alpha\beta\gamma)$, which contradicts our assumption on the minimality of $N\alpha$. This completes the proof.   □

8.3   The Theorems of Størmer and Ko Chao

Consider units $\varepsilon = t + u\sqrt{m}$ in $\mathbb{Z}[\sqrt{m}\,]$; if $u = 1$, then ε is fundamental in the sense that it is not a nontrivial power of another unit in $\mathbb{Z}[\sqrt{m}\,]$. The unit $3 + 2\sqrt{2}$ is the square of $1 + \sqrt{2}$, but all other units with $u = 2$ are also fundamental in $\mathbb{Z}[\sqrt{m}\,]$.

Størmer [119] has proved a general result that guarantees that certain units $t + u\sqrt{m}$ are not powers of other units.

Theorem 8.4 *Let m be a positive integer, and assume that m is not a square. Let $t + u\sqrt{m}$ be the unit corresponding to the minimal positive solution (t, u) of the Pell equation $t^2 - my^2 = e = \pm 1$, and define natural numbers t_n, u_n by*

$$t_n + u_n\sqrt{m} = (t + u\sqrt{m})^n. \tag{8.4}$$

If each prime dividing u_n also divides m, then $n = 1$, with the single exception $m = 2$ and $n = 2$, where $3 + 2\sqrt{2} = (1 + \sqrt{2})^2$.

Before we prove this result, we formulate an obvious corollary:

Corollary 8.5 *Given a positive integer m, the equation $t^2 - mu^2 = 1$ has at most one solution (t, u) with the property that every prime dividing u also divides m.*

This is clear since the exception $1 + \sqrt{2}$ in Størmer's Theorem has norm -1.

For the proof of Størmer's Theorem, we assume that u_n has the property that each of its prime divisors divides m. Equation (8.4), with $t = t_1$ and $u = u_1$, tells us that

$$u_n = nt_1^{n-1}u_1 + \binom{n}{3}t_1^{n-3}u_1^3 m + \ldots,$$

which implies $t_1 \mid t_n$. Write $t_n = s_n t_1$ and $mu_1^2 = t_1^2 - e = A$. Since m is a nonsquare > 1, so is A. Plugging these numbers into (8.4), we obtain

$$t_n + s_n\sqrt{A} = (t_1 + \sqrt{A})^n.$$

We know that each prime divisor of s_n divides A.

If $n > 1$, let p denote one of these prime divisors and set $n = pk$. Then

$$t_n + s_n\sqrt{A} = (t_p + s_p\sqrt{A})^k,$$

and hence

$$s_n = kt_p^{k-1}s_p + \binom{k}{3}t_p^{k-3}s_p A + \ldots.$$

Thus s_n is divisible by s_p, and hence each prime dividing s_p will divide A. But since

$$s_p = pt_1^{p-1} + \binom{p}{3} t_1^{p-3} A + \ldots, \tag{8.5}$$

we see that each prime dividing s_p divides pt_1^{p-1}. The equation $t_1^2 = A + e$ shows that t_1 and A are coprime, and hence the only prime dividing s_p is p. Since $s_p > 1$, we can write $s_p = p^r$ for some integer $r \geq 1$. Plugging this into (8.5), we obtain

$$pt_1^{p-1} + \binom{p}{3} t_1^{p-3} A + \ldots = p^r. \tag{8.6}$$

We now distinguish several cases.

$p = 2$ Here $(t + u\sqrt{m})^2 = t^2 + mu^2 + 2tu\sqrt{m}$; since every prime divisor of $2tu$ divides m, this is true for every prime divisor of t. But if $p \mid t$ divides m, then p also divides $t^2 - mu^2 = \pm 1$, which is impossible. Thus $t = 1$, and the only unit of the form $1 + \sqrt{m}$ is the unit $1 + \sqrt{2}$.

$p = 3$ In this case, $s_3 = 3s_1^2 + A = 3^r$ together with the fact that $A = t_1^2 - e$ implies $4t_1^2 - e = 3^r$.

If $e = -1$, then $e^r = 4t_1^2 + 1$ is a sum of coprime squares and cannot be divisible by 3. Thus $e = 1$, and we have

$$3^r = 4t_1^2 - 1 = (2t_1 - 1)(2t_1 + 1).$$

Since the factors are coprime, this is only possible if $2t_1 - 1 = 1$, which implies $t_1 = 1, r = 1$ and then $A = 0$, which is impossible.

$p > 3$ Here all the terms in (8.6) except possibly the first one are divisible by p^2 since $p \mid A$. Since t_1 and A are coprime, the first term pt_1^{p-1} is not divisible by p^2, and this implies $r = 1$. Dividing (8.6) by p yields

$$t_1^{p-1} + \frac{(p-1)(p-2)}{6} t_1^{p-3} A + \ldots = 1,$$

which is impossible since the left side is clearly > 1.

Størmer proved his result in connection with the numerical computation of logarithms; to this end, he was looking for pairs $(N, N + 1)$ of numbers such that both N and $N + 1$ are divisible only by a given finite set of prime numbers, and he showed how to construct such pairs by solving a finite set of Pell equations.

8.3.1 Application to Catalan's Equation

We now apply Størmer's Theorem to the equation $x^2 - y^q = 1$, where q is an odd integer. For proving that there are no integral solutions with $x > 1$, it is sufficient to prove the claim in the case where q is an odd prime number.

The analog of the following result[2] for the general Catalan equation plays a central role in Mihailescu's proof:

Theorem 8.6 (Nagell) *If there are positive integers x, y with $x^2 - y^q = 1$ for some odd prime number q, then $2 \mid y$ and $q \mid x$.*

For proving the first claim, we write the equation in the form $y^q = x^2 - 1 = (x - 1)(x + 1)$. Then $\gcd(x - 1, x + 1) \mid 2$; if y is odd, then x is even and the two factors are coprime. But then $x - 1 = a^q$ and $x + 1 = b^q$ must be q-th powers, hence $b^q - a^q = 2$, which is impossible. This contradiction shows that y must be even as claimed.

For the second claim, we write the equation in the form

$$x^2 = y^q + 1 = \frac{y^q + 1}{y + 1}(y + 1).$$

Observe that $\frac{y^q + 1}{y + 1} = Q_q(y, -1)$ in the notation of Sect. 3.5. By (3.18), we have $\gcd(Q_q(y, -1), y + 1) \mid q$. Thus if $q \nmid y$, the factors are coprime and must both be squares, i.e., there exist natural numbers a and b with

$$y + 1 = a^2, \quad \frac{y^q + 1}{y + 1} = b^2, \quad x = ab.$$

Now $a^2 - y = 1$; if y is a square, then $y = 0$, which is impossible. Thus y is not a square and $\varepsilon = a + \sqrt{y}$ is the fundamental unit of the order $\mathbb{Z}[\sqrt{y}]$. But the equation $x^2 - y^q = 1$ implies $\eta = x + y^{\frac{q-1}{2}}\sqrt{y}$ also is a unit in $\mathbb{Z}[\sqrt{y}]$, so we must have $\eta = \varepsilon^n$. By Størmer's Theorem, we must have $n = 1$, hence $x = a$, $b = 1$, and finally $q = 1$, which is impossible.

Now we give Chein's proof of the following theorem of Ko Chao:

Theorem 8.7 *The equation $x^2 - y^q = 1$, where $q \geq 5$ is prime, does not have any solutions in nonzero integers.*

Nagell's Theorem 8.6 tells us that x is odd. If $x \equiv 3 \bmod 4$, then $y^q = x^2 - 1 = (x - 1)(x + 1)$ implies the existence of natural numbers a and b with

$$x + 1 = 2^{q-1}a^q \quad \text{and} \quad x - 1 = 2b^q.$$

---

[2] See Theorem 8.14 below.

Observe that $a < b$ since $a^q = (b^q + 1)/2^{q-2} < b^q$. Now

$$(b^2 + 2a)\frac{b^{2q} + (2a)^q}{b^2 + 2a} = b^{2q} + (2a)^q = \left(\frac{x-1}{2}\right)^2 + 2(x+1) = \left(\frac{x+3}{2}\right)^2.$$

By Theorem 8.6, we have $q \mid x$. Since $q \neq 3$, the term $\frac{x+3}{2}$ is not divisible by q. Since the factors on the left are coprime, they are squares. But $b + 2a$ lies between two consecutive squares because $b^2 < b^2 + 2a < b^2 + 2b < (b+1)^2$. This is a contradiction.

If $x \equiv 1 \bmod 4$, then $x - 1 = 2^{q-1}a^q$ and $x + 1 = 2b^q$, and hence we obtain in a similar way

$$b^{2q} - (2a)^q = \left(\frac{x-3}{2}\right)^2,$$

and the rest of the proof carries over almost word for word.

8.4 Euler's Equation via Pure Cubic Number Fields

We now return to Euler's equation $x^2 = y^3 + 1$ and determine its integral solutions. We begin by writing our equation in the form $y^3 = x^2 - 1 = (x-1)(x+1)$; a common divisor of $x + 1$ and $x - 1$ divides their difference 2, and hence there are two possibilities:

1. x is even. Then $\gcd(x + 1, x - 1) = 1$, and according to Corollary 4.13, there exist integers $a, b \in \mathbb{Z}$ such that $x + 1 = \pm a^3$ and $x - 1 = \pm b^3$. By pulling $-1 = (-1)^3$ into the cube, we may omit the signs and obtain $x + 1 = a^3$ and $x - 1 = b^3$. Subtracting these equations from each other yields $2 = a^3 - b^3 = (a - b)(a^2 + ab + b^2)$; thus $a - b$ divides 2.
   If $a - b = \pm 1$, then

   $$\pm 2 = a^2 + ab + b^2 = (b \pm 1)^2 + b(b \pm 1) + b^2 = 3b^2 \pm 3b + 1.$$

   Solving these quadratic equations yields a contradiction as the solutions are not integers.
   If $a - b = \pm 2$, on the other hand, then we find

   $$\pm 1 = a^2 + ab + b^2 = (b \pm 2)^2 + b(b \pm 2) + b^2 = 3b^2 \pm 6b + 4,$$

   and now we obtain the unique solution $b = -1$, $a = 1$, $x = 0$, and $y = -1$.
2. x is odd. Then $\gcd(x + 1, x - 1) = 2$, and according to Corollary 4.13, there exist integers $a, b \in \mathbb{Z}$ with $x + 1 = 2a^3$ and $x - 1 = 4b^3$, where the signs may be omitted as before (the possibility $x + 1 = 4a^3$ and $x - 1 = 2b^3$ can be reduced

to the first: simply replace x by $-x$). In a similar way as above, we now find the
equation $1 = a^3 - 2b^3$.

We will show below that the only integral solutions of this equation are
given by $(a, b) = (1, 0)$ and $(-1, -1)$. These lead to the solutions $(x, y) = (\pm 1, 0), (\pm 3, 2)$ of the original equation.

We will attack the equation $a^3 - 2b^3 = 1$ directly by writing it in the form

$$1 = (a - b\sqrt[3]{2})(a^2 + \sqrt[3]{2}\,ab + \sqrt[3]{4}\,b^2)$$

and observing that $a - b\sqrt[3]{2}$ is a unit in the ring $\mathbb{Z}[\sqrt[3]{2}]$. It can be shown that
$R^\times = \langle -1, 1 - \sqrt[3]{2}\rangle$, and the claim then boils down to showing that $\pm(1 - \sqrt[3]{2})^n = a - b\sqrt[3]{2}$ implies $|n| \leq 1$ (in general, this power will have the form $r + s\sqrt[3]{2} + t\sqrt[3]{4}$
for some $t \neq 0$).

The calculations[3] will be performed in the pure cubic number field $\mathbb{Q}(\sqrt[3]{2})$. Its
ring of integers is

$$\mathbb{Z}[\sqrt[3]{2}] = \{a + b\sqrt[3]{2} + c\sqrt[3]{4} : a, b, c \in \mathbb{Z}\}.$$

This ring is Euclidean with respect to the absolute value of the norm, and hence it
has unique factorization. The solution of the equation $a^3 + 2b^3 = 1$ that we will
give does not use any of this: All we need is the fact that the units of the ring $\mathbb{Z}[\sqrt[3]{2}]$
are generated by -1 and $\sqrt[3]{2} - 1$, and we will prove this below.

8.4.1   Units in Pure Cubic Number Fields

A pure cubic number field is a number field of the form $K = \mathbb{Q}(\sqrt[3]{m})$ for some
integer $m > 0$, which we may assume to be cubefree, i.e., not divisible by a cube
$\neq \pm 1$. The elements of K have the form $x + y\sqrt[3]{m} + z\sqrt[3]{m^2}$ with $x, y, z \in \mathbb{Q}$. We
will not determine the ring of integers of K and are content with observing that if
$m = ab^2$ for squarefree integers $a, b \in \mathbb{N}$, then $\sqrt[3]{m^2} = b\sqrt[3]{a^2b}$. If $r, s, t \in \mathbb{Z}$, then
the elements $r + s\sqrt[3]{ab^2} + t\sqrt[3]{a^2b}$ belong to the ring of integers in K.

We also need to determine the norm of an element $\alpha = r + s\sqrt[3]{ab^2} + t\sqrt[3]{a^2b}$,
which is defined as the product $N\alpha = \alpha\alpha'\alpha''$ of the conjugates of α, where

$$\alpha' = r + s\rho\sqrt[3]{ab^2} + t\rho^2\sqrt[3]{a^2b} \quad \text{and} \quad \alpha'' = r + s\rho^2\sqrt[3]{ab^2} + t\rho\sqrt[3]{a^2b},$$

---

[3] We will follow Nagell's publication [99] (see also [92]).

and where $\rho = \frac{-1+\sqrt{-3}}{2}$. A straightforward calculation yields

$$N\alpha = r^3 + ab^2s^3 + a^2bt^3 - 3abrst.$$

As in quadratic number fields, units have norm ± 1, and integral elements with norm ± 1 are units. In particular, if (x, y) is an integral solution of the Diophantine equation $x^3 + dy^3 = 1$, then $x + y\sqrt[3]{d}$ is a unit in $\mathbb{Z}[\sqrt[3]{d}]$. According to a theorem due to Dirichlet, this ring has a fundamental unit ε with the property that all units can be written in the form $\eta = \pm\varepsilon^n$ for some $n \in \mathbb{Z}$. The nontrivial part of this theorem claims that the equation

$$r^3 + ab^2s^3 + a^2bt^3 - 3abrst = 1$$

has solutions, whereas the assertion that each unit can be written up to sign as a power of the fundamental unit follows as in the real quadratic case by studying absolute values $|r + s\sqrt[3]{ab^2} + t\sqrt[3]{a^2b}|$.

Let us determine the fundamental unit in the case we are mainly interested in, namely for the field $\mathbb{Q}(\sqrt[3]{2})$. Here $\sqrt[3]{2} - 1$ is a unit since

$$(\sqrt[3]{2} - 1)(1 + \sqrt[3]{2} + \sqrt[3]{4}) = 1.$$

If we interpret $\sqrt[3]{2}$ as a real number, then $\varepsilon = 1 + \sqrt[3]{2} + \sqrt[3]{4} > 1$. Now we claim the following:

Lemma 8.8 *Let $\varepsilon = 1 + \sqrt[3]{2} + \sqrt[3]{4}$. Then each unit $\eta > 1$ in $\mathbb{Z}[\sqrt[3]{2}]$ has the form $\eta = \varepsilon^n$ for an integer $n \geq 1$.*

Proof The units ε^n all have value > 1 for $n \geq 1$. If there is a unit $\eta > 1$ not of this form, then η lies between two powers of ε:

$$\varepsilon^m < \eta < \varepsilon^{m+1}.$$

But then $\eta_1 = \eta\varepsilon^{-m}$ is a unit with $1 < \eta_1 < \varepsilon$. Since

$$1 = |\eta_1\eta_1'\eta_1''| = \eta_1|\eta_1'|^2,$$

this implies

$$\frac{1}{\sqrt{\varepsilon}} < |\eta_1'| < 1.$$

If we write $\eta_1 = r + s\sqrt[3]{2} + t\sqrt[3]{4}$, then the triangle inequality shows that

$$3|r| = |\eta_1 + \eta_1' + \eta_1''| \leq |\eta_1| + |\eta_1'| + |\eta_1''| \leq \varepsilon + 2 < 5.9,$$

and hence $|r| \leq 1$. Similarly, we find

$$3|s|\sqrt[3]{2} = |\eta_1 + \rho^2\eta_1' + \rho\eta_1''| < 5.9,$$

and thus $|s| \leq 1$, as well as

$$3|t|\sqrt[3]{4} = |\eta_1 + \rho\eta_1' + \rho^2\eta_1''| < 5.9,$$

and therefore $|t| \leq 1$. Going through all possible values then yields the desired contradiction. $\qquad\square$

If $0 < \eta < 1$, then $1/\eta = \varepsilon^n$ for some $n \geq 1$, and hence every positive unit has the form $\eta = \varepsilon^n$ for some $n \in \mathbb{Z}$. Finally, if $\eta < 0$, then $-\eta$ must be a power of ε. We have shown the following:

Proposition 8.9 *Each unit $\varepsilon \in \mathbb{Z}[\sqrt[3]{2}]$ can be written uniquely in the form*

$$\eta = (-1)^m \varepsilon^n$$

for $m \in \{0, 1\}$ and $n \in \mathbb{Z}$, where $\varepsilon = 1 + \sqrt[3]{2} + \sqrt[3]{4}$ is the fundamental unit of $\mathbb{Z}[\sqrt[3]{2}]$.

This statement remains correct if we replace ε by $\varepsilon^{-1} = \sqrt[3]{2} - 1$.

8.4.2 The Equation $x^3 + 2y^3 = 1$

Let us first consider the equation

$$(1 - \sqrt[3]{2})^n = x + y\sqrt[3]{2}. \tag{8.7}$$

Expanding the left hand side using the binomial theorem and comparing coefficients, we obtain

$$x = 1 - 2\binom{n}{3} + 4\binom{n}{6} - 8\binom{n}{9} + \cdots$$

$$-y = \binom{n}{1} - 4\binom{n}{4} + 4^2\binom{n}{7} \mp$$

$$0 = \binom{n}{2} - 4\binom{n}{5} + 4^2\binom{n}{8} \mp \cdots$$

The last equation implies that $\binom{n}{2}$ must be divisible by 4, which happens if and only if $n \equiv 0, 1 \bmod 4$.

We now assume $n \geq 2$ (and thus $n \geq 4$); dividing the last equation through by $\binom{n}{2}$, we find

$$-1 = \sum_{k \geq 1} \binom{n-2}{3k} \frac{2(-2)^k}{(3k+1)(3k+2)}.$$

Since $-2 \equiv 1 \bmod 3$ and $(3k+1)(3k+2) \equiv 2 \bmod 3$, reduction modulo 3 yields the congruence

$$1 + \binom{n-2}{3} + \binom{n-2}{6} + \ldots \equiv 0 \bmod 3.$$

The following lemma shows that this is impossible.

Lemma 8.10 *For each positive integer m, we have*

$$\binom{m}{0} + \binom{m}{3} + \binom{m}{6} + \ldots \not\equiv 0 \bmod 3.$$

Proof If we set

$$S_0 = \binom{m}{0} + \binom{m}{3} + \binom{m}{6} + \ldots,$$

$$S_1 = \binom{m}{1} + \binom{m}{4} + \binom{m}{7} + \ldots,$$

$$S_2 = \binom{m}{2} + \binom{m}{5} + \binom{m}{8} + \ldots,$$

then we find

$$S_0 + S_1 + S_2 = 2^m \equiv (-1)^m \bmod 3$$

as well as

$$S_1 = \binom{m}{0}\frac{m}{1} + \binom{m}{3}\frac{m-3}{4} + \ldots \equiv \binom{m}{0} \cdot m + \binom{m}{3} \cdot m + \ldots \equiv m S_0 \bmod 3,$$

$$S_2 = \binom{m}{1}\frac{m-1}{2} + \binom{m}{4}\frac{m-4}{5} + \ldots$$

$$\equiv \binom{m}{1} \cdot (1-m) + \binom{m}{4} \cdot (1-m) + \ldots \equiv -m S_1 + S_1 \bmod 3,$$

and hence

$$(-1)^m \equiv S_0 + S_1 + S_2 \equiv (1 + 2m - m^2)S_0 \bmod 3.$$

The claim follows. □

For negative exponents,

$$(\sqrt[3]{2} - 1)^{-1} = 1 + \sqrt[3]{2} + \sqrt[3]{4}$$

shows that (8.7) boils down to the equation

$$(1 + \sqrt[3]{2} + \sqrt[3]{4})^n = x + y\sqrt[3]{2}.$$

This equation cannot hold for $n \geq 1$ since the coefficient of $\sqrt[3]{4}$ in the multinomial development of the left side only contains positive summands. We have proved the following result:

Proposition 8.11 *The only integral solutions of the equation (8.7) are $n = 0$ and $n = 1$.*

Now if $x^3 + 2y^3 = 1$, then $x + y\sqrt[3]{2}$ is a unit in $\mathbb{Z}[\sqrt[3]{2}]$ since $N(x + y\sqrt[3]{2}) = x^3 + 2y^3$. Thus $x + y\sqrt[3]{2} = \pm(1 - \sqrt[3]{2})^n$ according to Proposition 8.9, and now Proposition 8.11 implies that $n = 0$ or $n = 1$. Therefore $(x, y) = (1, 0)$ and $(x, y) = (-1, 1)$ are the only integral solutions of the equation $x^3 + 2y^3 = 1$.

8.4.3 The Theorem of Delaunay and Nagell

For solving the equation $x^3 + dy^3 = 1$, we have to study units of the form $x + y\sqrt[3]{d}$ in pure cubic number fields. A theorem of Boris Nikolaevic Delaunay[4] and Trygve Nagell essentially states that with explicitly given exceptions, the unit $x + y\sqrt[3]{d}$ is the fundamental unit of the field $\mathbb{Q}(\sqrt[3]{d})$, so that the equation has at most one integral solution.

Detailed proofs can be found in Nagell [99] and Mordell [97], as well as in Leveque [92, S. 104ff]. An elegant proof due to Thoralf Skolem of the theorem that $x^3 + dy^3 = 1$ has at most one solution in nonzero integers is given in Cassels [15, Thm. 10.1]. For a somewhat surprising connection with elliptic curves, see [84].

---

[4] Another often used transliteration is Delone; Delaunay is the French variant. In 1915, Delaunay published his theorem in Russian, but it became known in the West only through a publication in French in 1920.

8.5 Mihailescu's Proof

So far we have presented the solution of the Catalan equation $x^p - y^q = 1$ when $p = 2$ or $q = 2$. It would have been possible to cover a few more results, in particular Nagell's solution of the equations $x^3 - y^q = 1$ and $x^p - y^3 = 1$. For proving that there are no nontrivial integral solutions of these equations, he wrote them in the form $y^q = (x - 1)(x^2 + x + 1)$ and $x^p = (y + 1)(1 - y + y^2)$ and had to solve the equations $x^2 + x + 1 = y^q$ and $x^2 + x + 1 = 3y^q$. In order to give the readers an idea of the complexity of these investigations, we will present his results:

Theorem 8.12 *The integral solutions of the Diophantine equation* $x^2 + x + 1 = y^q$ *are*

- $(x, y) = (0, \pm 1)$, $(-1, \pm 1)$ *if q is even;*
- $(x, y) = (0, 1)$, $(-1, 1)$ *if q $\neq$ 3 is odd;*
- $(x, y) = (0, 1)$, $(-1, 1)$, $(18, 7)$, $(-19, 7)$ *if q = 3.*

The corresponding result for the equation $x^2 + x + 1 = 3y^q$ is as follows:

Theorem 8.13 *The integral solutions of* $x^2 + x + 1 = 3y^q$ *for q $\geq$ 3 are given by*

- $(x, y) = (1, \pm 1)$, $(2, \pm 1)$ *if q is even;*
- $(x, y) = 1, 1)$, $(-2, 1)$ *if q is odd.*

Another result with an elementary (but technical) proof is the following observation by Cassels:

Theorem 8.14 *If* $x^p - y^q = \pm 1$ *for nonzero integers x, y and odd prime exponents p, q, then p $\mid$ y and q $\mid$ x.*

Using the action of the Galois group on the class group of cyclotomic number fields,[5] in particular a result called Stickelberger's Theorem, Mihailescu was able to strengthen Cassels' result:

Theorem 8.15 *If* $x^p - y^q = \pm 1$ *for nonzero integers x, y and odd prime exponents p, q, then* $p^2 \mid y$ *and* $q^2 \mid x$.

This result then quickly implies the following:

Corollary 8.16 *If* $x^p - y^q = \pm 1$ *for nonzero integers x, y and odd prime exponents p, q, then (p, q) is a Wieferich pair, i.e., p and q satisfy the congruences*

$$p^{q-1} \equiv 1 \bmod q^2 \quad and \quad q^{p-1} \equiv 1 \bmod p^2.$$

---

[5] We will give an example of the strength of such investigations in Chap. 9.

These are strong conditions; there are only 7 Wieferich pairs known; the smallest are $(2, 1093)$ and $(83, 4871)$.

The full proof (presented in the books [12] and [113]) uses a more detailed analysis of Stickelberger's method. The state of the art before Mihailescu's proof is presented in Ribenboim's book [108], where the full proofs of Nagell's results on the equations $x^3 - y^q = 1$ can be found.

8.5.1 Summary

In this chapter we have proved some special cases of Catalan's conjecture that were accessible with elementary methods.

8.6 Exercises

8.1. (Nagell) Show that if $x^2 - y^q = 1$ has a nontrivial solution for primes $q \geq 5$, then $q \equiv 1 \bmod 8$.

8.2. Show that the sums

$$S_0 = \binom{m}{0} + \binom{m}{3} + \binom{m}{6} + \ldots$$

satisfy

$$S_0 \equiv \begin{cases} 1 \bmod 3 & \text{if } m \equiv 0, 1, 2 \bmod 6, \\ 2 \bmod 3 & \text{if } m \equiv 3, 4, 5 \bmod 6. \end{cases}$$

8.3. Show that

$$\binom{m}{0} + \binom{m}{2} + \binom{m}{4} + \ldots = \binom{m}{1} + \binom{m}{3} + \binom{m}{5} + \ldots = 2^{m-1}.$$

8.4. Show that

$$\binom{m}{0} + \binom{m}{3} + \binom{m}{6} + \ldots = \begin{cases} \frac{2^m + (-1)^m \cdot 2}{3} & \text{if } m \equiv 0 \bmod 3, \\ \frac{2^m - (-1)^m}{3} & \text{if } m \equiv 1, 2 \bmod 3. \end{cases}$$

These equations provide us with a new proof that $3 \nmid S_0$.

8.5. Show that

$$
\binom{m}{0} + \binom{m}{4} + \binom{m}{8} + \ldots =
\begin{cases}
2^{m-2} + (-1)^{\frac{m}{4}} 2^{\frac{m-2}{2}} & \text{if } m \equiv 0 \bmod 4, \\
2^{m-2} + (-1)^{\frac{m-1}{4}} 2^{\frac{m-3}{2}} & \text{if } m \equiv 1 \bmod 4, \\
2^{m-2} & \text{if } m \equiv 2 \bmod 4, \\
2^{m-2} + (-1)^{\frac{m+1}{4}} 2^{\frac{m-3}{2}} & \text{if } m \equiv 3 \bmod 4.
\end{cases}
$$

8.6. Let $S = \{p_1, \ldots, p_n\}$ be a finite set of prime numbers. An S-smooth integer is an integer N all of whose prime factors are contained in S. Prove Størmer's Theorem: The equation $x - y = 1$ has at most 3^n solutions in S-smooth integers x and y.

8.7. Show that the only integral solutions of the equation $x^2 + x + 1 = 3y^2$ (compare Theorem 8.13) are given by (x_m, y_m) with

$$
x_m = \pm \frac{\sqrt{3}}{4} \left[(2 + \sqrt{3})^{2n+1} - (2 - \sqrt{3})^{2n+1} \right] - \frac{1}{2}
$$

$$
y_m = \pm \frac{1}{4} \left[(2 + \sqrt{3})^{2n+1} - (2 - \sqrt{3})^{2n+1} \right].
$$

Chapter 9
Ambiguous Ideal Classes and Quadratic Reciprocity

It is quite difficult to determine class numbers, even in the simplest case of quadratic number fields, for fields with large discriminant. It is, however, possible to make several rather precise statements concerning the parity of class numbers of quadratic number fields. The theory behind these statements is called genus theory and goes back to Gauss, who worked with quadratic forms rather than quadratic number fields. Genus theory may be generalized to cyclic extensions, and in fact the question we will answer is how the Galois group of an extension acts on the ideal classes. In this chapter we will only scratch the surface of genus theory by proving the ambiguous class number formula.

The essential idea behind the proof is to reduce the action of the Galois group on ideal classes to the action on ideals, then on principal ideals and finally on elements, where everything can be done explicitly. Once more we will be studying a difficult object, namely the class group, by studying homomorphisms into simpler structures.

9.1 Ambiguous Ideal Classes

Let A be a finite abelian group. Then A can be written as a direct sum of cyclic groups, say $A = A_1 \oplus \cdots \oplus A_n$. If A is a finite 2-group, i.e., a group whose order is a power of 2, then the 2-rank of A is the number n of cyclic components. Since it is easy to see that $A/A^2 \simeq A_1/A_1^2 \oplus \cdots \oplus A_n/A_n^2$, and since $A_j/A_j^2 \simeq \mathbb{Z}/2\mathbb{Z}$ for cyclic groups A_j, the 2-rank of A is n if and only if $\#A/A^2 = 2^n$.

The determination of the order of the quotient group $\mathrm{Cl}(k)/\mathrm{Cl}(k)^2$, i.e., of the 2-rank of the ideal class group, goes back to Gauss, who solved this problem in the language of binary quadratic forms. It is almost impossible to miss the central questions of this theory when studying the operation of the Galois group $G = \{1, \sigma\}$ of $k/\mathbb{Q}$ on the ideal class group.

© The Author(s), under exclusive license to Springer Nature Switzerland AG 2021
F. Lemmermeyer, *Quadratic Number Fields*, Springer Undergraduate
Mathematics Series, https://doi.org/10.1007/978-3-030-78652-6_9

For an ideal class $c = [\mathfrak{a}]$ we set $c^\sigma = [\mathfrak{a}^\sigma]$; of course we have to show that this action is well defined (see Exercise 9.1). Clearly an ideal class c and its conjugate c^σ always have the same order. Moreover, since $c \cdot c^\sigma = [\mathfrak{a}][\mathfrak{a}^\sigma] = [(N\mathfrak{a})] = (1)$ is the principal class, $c^\sigma = c^{-1}$ is always the inverse class of c.

We call an ideal class $c \in \mathrm{Cl}(k)$ ambiguous if $c^\sigma = c$. Similarly, an ideal $\mathfrak{a}$ is called ambiguous, if $\mathfrak{a}^\sigma = \mathfrak{a}$.

Lemma 9.1 *The nontrivial automorphism* $\sigma : \sqrt{m} \to -\sqrt{m}$ *of* $k = \mathbb{Q}(\sqrt{m})$ *acts as* -1 *on the class group* $\mathrm{Cl}(k)$. *In particular, an ideal class c is ambiguous if and only if* $c^2 = 1$.

Proof We have already seen that $\mathfrak{a}^{1+\sigma} = \mathfrak{a}\mathfrak{a}' = (N\mathfrak{a})$ is principal, and that this implies that $c^\sigma = c^{-1}$.

If c is ambiguous, i.e., if $c = c^\sigma$, then $c^2 = c^{1+\sigma} = 1$. Conversely, if $c^2 = 1$, then $c^\sigma = c^{-1} = c$.   □

If k is a number field with class number 2, then the nontrivial ideal class c is always ambiguous. For $k = \mathbb{Q}(\sqrt{-5})$, the nontrivial ideal class is generated by the prime ideal $(2, 1+\sqrt{-5})$; since this ideal is ambiguous because of $(2, 1+\sqrt{-5})^\sigma = (2, 1-\sqrt{-5}) = (2, 1+\sqrt{-5})$. The ideal class c is also generated by $(3, 1+\sqrt{-5})$, and here $(3, 1+\sqrt{-5})^\sigma = (3, 1-\sqrt{-5}) \neq (3, 1+\sqrt{-5})$. In $\mathbb{Q}(\sqrt{-5})$, each ideal class contains an ambiguous ideal (the principal class contains the ambiguous ideal (1)), as well as many non-ambiguous ideals.

For ideal class groups of order 4, the number of ambiguous classes determines the structure. If $\mathrm{Cl}(k) \simeq \mathbb{Z}/2\mathbb{Z} \oplus \mathbb{Z}/2\mathbb{Z}$ is elementary abelian, then the number of ambiguous ideal classes is 4 since in this case, every ideal class is ambiguous. If $\mathrm{Cl}(k) \simeq \mathbb{Z}/4\mathbb{Z}$, on the other hand, then the two classes with order 4 are not ambiguous, whereas the class with order 2 and the principal class are ambiguous, Thus there are only 2 ambiguous ideal classes in this case.

If an ideal $\mathfrak{a}$ is ambiguous, then so is the ideal class $c = [\mathfrak{a}]$ it generates; the converse is not true in general: Since $k = \mathbb{Q}(\sqrt{34})$ has class number 2, the ideal class c of order 2 is ambiguous. This ideal class is not generated by an ambiguous ideal for the simple reason that all ambiguous ideals in k are principal. As we will see below, each ambiguous ideal is a product of ramified prime ideals and ideals generated by ordinary integers. But in k we have $(2, \sqrt{34}) = (6 + \sqrt{34})$ and $(17, \sqrt{34}) = (17 + 3\sqrt{34})$.

The ambiguous ideal classes form a group $\mathrm{Am}(k)$, in which the ideal classes generated by ambiguous ideals form a subgroup, namely the group $\mathrm{Am_{st}}(k)$ of strongly ambiguous ideal classes. Our goal is determining the structure of the group $\mathrm{Am}(k)$.

This will allow us to deduce information about the elements of order 2 in the class group. In fact, since $c^\sigma = c^{-1}$ we have $c^{1-\sigma} = c^2$, and therefore the homomorphism $c \mapsto c^{1-\sigma}$ maps the class group $\mathrm{Cl}(k)$ of a quadratic number field k to the group $\mathrm{Cl}(k)^{1-\sigma} = \mathrm{Cl}(k)^2$ of ideal classes that are squares, and this homomorphism is onto. Its kernel consists of the ideal classes c with $c^{1-\sigma} = 1$, i.e., of the ambiguous ideal classes. This implies that the order of the group $\text{Am}(k)$ of ambiguous ideal classes is equal to the order of $Cl(k)/Cl(k)^2$:

Proposition 9.2 *Let k be a quadratic number field. Then*

$$\#Cl(k)/Cl(k)^2 = \#\text{Am}(k),$$

and, in particular, the class number of k is odd if and only if the number of ambiguous ideal classes is 1.

Actually, since both groups are elementary abelian, equal cardinality implies isomorphism. The last claim follows from the observation that squaring is an isomorphism on a finite group if and only if it has odd order.

9.1.1 Exact Sequences

The calculations below are far easier to digest by using exact sequences. A short sequence of abelian groups A, B, C consists of group homomorphisms $\alpha : A \longrightarrow B$ and $\beta : B \longrightarrow C$, which are composed as follows:

$$1 \longrightarrow A \xrightarrow{\ \alpha\ } B \xrightarrow{\ \beta\ } C \longrightarrow 1. \tag{9.1}$$

The map $1 \longrightarrow A$ (which is often denoted by $0 \longrightarrow A$ if A is written additively) sends the element of the trivial group $\{1\}$ to the neutral element of A. Similarly, $C \longrightarrow 1$ is the homomorphism sending each element of C to the element of the trivial group $\{1\}$.

A sequence of abelian groups is called exact if the kernel of each map in the sequence is equal to the image of the preceding map (if there is one). Thus the sequence (9.1) is exact if and only if the following conditions are satisfied:

- $\ker \alpha = \text{im } (1 \longrightarrow A) = \{1\}$; in other word, α must be injective;
- $C = \ker(C \longrightarrow 1) = \text{im } \beta$; in other words, β must be surjective;
- $\ker \beta = \text{im } \alpha$.

Essentially, this short exact sequence contains the same information as the homomorphism theorem $C \simeq B/\text{im } A$, but it has the advantage that all the maps occur explicitly in the diagram. Perhaps this advantage will only become clear by studying homological algebra more carefully. One goal of this chapter is showing that this is a useful thing to do for those who are interested in algebraic number theory.

The proof of Proposition 9.2 consisted in verifying the exactness of the sequence

$$1 \longrightarrow \text{Am}(k) \longrightarrow Cl(k) \xrightarrow{\ 1-\sigma\ } Cl(k)^2 \longrightarrow 1.$$

The definitions of principal ideals and the ideal class group $\mathrm{Cl}(k)$ of a number field k provide us with two exact sequences, namely

$$1 \longrightarrow E_k \longrightarrow k^\times \longrightarrow H_k \longrightarrow 1,$$

$$1 \longrightarrow H_k \longrightarrow I_k \longrightarrow \mathrm{Cl}(k) \longrightarrow 1,$$

where E_k is the unit group, H_k the group of (fractional) principal ideals $\neq (0)$, and I_k the group of all fractional ideals $\neq (0)$.

9.1.2   Ambiguous Ideal Classes

The group $\mathrm{Am_{st}}(k)$ of strongly ambiguous ideal classes is, by definition, equal to $\mathrm{Am_{st}}(k) = AH/H \simeq A/A \cap H$, where A denotes the group of nonzero ambiguous ideals and H the group of nonzero principal ideals. Clearly $A \cap H = H^G$ is the group of ambiguous principal ideals, and so we have $\mathrm{Am_{st}}(k) \simeq A/H^G$. This observation gives us the exact sequence

$$1 \longrightarrow H^G \overset{\iota}{\longrightarrow} A \longrightarrow \mathrm{Am_{st}}(k) \longrightarrow 1.$$

The group P of all fractional ideals (a) with $a \in \mathbb{Q}^\times$ is a subgroup of both H^G and A; this allows us to modify the exact sequence slightly and turn it into

$$1 \longrightarrow H^G/P \overset{\iota}{\longrightarrow} A/P \overset{\pi}{\longrightarrow} \mathrm{Am_{st}}(k) \longrightarrow 1. \qquad (9.2)$$

Since $\mathrm{Am}(k)$ is elementary abelian, i.e., since $c^2 = 1$ for each ambiguous ideal class c, for determining the structure of $\mathrm{Am}(k)$ and $\mathrm{Am_{st}}(k)$ it is sufficient to compute the orders of these groups. The exact sequence (9.2) is a first step in this direction. The next steps consist in the computation of the order of H^G/P and of A/P. Before we do so we present a simple but very effective tool.

9.1.3   Hilbert's Theorem 90

Hilbert's Theorem 90 (in Hilbert's report on algebraic numbers, his famous *Zahlbericht*, the theorems were numbered, and this one had the number 90) comes in two versions, one for elements and one for ideals.

Theorem 9.3 (Hilbert's Theorem 90 for Elements) *Let k be a quadratic number field and $\alpha \in k^\times$. Then $N\alpha = 1$ if and only if α has the form $\alpha = \beta^{1-\sigma}$. Here β is determined uniquely up to rational factors.*

Equivalent formulations of Hilbert's Theorem 90 are the following:

1. There is an exact sequence

$$1 \longrightarrow (k^\times)^{1-\sigma} \longrightarrow k^\times \xrightarrow{\;N\;} k^\times,$$

   where N denotes the norm map $N_{k/\mathbb{Q}} : k^\times \longrightarrow \mathbb{Q}^\times$.
2. The group $k^\times[N]/(k^\times)^{1-\sigma}$ is trivial. Here $k^\times[N]$ denotes the kernel of the norm map $N : k^\times \longrightarrow \mathbb{Q}^\times$.

Proof The proof of "$\Longleftarrow$" is trivial. Assume therefore that $N\alpha = 1$. If $\alpha = -1$, we set $\beta = \sqrt{m}$; if $\alpha \neq -1$, we set $\beta = \alpha^\sigma + 1$; then $\beta^{\sigma-1} = \frac{\alpha+1}{\alpha'+1} = \frac{\alpha(\alpha+1)}{\alpha\alpha'+\alpha} = \frac{\alpha(\alpha+1)}{1+\alpha} = \alpha$. $\qquad\square$

The corresponding result for ideals is

Theorem 9.4 (Hilbert's Theorem 90 for Ideals) *If $\mathfrak{a}$ is a fractional ideal[1] in $\mathcal{O}_k$, then we have $N\mathfrak{a} = (1)$ if and only if $\mathfrak{a}$ has the form $\mathfrak{a} = \mathfrak{b}^{\sigma-1}$ for some (integral) ideal $\mathfrak{b}$.*

Proof As in the case of elements, the proof of "$\Longleftarrow$" is trivial. Assume therefore that $N\mathfrak{a} = 1$ (hence $\mathfrak{a} = \mathfrak{c}\mathfrak{d}^{-1}$ is the quotient of two integral ideals $\mathfrak{c}$ and $\mathfrak{d}$ with the same norm). By the uniqueness of prime ideal factorization we may assume that $\mathfrak{c}$ and $\mathfrak{d}$ are coprime. This immediately implies that $\mathfrak{c}$ and $\mathfrak{d}$ are not divisible by any inert prime ideals: If, for example, we had $(q) \mid \mathfrak{c}$, then q^2 would occur in the factorization of $N\mathfrak{d}$, hence $\mathfrak{d}$ would also be divisible by (q), and this contradicts our assumption that $\mathfrak{c}$ and $\mathfrak{d}$ are coprime. For the same reason, no ramified prime ideals can divide $\mathfrak{c}$. Thus $\mathfrak{c}$ and $\mathfrak{d}$ are products of split prime ideals. If $\mathfrak{c} = \mathfrak{p}_1^{e_1} \cdots \mathfrak{p}_r^{e_r}$ is the prime ideal factorization of $\mathfrak{c}$, then we must have $N\mathfrak{c} = p_1^{e_1} \cdots p_r^{e_r} = N\mathfrak{d}$. Since $\mathfrak{c}$ and $\mathfrak{d}$ are coprime, none of the $\mathfrak{p}_j$ can divide $\mathfrak{d}$, hence the only possibility is that $\mathfrak{d} = \mathfrak{p}_1'^{e_1} \cdots \mathfrak{p}_r'^{e_r} = \mathfrak{c}'$. But then $\mathfrak{a} = \mathfrak{c}\mathfrak{d} = \mathfrak{d}'\mathfrak{d}^{-1} = \mathfrak{d}^{\sigma-1}$. $\qquad\square$

9.2   The Ambiguous Class Number Formula

As a warm-up we construct a few exact sequences involving the following groups:

- $E = \mathcal{O}_k^\times$ is the unit group of $\mathcal{O}_k$;
- $E[N] = \{\varepsilon \in E : N_{k/\mathbb{Q}}(\varepsilon) = 1\}$ is the kernel of the norm map on the unit group, that is, the subgroup of units with norm $+1$;
- $E^{1-\sigma} = \{\varepsilon^{1-\sigma} : \varepsilon \in E\}$;
- $H^G = \{(\alpha) : (\alpha)^\sigma = (\alpha)\}$ is the group of ambiguous principal ideals;
- $P = \{(a) : a \in \mathbb{Q}^\times\}$ is the subgroup of A consisting of all nonzero ideals generated by rational numbers.

---

[1] For integral ideals, the statement is trivial since then $N\mathfrak{a} = (1)$ is equivalent to $\mathfrak{a} = (1)$.

Now we claim

Proposition 9.5 *There is an exact sequence*

$$1 \longrightarrow E^{1-\sigma} \longrightarrow E[N] \overset{\lambda}{\longrightarrow} H^G/P \longrightarrow 1.$$

Proof The map $E^{1-\sigma} \longrightarrow E[N]$ is the inclusion map: Each unit $\varepsilon^{1-\sigma}$ has norm 1 and thus is an element of $E[N]$. For constructing $\lambda : E[N] \longrightarrow H^G/P$ assume that $\varepsilon \in E[N]$, i.e., $N\varepsilon = 1$. By Hilbert's Theorem 90 there is an $\alpha \in k^\times$ such that $\varepsilon = \alpha^{1-\sigma}$; clearly $(\alpha) \in H^G$ since $(\alpha)^\sigma = (\alpha^\sigma) = (\varepsilon\alpha) = (\alpha)$. The map $\varepsilon \mapsto (\alpha)$ is not well defined, however, since with α each element αa for any $a \in \mathbb{Q}^\times$ has the property $(\alpha a)^{1-\sigma} = \varepsilon$. For this reason we set $\lambda(\varepsilon) = (\alpha)P$, and this map now is well defined. Clearly $\varepsilon \in \ker\lambda$ if and only if $\lambda(\varepsilon) = P$; this is equivalent to $(\alpha) = (a)$, i.e., to $\alpha = a\eta$ for some unit η. This implies $\varepsilon = \alpha^{1-\sigma} = \eta^{1-\sigma}$, which shows that $\ker\lambda = E^{1-\sigma}$.

The surjectivity of λ is clear: If (α) is ambiguous, then $(\alpha)^\sigma = (\alpha)$ and thus $\varepsilon\alpha^\sigma = \alpha$ for some unit ε, hence $\varepsilon = \alpha^{1-\sigma}$. □

The content of this proposition may also be expressed by the isomorphism

$$E[N]/E^{1-\sigma} \simeq H^G/P.$$

The quotient group $H^{-1}(G, E) = E[N]/E^{1-\sigma}$ is a cohomology group. We have come across such a group already in Hilbert's Theorem 90, which says that $H^{-1}(G, k^\times) = k^\times[N]/(k^\times)^{1-\sigma} = 1$. Hilbert's Theorem 90 for ideals claims accordingly that $H^{-1}(G, I_k) = I_k[N]/I_k^{1-\sigma} = 1$, where I_k denotes the group of nonzero fractional ideals in a quadratic number field. Such cohomology groups for cyclic Galois groups $G = \langle\sigma\rangle$ are all over the place in class field theory, the theory of abelian extensions of number fields.

Galois cohomology[2] gives the exact sequence in Proposition 9.5 in the other direction (Exercise 9.10).

The order of the group $E[N]/E^{1-\sigma}$ can be determined quickly. If $\Delta < 0$, then E consists only of roots of unity with norm 1. Thus $\varepsilon^\sigma = \varepsilon^{-1}$, hence $E^{1-\sigma} = E^2$ and

---

[2]Those who are familiar with the first principles of cohomology get the sequence for free: The trivial sequence

$$1 \longrightarrow E \longrightarrow k^\times \longrightarrow H \longrightarrow 1,$$

in which H denotes the group of nonzero fractional principal ideals, provides the long exact sequence

$$1 \longrightarrow E^G \longrightarrow (k^\times)^G \longrightarrow H^G \longrightarrow H^1(G, E) \longrightarrow H^1(G, k^\times),$$

from which the claim follows using Hilbert's Theorem 90 ($H^1(G, k^\times) = 1$), the periodicity $H^1(G, A) \simeq H^{-1}(G, A)$ for cyclic groups G, as well as $(k^\times)^G = \mathbb{Q}^\times$, $E^G = \{\pm 1\}$ and $\mathbb{Q}^\times/E^G \simeq P$.

$E[N]/E^{1-\sigma} = E/E^2 \simeq \mathbb{Z}/2\mathbb{Z}$. If $\Delta > 0$, then let ε denote the fundamental unit. If $N\varepsilon = +1$, then again $E[N] = E$ and $E^{1-\sigma} = E^2$, hence $E[N]/E^{1-\sigma} = E/E^2 = \langle -1, \varepsilon \rangle / \langle \varepsilon^2 \rangle \simeq (\mathbb{Z}/2\mathbb{Z})^2$. If $N\varepsilon = -1$, on the other hand, then $E[N] = \langle -1, \varepsilon^2 \rangle$ and $E^{1-\sigma} = E^2 = \langle \varepsilon^2 \rangle$, hence $E[N]/E^{1-\sigma} = E/E^2 \simeq \mathbb{Z}/2\mathbb{Z}$.

Lemma 9.6 *Let k be a quadratic number field whose unit group E is generated by the fundamental unit ε (and -1). Then*

$$H^{-1}(G, E) = E[N]/E^{1-\sigma} \simeq \begin{cases} \mathbb{Z}/2\mathbb{Z}, & \text{if } d < 0, \\ \mathbb{Z}/2\mathbb{Z}, & \text{if } d > 0, N\varepsilon = -1, \\ (\mathbb{Z}/2\mathbb{Z})^2, & \text{if } d > 0, N\varepsilon = +1. \end{cases}$$

It remains to determine the order of A/P. To this end we will use the following lemma.

Lemma 9.7 *An ideal $\mathfrak{a}$ is ambiguous if and only if $\mathfrak{a}$ is the product of ramified prime ideals and an ideal (a) with $a \in \mathbb{Q}^\times$. More exactly we have*

$$A/P \simeq (\mathbb{Z}/2\mathbb{Z})^t,$$

where t is the number of primes that ramify in $k/\mathbb{Q}$, in other words, the number of distinct prime factors of the discriminant of k.

Proof We may assume that $\mathfrak{a}$ is an integral ideal (otherwise we multiply it by a suitable rational integer). Among all decompositions $\mathfrak{a} = (a)\mathfrak{b}$ with an integral ideal $\mathfrak{b}$ we pick one in which $a \in \mathbb{N}$ is maximal.

Let $\mathfrak{p}$ denote a prime ideal with $\mathfrak{p}^\sigma \neq \mathfrak{p}$; if $\mathfrak{p}$ divides $\mathfrak{b}$, then we must have $\mathfrak{p}^\sigma \mid \mathfrak{b}$. In fact by applying σ to $\mathfrak{p} \mid \mathfrak{a}$ we see that $\mathfrak{p}^\sigma \mid \mathfrak{b}^\sigma = \mathfrak{b}$. Thus $(p) \mid \mathfrak{b}$, where $(p) = \mathfrak{p}\mathfrak{p}^\sigma$, which contradicts the maximality of a. This shows that $\mathfrak{b}$ is not divisible by a split prime ideal.

For the same reason, $\mathfrak{b}$ is not divisible by any inert prime ideal (p). Thus $\mathfrak{b}$ is a product of ramified prime ideals. If $\mathfrak{p}$ is such a prime ideal, then $\mathfrak{p}^2 = (p)$, and the maximality of a implies that we can write $\mathfrak{a}$ uniquely in the form

$$\mathfrak{a} = (a) \prod \mathfrak{p}_j^{e_j},$$

where $\mathfrak{p}_j$ runs through the ramified prime ideals and where $e_j \in \{0, 1\}$. Now we set

$$\phi : A/P \longrightarrow (\mathbb{Z}/2\mathbb{Z})^t : (a) \prod \mathfrak{p}_j^{e_j} \longmapsto (e_1, \ldots, e_t)$$

and show that ϕ is a group isomorphism, which is left as an exercise. $\square$

If we collect everything, then the exact sequence (9.2) now implies

Corollary 9.8 *In the quadratic number field k with discriminant Δ and fundamental unit ε we have*

$$\# \mathrm{Am}_{\mathrm{st}}(k) = \begin{cases} 2^{t-1} & \text{if } \Delta < 0, \\ 2^{t-1} & \text{if } \Delta > 0, N\varepsilon = -1, \\ 2^{t-2} & \text{if } \Delta > 0, N\varepsilon = +1, \end{cases}$$

where t denotes the number of primes that ramify in k.

Thus it remains only to determine the difference between the group of ambiguous ideal classes $\mathrm{Am}(k)$ and that of strictly ambiguous ideal classes $\mathrm{Am}_{\mathrm{st}}(k)$:

Proposition 9.9 *There is an exact sequence*

$$1 \longrightarrow \mathrm{Am}_{\mathrm{st}}(k) \longrightarrow \mathrm{Am}(k) \overset{\mu}{\longrightarrow} (E_{\mathbb{Q}} \cap Nk^{\times})/NE_k \longrightarrow 1.$$

In particular, $\mathrm{Am}(k) = \mathrm{Am}_{\mathrm{st}}(k)$ except when -1 is the norm of an element, but not of a unit. In this case, $\# \mathrm{Am}(k) = 2 \cdot \# \mathrm{Am}_{\mathrm{st}}(k)$.

Proof Let $c = [\mathfrak{a}]$ be ambiguous. Then $\mathfrak{a}^\sigma \sim \mathfrak{a}$, hence $\mathfrak{a}^\sigma = \alpha\mathfrak{a}$. Taking norms yields $(N\alpha) = (1)$, that is $N\alpha = \pm 1 \in E_{\mathbb{Q}} \cap Nk^{\times}$. We set $\mu(c) = N\alpha \cdot NE_k$ and claim that μ is well defined. In fact if we start from $c = [\mathfrak{b}]$, then $\mathfrak{b} = \gamma\mathfrak{a}$, and $\mathfrak{b}^\sigma = \gamma^\sigma \mathfrak{a}^\sigma = \gamma^\sigma \alpha\mathfrak{a} = \gamma^{\sigma-1}\alpha\mathfrak{b}$ shows that $N(\gamma^{\sigma-1}\alpha) \cdot NE_k = N\alpha \cdot NE_k$ since elements of the form $\gamma^{\sigma-1}$ have norm 1. Thus μ is well defined.

If $c \in \ker \mu$, then $N\alpha = N\eta$, d.h. $N(\alpha\eta) = 1$. According to Hilbert's Theorem 90, we have $\alpha\eta = \beta^{1-\sigma}$, and now $\mathfrak{a}^\sigma = \alpha\mathfrak{a}$ implies $(\beta\mathfrak{a})^\sigma = (\beta)\mathfrak{a}$. Thus $\mathfrak{b} = \beta\mathfrak{a}$ is an ambiguous ideal equivalent to $\mathfrak{a}$, and therefore $c = [\mathfrak{b}]$ is strongly ambiguous. Conversely, strongly ambiguous ideal classes are clearly contained in $\ker \mu$.

In order to prove the surjectivity of μ we have to show that $-1NE_k$ lies in the image of μ if -1 is the norm of an element from k. Assume therefore that $N\alpha = -1$ for $\alpha = x + y\sqrt{m}$. Then $x^2 - my^2 = -1$, hence -1 is a quadratic residue modulo each odd prime divisor p of m. We know from elementary number theory (or from the arithmetic of Gaussian integers) that this holds if and only if $m = a^2 + b^2$ is a sum of two squares; here we may assume that a is odd. Now we verify that $\mathfrak{a} = (a, b + \sqrt{m})$ generates an ambiguous ideal class $c = [\mathfrak{a}]$, and that $\mu(c) = -1$. In fact we have

$$\mathfrak{a}^2 = (a^2, ab + a\sqrt{m}, b^2 + 2b\sqrt{m} + m)$$
$$= (a^2, ab + a\sqrt{m}, 2b^2 + 2b\sqrt{m})$$
$$= (a^2, a(b + \sqrt{m}), 2b(b + \sqrt{m})) = (a^2, b + \sqrt{m}) = (b + \sqrt{m})$$

because of $\gcd(a^2, 2b) = 1$ and $(b + \sqrt{m})(b - \sqrt{m}) = b^2 - m = -a^2$. Thus $\mathfrak{a}^2 = (b + \sqrt{m})$ and $\mathfrak{a}\mathfrak{a}^\sigma = N\mathfrak{a} = (a)$, and therefore $\mathfrak{a}^{1-\sigma} = \mathfrak{a}^2/\mathfrak{a}^{1+\sigma} = \frac{1}{a}(b + \sqrt{m})$. This yields $\mu(c) = N(\frac{b+\sqrt{m}}{a}) = \frac{b^2-m}{a^2} = -1$ as claimed.   □

The group $(E_{\mathbb{Q}} \cap Nk^\times)/NE_k$ is small, because even $E_{\mathbb{Q}} = \{\pm 1\}$ has only two elements. In fact we have $(E_{\mathbb{Q}} \cap Nk^\times)/NE_k = 1$ unless -1 is the norm of an element from k or if $N\varepsilon = -1$, and $(E_{\mathbb{Q}} \cap Nk^\times)/NE_k \simeq \mathbb{Z}/2\mathbb{Z}$ if -1 is the norm of an element, but not the norm of a unit. As we just have seen, -1 is the norm of an element if and only if $\Delta = \square + \square$ is the sum of two squares. Thus we have

Theorem 9.10 (Ambiguous Class Number Formula) *In quadratic number fields k with discriminant Δ and fundamental unit ε we have*

$$\# \operatorname{Am}(k) = \begin{cases} 2^{t-2}, & \text{if } \Delta > 0, N\varepsilon = +1, \ \Delta \neq \square + \square, \\ 2^{t-1}, & \text{otherwise.} \end{cases}$$

where t denotes the number of primes ramified in k.

Examples

Δ	t	$N\varepsilon$	$\square + \square$	$\# \operatorname{Am}_{st}(k)$	$\# \operatorname{Am}(k)$
8	1	-1	$1^2 + 1^2$	1	1
10	2	-1	$1^2 + 3^2$	2	2
12	2	$+1$	no	1	1
30	3	$+1$	no	2	2
34	2	$+1$	$3^2 + 5^2$	1	2
-30	3	$-$	no	4	4

As an additional consequence of the ambiguous class number formula we claim:

Corollary 9.11 *The class number of the quadratic number field with discriminant Δ is odd if and only if we are in one of the following cases; there p denotes prime numbers $\equiv 1 \bmod 4$ and q, q' prime numbers $\equiv 3 \bmod 4$:*

(1) Δ *is a prime discriminant, i.e.,* $\Delta = -4, \pm 8, p, -q$;
(2) Δ *is a product of two negative prime discriminants:* $\Delta = 4q$, $\Delta = 8q$ *or* $\Delta = qq'$.

Proof The class number of k is even if and only if $\# \operatorname{Am}(k) \neq 1$, thus if the number of ambiguous ideal classes is even. The other claims follow directly from the ambiguous class number formula.   □

9.3 The Quadratic Reciprocity Law

The quadratic reciprocity law is a corollary of Corollary 9.11. We begin by proving the two supplementary laws.

Theorem 9.12 (First Supplementary Law) *For all odd prime numbers* p, *the following assertions are equivalent:*

(1) $(\frac{-1}{p}) = +1$, *i.e., the congruence* $x^2 \equiv -1$ mod p *is solvable.*
(2) $p = a^2 + 4b^2$ *is sum of two squares.*
(3) *We have* $p \equiv 1$ mod 4.

The equivalence of (1) and (3) may also be expressed by the equation

$$\left(\frac{-1}{p}\right) = (-1)^{\frac{p-1}{2}}.$$

Proof (1) $\Longrightarrow$ (2): If $(-1/p) = 1$, then p splits in $k = \mathbb{Q}(i)$. Multiplying through by i we may assume that the coefficient of i is even. Thus $p = (a + 2bi)(a - 2bi)$, and taking the norm yields $p = a^2 + 4b^2$.

(2) $\Longrightarrow$ (3): Since p and a are odd, $p = a^2 + 4b^2$ implies $p \equiv 1$ mod 4.

(3) $\Longrightarrow$ (1): If $p \equiv 1$ mod 4, then $(\frac{-1}{p}) = (-1)^{(p-1)/2} = 1$ according to Euler's Criterion.   □

Similarly we can prove

Theorem 9.13 (Second Supplementary Law) *For all odd prime numbers* p, *the following assertions are equivalent:*

(1) $(\frac{2}{p}) = +1$, *i.e., the congruence* $x^2 \equiv 2$ mod p *is solvable.*
(2) *We have* $p = e^2 - 2f^2$ *for integers* $e, f \in \mathbb{Z}$.
(3) *We have* $p \equiv \pm 1$ mod 8.

The equivalence of (1) and (3) can also be expressed by the equation

$$\left(\frac{2}{p}\right) = (-1)^{\frac{p^2-1}{8}}.$$

Proof (1) $\Longrightarrow$ (2): If $(\frac{2}{p}) = +1$, then p splits in $\mathbb{Q}(\sqrt{2})$, and we have $\pm p = x^2 - 2y^2$; multiplying $x + y\sqrt{2}$, if necessary, by the unit $1 + \sqrt{2}$ we can make sure that $p = e^2 - 2y^2$.

(2) $\Longrightarrow$ (3): Reduction modulo 8 yields $p \equiv \pm 1$ mod 8 in all cases.

(3) $\Longrightarrow$ (1): Let h denote the class number of $k = \mathbb{Q}(\sqrt{p})$, which is odd by Corollary 9.11. If $p \equiv \pm 1$ mod 8, then 2 splits in $k/\mathbb{Q}$, hence $2\mathcal{O}_k = \mathfrak{p}\mathfrak{p}'$ for prime ideals $\mathfrak{p}, \mathfrak{p}'$. Since $\mathfrak{p}^h = \frac{1}{2}(x + y\sqrt{p})$ is a principal ideal, taking the norm yields $x^2 - py^2 = \pm 4 \cdot 2^h$. Reduction modulo p shows that $\pm 2^h$ and thus ± 2 is a quadratic residue modulo p; the claim now follows from the first supplementary law.   □

The quadratic reciprocity law for odd prime numbers is the content of the following theorem.

Theorem 9.14 (Quadratic Reciprocity Law) *If p and q are odd primes, then*

$$\left(\frac{p}{q}\right)\left(\frac{q}{p}\right) = (-1)^{\frac{p-1}{2}\frac{q-1}{2}}.$$

Proof We first discuss the case where one of the primes, say p, is congruent to $1 \bmod 4$. We will show that in this case we have $\left(\frac{p}{q}\right) = +1 \iff \left(\frac{q}{p}\right) = +1$.

Since $\left(\frac{p}{q}\right) = +1$, the prime q splits in $k = \mathbb{Q}(\sqrt{p})$. Thus $q\mathcal{O}_k = \mathfrak{q}\mathfrak{q}'$ and $\mathfrak{q}^h = \frac{1}{2}(x + y\sqrt{p})$ is a principal ideal, where h is the class number of k, which is odd by Corollary 9.11. Taking the norm yields $\pm 4q^h = x^2 - py^2$. This in turn provides us with the congruence $\pm 4q^h \equiv x^2 \bmod p$, and then $\left(\frac{-1}{p}\right) = +1$ implies $\left(\frac{q}{p}\right) = +1$ as claimed.

If $\left(\frac{q}{p}\right) = +1$, on the other hand, then we use the number field $k = \mathbb{Q}(\sqrt{q})$, which also has odd class number h. Again the fact that p splits in $\mathcal{O}_k$ yields the equation $\pm 4p^h = x^2 - qy^2$ and thus $\left(\frac{\pm p}{q}\right) = +1$. Since either $q \equiv 1 \bmod 4$ and $\left(\frac{-1}{q}\right) = +1$ or $q \equiv 3 \bmod 4$ and the sign is necessarily positive (Exercise 9.15), we obtain $\left(\frac{p}{q}\right) = +1$.

Finally assume that $p \equiv q \equiv 3 \bmod 4$. Consider the field $k = \mathbb{Q}(\sqrt{pq})$. According to Corollary 9.11, the class number h of k is odd. Thus the prime ideal $\mathfrak{p} = (p, \sqrt{pq})$ above p must be principal: In fact we have $\mathfrak{p}^2 \sim (1)$ and $\mathfrak{p}^h \sim 1$, and since $h = 2j+1$ we get $\mathfrak{p} = \mathfrak{p}^{h-2j} \sim (1)$. Assume therefore that $\mathfrak{p} = \frac{1}{2}(x+y\sqrt{pq})$. Then $\pm 4p = x^2 - pqy^2$, hence $x = pz$ and $\pm 4 = pz^2 - qy^2$. If the positive sign holds, then reduction modulo q and p shows that $\left(\frac{p}{q}\right) = +1$ and $\left(\frac{q}{p}\right) = -1$. If the negative sign holds, then we find accordingly that $\left(\frac{p}{q}\right) = -1$ and $\left(\frac{q}{p}\right) = +1$. This completes the proof. $\square$

9.3.1 Summary

In this chapter we have proved the ambiguous class number formula for quadratic number fields, and derived the quadratic reciprocity as a corollary.

9.4 Exercises

9.1. Show that the operation $[\mathfrak{a}]^\sigma = [\mathfrak{a}^\sigma]$ on the ideal class group of a quadratic number field is well defined, i.e., that $[\mathfrak{a}] = [\mathfrak{b}]$ implies $[\mathfrak{a}^\sigma] = [\mathfrak{b}^\sigma]$.

9.2. Show that the ideal class of order 2 in $\mathbb{Q}(\sqrt{10})$ contains the ideals $(2, \sqrt{10})$, $(3, 1 + \sqrt{10})$ and $(5, 1 + \sqrt{10})$. Which of these ideals are ambiguous?

9.3. Let $p \equiv 5 \bmod 8$ be prime. Show that the ideal class $\mathbb{Q}(\sqrt{2p})$ generated by the ambiguous ideal $(2, \sqrt{2p})$ has order 2.

9.4. Let $p \equiv 1 \bmod 8$ be prime. Show that the ambiguous ideal $(2, \sqrt{2p})$ is principal in $k = \mathbb{Q}(\sqrt{2p})$ if and only if the norm of the fundamental unit in k is $+1$.

In this case write $2p = a^2 + b^2$ with $a > b > 0$ and show that the ideal $\mathfrak{a} = (a, b + \sqrt{m})$ generates an ambiguous ideal class of order 2.

9.5. Show that if k is a quadratic number field with class number 2, then $\mathrm{Am}(k) = \mathrm{Cl}(k)$.

9.6. Show that if k is a quadratic number field with odd class number, then $\mathrm{Am}(k) = 1$.

9.7. Show: If A and B are subgroups of an abelian group, then $AB/B \simeq A/A \cap B$. Hint: Show that $A \cap B$ is the kernel of the natural map $A \longrightarrow AB/B$.

9.8. Show that the inclusion $\iota : H^G/P \longrightarrow A/P$ in (9.2) is injective and that the map $\pi : A/P \longrightarrow \mathrm{Am}_{\mathrm{st}}(k)$ defined by $\pi(\mathfrak{a}P) = [\mathfrak{a}]$ is well defined and surjective. Also show that $\ker \pi = \mathrm{im}\, \iota$.

9.9. (O. Taussky) Solve the Pythagorean equation $x^2 + y^2 = z^2$ using Hilbert's Theorem 90. Hint: $\alpha = \frac{x+yi}{z} \in \mathbb{Q}(i)$ satisfies the equation $N\alpha = 1$. Write $\alpha = \frac{m+ni}{m-ni}$ and rationalize the denominator.

Generalize this exercise to all equations of the form $x^2 - my^2 = z^2$ for squarefree values $m \in \mathbb{Z} \setminus \{0, 1\}$.

9.10. Show that there is an exact sequence

$$1 \longrightarrow P \longrightarrow H^G \longrightarrow E[N]/E^{1-\sigma} \longrightarrow 1,$$

where H denotes the group of nonzero principal ideals.

9.11. Let $m = a^2 + b^2$ be a sum of two squares. Then the ideals $(a, b + \sqrt{m})$ do not necessarily lie in the same ideal class for each choice of a and b. Verify this for $m = 10 = 1^2 + 3^2 = 3^2 + 1^2$.

For the distribution of these ideals over the ideal classes see [82] and [9].

9.12. Let p be a ramified prime in $k = \mathbb{Q}(\sqrt{m})$, where $m \neq \pm p$, and assume that the prime ideal $\mathfrak{p}$ above p is principal, say $\mathfrak{p} = (\pi)$. Show that $\varepsilon = \frac{1}{p}\pi^2$ is a unit in $\mathcal{O}_k$, and that neither ε nor $-\varepsilon$ is a square. Generalize this to products of ramified prime ideals. Use this to compute the fundamental unit of $\mathbb{Q}(\sqrt{30})$.

9.13. Let $p \equiv 1 \bmod 4$ be a prime number. Show that $\left(\frac{q}{p}\right) = +1$ implies $\left(\frac{p}{q}\right) = +1$.

Hint: Use $k = \mathbb{Q}(\sqrt{q^*})$ for $q^* = \left(\frac{-1}{q}\right)q$ instead of $\mathbb{Q}(\sqrt{q})$.

9.14. Show that the solvability of the Pell equation implies that the norm of the fundamental unit ε_p of $\mathbb{Q}(\sqrt{p})$ for primes p is equal to $N\varepsilon_p = -\left(\frac{-1}{p}\right)$.

Show also that $N\varepsilon_{pq} = -1$ if $p \equiv q \equiv 1 \bmod 4$ are primes with $\left(\frac{p}{q}\right) = -1$.

9.15. Let $\pm 4p^h = x^2 - qy^2$ for prime numbers $p \equiv 1 \bmod 4$ and $q \equiv 3 \bmod 4$. Show that x and y are both even, and that the plus sign must hold.

9.16. Let $k = \mathbb{Q}(\sqrt{m})$ be a quadratic number field with fundamental unit ε_m. Show: If $N\varepsilon_m = +1$, then there is an ambiguous principal ideal $\mathfrak{a} = (\alpha)$ with $\mathfrak{a} \neq (1), (\sqrt{m})$.

Hint: By Hilbert's Theorem 90 we have $\varepsilon = \alpha'/\alpha$.

9.17. Show that the norm of the fundamental unit ε of $\mathbb{Q}(\sqrt{p})$ is negative if $p \equiv 1 \bmod 4$ is prime.

Hint: Use the preceding exercise.

9.18. The idea behind Kummer's ideal numbers was the construction of ring homomorphisms $\mathcal{O}_k \longrightarrow \mathbb{F}_q$ of the ring of integers of number fields into finite fields. Restrict these homomorphisms to the multiplicative group, that is, consider the group homomorphism $\psi : \mathcal{O}_k^\times \longrightarrow \mathbb{F}_q^\times$. Find examples of real quadratic number fields and primes q for which this homomorphism is trivial, or where it is surjective.

9.19. Let p be a prime number such that $\left(\frac{10}{p}\right) = +1$. Then there exist two possibilities:

1. $\left(\frac{2}{p}\right) = \left(\frac{5}{p}\right) = +1$; in this case $p = x^2 - 10y^2$.
2. $\left(\frac{2}{p}\right) = \left(\frac{5}{p}\right) = -1$; in this case $\pm 2p = X^2 + 10y^2$ and, using $X = 2x$, $\pm p = 2x^2 - 5y^2$.

Show that this implies that each element $x + y\sqrt{10} \in \mathbb{Z}[\sqrt{10}]$ can be written uniquely as a product of a unit and irreducible elements of the form $a + b\sqrt{10}$ or $c\sqrt{2} + d\sqrt{5}$.

9.20. Let $q \equiv 3 \bmod 8$ be a prime number. Show that the class number of $\mathbb{Q}(\sqrt{2q})$ is odd and deduce that the equation $2x^2 - qy^2 = -1$ is solvable. Deduce that $\left(\frac{2}{q}\right) = -1$.

Chapter 10
Quadratic Gauss Sums

In Chap. 3 we have already pointed out the importance of Euler's Modularity Theorem, which is not an isolated curiosity but a part of a whole family of related modularity theorems (most of which are beyond the scope of this book). Here we will apply the method of generating functions, which we have used in our proof of Binet's formula (2.5) for Fibonacci numbers in Sect. 2.5, to the investigation of problems connected with Euler's Modularity Theorem.

10.1 Dirichlet Characters

At the heart of the notion of a Dirichlet character is the idea of studying algebraic structures by constructing (and investigating) homomorphisms into simpler structures. Characters map groups to groups of complex numbers, and the multiplicative group of complex numbers is simple in the sense that its finite subgroups are cyclic.

An example of a character is the Legendre symbol $(\frac{\cdot}{p})$, which is a group homomorphism from the coprime residue class group $(\mathbb{Z}/p\mathbb{Z})^{\times}$, where p is an odd prime number, to the subgroup $\{-1, +1\}$ of the complex numbers. More generally, a *Dirichlet character* χ *defined modulo m* is a group homomorphism

$$\chi : (\mathbb{Z}/m\mathbb{Z})^{\times} \longrightarrow \mathbb{C}^{\times}$$

assigning complex numbers to all the coprime residue classes modulo m. Since $(\mathbb{Z}/m\mathbb{Z})^{\times}$ is a finite abelian group of order $n = \phi(m)$, the Theorem of Euler-Fermat $a^n \equiv 1 \bmod m$ implies $1 = \chi(1) = \chi(a^n) = \chi(a)^n$, from which we can read off that the image of a Dirichlet character is an n-th root of unity. The Dirichlet character χ is called a *quadratic Dirichlet character* if χ only attains the values $+1$ and -1.

Example There exist three nontrivial Dirichlet characters defined modulo 8. For positive representatives a of the coprime residue classes modulo 8 these may be

© The Author(s), under exclusive license to Springer Nature Switzerland AG 2021
F. Lemmermeyer, *Quadratic Number Fields*, Springer Undergraduate
Mathematics Series, https://doi.org/10.1007/978-3-030-78652-6_10

defined by the Legendre symbols

$$\chi_{-4}(a) = \left(\frac{-1}{a}\right), \quad \chi_8(a) = \left(\frac{2}{a}\right), \quad \chi_{-8}(a) = \left(\frac{-2}{a}\right).$$

The fact that all Dirichlet characters modulo 8 are quadratic characters reflects the structure of $(\mathbb{Z}/8\mathbb{Z})^\times$: Since this group is elementary abelian (the square of each coprime residue class a modulo 8 is the unit element), the same must be true for the values $\chi(a)$, which attain only the values $+1$ or -1. We can define these characters also by the following table:

$a \bmod 8$	1	3	5	7
χ_{-4}	$+1$	-1	$+1$	-1
χ_8	$+1$	-1	-1	$+1$
χ_{-8}	$+1$	$+1$	-1	-1

In many cases it is necessary to extend these Dirichlet characters χ from residue classes modulo m to all natural numbers by setting

$$\chi(a) = \begin{cases} \chi(a + m\mathbb{Z}) & \text{if } \gcd(a, m) = 1, \\ 0, & \text{if } \gcd(a, m) \neq 1. \end{cases}$$

This extension clearly has the property that $\chi(a + m) = \chi(a)$ for all natural numbers a.

Example The Dirichlet character χ_{-4} defined modulo 4 by $\chi_{-4}(1 + 4\mathbb{Z}) = 1$ and $\chi_{-4}(3 + 4\mathbb{Z}) = -1$ may be extended to all natural numbers by setting $\chi_{-4}(2n) = 0$ and $\chi_{-4}(2n + 1) = (-1)^n$. For odd integers $a \in \mathbb{N}$ we then have $\chi_{-4}(a) = \left(\frac{-4}{a}\right) = \left(\frac{-1}{a}\right)$.

The extension to negative integers, when needed, must be done with care. For the Dirichlet character χ_{-4} we have $\chi(-1 + 4\mathbb{Z}) = \chi(3 + 4\mathbb{Z}) = -1$, whereas $\left(\frac{-1}{-1}\right) = +1$.

10.1.1 Primitive Characters

The Dirichlet character χ_{-4} may be interpreted in a natural way as a Dirichlet character ψ_8 defined modulo 8 by setting $\psi_8(1 + 8\mathbb{Z}) = \psi_8(5 + 8\mathbb{Z}) = 1$ and $\psi_8(3 + 8\mathbb{Z}) = \psi_8(7 + 8\mathbb{Z}) = -1$. We also say that ψ_8 factors over $\mathbb{Z}/4\mathbb{Z}$ since $\chi_{-4} \circ \pi(a) = \psi_8(a)$, where $\pi : (\mathbb{Z}/8\mathbb{Z})^\times \longrightarrow (\mathbb{Z}/4\mathbb{Z})^\times$ is the natural projection map that sends residue classes modulo 8 to residue classes modulo 4. This property becomes a little bit more impressive when stated via the commutativity of the diagram

$$
\begin{array}{ccc}
(\mathbb{Z}/8\mathbb{Z})^{\times} & \xrightarrow{\ \psi_8\ } & \{-1,+1\} \\
\pi \downarrow & & \downarrow \mathrm{id} \\
(\mathbb{Z}/4\mathbb{Z})^{\times} & \xrightarrow{\ \chi_{-4}\ } & \{-1,+1\}.
\end{array}
$$

A Dirichlet character χ defined modulo N is called *primitive* if there is no proper divisor N_1 of N such that χ is already defined modulo N_1. In this case we call N the *conductor* of χ. We have to show, however, that the conductor is well defined. To this end we have to show that if a Dirichlet character χ is defined modulo N_1 and modulo N_2, then χ is also defined modulo $\gcd(N_1, N_2)$ (Exercise 10.4).

In the following we will classify the primitive Dirichlet characters. This classification is deeper than it might first appear, and it will lead us into the heart of number theory, namely reciprocity laws, the notion of modularity, and class number formulas.

10.1.2   The Character Group of Finite abelian Groups

Let A be a finite abelian and multiplicatively written group. A character on A is a homomorphism $\chi : A \longrightarrow \mathbb{C}^{\times}$ into the multiplicative group of complex numbers. If n denotes the order of an element $a \in A$, then $\chi(a)^n = \chi(a^n) = \chi(1) = 1$ implies that the image $\chi(a)$ is an n-th root of unity. In particular, the values that χ attains are roots of unity. Characters on $A = (\mathbb{Z}/n\mathbb{Z})^{\times}$ are the usual Dirichlet characters defined modulo n, which are our main objects of interest.

The set $X(A)$ of all characters on A is a group with respect to the multiplication defined by

$$
\chi_1\chi_2(a) = \chi_1(a) \cdot \chi_2(a).
$$

The group $X(A)$ is called the *character group* of A. Our goal is the determination of the algebraic structure of the character group; in fact we will find that the character group of a finite abelian group is isomorphic to the group itself, which means that we can read off all algebraic properties of such groups from their character group.

Lemma 10.1 *For finite abelian groups A and B we have*

$$
X(A \oplus B) \simeq X(A) \oplus X(B). \tag{10.1}
$$

If χ_A is a character on A and if χ_B is a character on B, then

$$\chi((a, b)) = \chi_A(a) \cdot \chi_B(b)$$

defines a character on $A \oplus B$. Conversely, a character χ on $A \oplus B$ defines characters on A and B by restriction, that is, by setting $\chi_A(a) = \chi(a, 1)$ and $\chi_B(b) = \chi(1, b)$. The map $\chi \mapsto (\chi_A, \chi_B)$ defines a homomorphism

$$\lambda : X(A \oplus B) \longrightarrow X(A) \oplus X(B),$$

which is surjective by what we have already said, and whose kernel consists of all characters χ of $A \oplus B$ for which we have $\chi(1, b) = \chi(a, 1) = 1$. But this implies $\chi(a, b) = \chi(a, 1) \cdot \chi(1, b) = 1$ for all $a \in A$ and $b \in B$, hence χ is the trivial character, and λ is injective.

Next we show

Proposition 10.2 *For each finite abelian group A we have $X(A) \simeq A$.*

Each finite abelian group can be written as a direct product of cyclic groups. According to Lemma 10.1 it is therefore sufficient to prove the claim for cyclic groups.

If $A = \langle g \rangle$ is cyclic, then each character $\chi \in X(A)$ is completely determined by the value $\chi(g)$. If n denotes the order of A (and thus also of g) and if ζ_n is a primitive n-th root of unity, then $\omega(g) = \zeta_n$ defines a character $\omega \in X(A)$ with the property that each χ can be written as a power of ω. Thus $X(A) = \langle \omega \rangle$ is cyclic with the same order as A, and the homomorphism that sends $a = g^m \in A$ to $\chi = \omega^m \in X(A)$ provides us with an isomorphism between A and $X(A)$.

Let us denote the subgroup of quadratic characters on A, that is, the characters that attain only the values ± 1, by $X_2(A)$. Then again $X_2(A \oplus B) \simeq X_2(A) \oplus X_2(B)$. Moreover we have

Proposition 10.3 *For each finite abelian group A we have $X_2(A) \simeq A/A^2$.*

The proof is similar to the one above: It is sufficient to prove the claim for cyclic groups. If $A = \langle g \rangle$ and if the order n of A (and g) is odd, then $A/A^2 = 1$ and $X_2(A) = 1$, since $1 = \chi(1) = \chi(g)^n = \chi(g)^n$ and $\chi(g) = \pm 1$ imply $\chi(g) = 1$. If n is even, then $\chi_0(g) = 1$ and $\chi_1(g) = -1$ define the only two possible quadratic characters on A, and then $A/A^2 \simeq \mathbb{Z}/2\mathbb{Z}$ and $X_2(A) \simeq \mathbb{Z}/2\mathbb{Z}$.

If B is a subgroup of the finite abelian group A, then each character χ_0 on A/B defines a character χ on A via

$$\chi(a) = \chi_0(aB).$$

Each character that is not already defined on a proper quotient of A is called *primitive*.

Lemma 10.4 *In the decomposition (10.1), a character $\chi = \chi_A \cdot \chi_B$ is primitive if and only if χ_A and χ_B are primitive.*

Proof If χ is not primitive, then there is a nontrivial subgroup $A_1 \oplus B_1$ of $A \oplus B$ such that χ is induced by a character on $A/A_1 \oplus B/B_1$. Then χ_A and χ_B are induced by characters on A/A_1 and B/B_1, and at least one of the subgroups A_1 or A_2 is nontrivial.

If χ_A is not primitive, then χ_A is induced by a character on A/A_1, hence χ is induced by a character on $(A \oplus B)/(A_1 \oplus 1)$. $\square$

Not every finite abelian group admits primitive quadratic characters; groups of odd order, for example, have only the trivial quadratic character. More generally we have

Lemma 10.5 *If $A = B \oplus U$ for a group U of odd order, then $X_2(A) \simeq X_2(B)$ and $B \simeq A/U$ (where we have identified $U \simeq 1 \oplus U$); in particular, A does not have any primitive quadratic character if U is nontrivial.*

This is clear since $X_2(U) = 1$.

10.1.3 Classification of Quadratic Dirichlet Characters

Now we can determine the primitive quadratic Dirichlet characters modulo N. We write $N = p_1^{a_1} \cdots p_t^{a_t}$; then

$$(\mathbb{Z}/N\mathbb{Z})^\times \simeq (\mathbb{Z}/p_1^{a_1}\mathbb{Z})^\times \oplus \cdots \oplus (\mathbb{Z}/p_t^{a_t}\mathbb{Z})^\times.$$

Next we know that for odd prime numbers p we have

$$(\mathbb{Z}/p^m\mathbb{Z})^\times \simeq \mathbb{Z}/(p-1)\mathbb{Z} \oplus \mathbb{Z}/p^{m-1}\mathbb{Z}.$$

According to Lemma 10.5 there exists a primitive quadratic Dirichlet character on $(\mathbb{Z}/p^m\mathbb{Z})^\times$ only if $m = 1$. Finally, $X_2((\mathbb{Z}/p\mathbb{Z})^\times)$ only consists of the trivial character and the quadratic Dirichlet character $\chi(a) = \left(\frac{a}{p}\right)$ defined by the Legendre symbol. Thus we have

Lemma 10.6 *Let p denote an odd prime number. Then there exist exactly two quadratic Dirichlet characters defined modulo p, namely the trivial character and the primitive quadratic Dirichlet character χ defined by $\chi(a) = \left(\frac{a}{p}\right)$. For $n \geq 2$ there does not exist any primitive quadratic Dirichlet character modulo p^n.*

In the case $p = 2$ each coprime residue class modulo 2^m can be written as a product of a power -1 and of 5, which shows that

$$(\mathbb{Z}/2^m\mathbb{Z})^\times \simeq \mathbb{Z}/2\mathbb{Z} \oplus \mathbb{Z}/2^{m-2}\mathbb{Z}.$$

With $A = (\mathbb{Z}/2^m\mathbb{Z})^\times$ we thus have $A/A^2 \simeq \mathbb{Z}/2\mathbb{Z} \oplus \mathbb{Z}/2\mathbb{Z}$, hence there are four quadratic characters modulo $2^m \geq 8$; apart from the trivial character these are the characters χ_{-4}, χ_8 and χ_{-8} defined above; here χ_{-4} is a primitive character modulo 4, the other two are primitive characters modulo 8.

Lemma 10.7 *There exist exactly four quadratic characters defined modulo 2^m for $m \geq 3$, namely the trivial character, as well as the characters χ_{-4} with conductor 4 and the characters χ_8 and χ_{-8} with conductor 8.*

Thus primitive quadratic Dirichlet characters exist only modulo 4, 8 and for odd prime numbers p. Because of Lemma 10.1 there is exactly one primitive quadratic Dirichlet character modulo N, where N is a product of such moduli. These integers N are exactly those positive integers that are, up to sign, discriminants of quadratic number fields: $N = |\Delta|$. The decomposition of $\Delta = \Delta_1 \cdots \Delta_t$ into prime discriminants corresponds to a decomposition $\chi = \chi_1 \cdots \chi_t$ of a primitive quadratic character χ into primitive quadratic characters defined modulo $N_j = |\Delta_j|$. According to Lemma 10.4 χ is primitive if and only if the components χ_j are primitive. Thus we have the following

Theorem 10.8 *There exists a bijection between primitive quadratic Dirichlet characters and discriminants of quadratic number fields.*

The fact that there is a bijection between the primitive quadratic Dirichlet characters and quadratic number fields suggests the question whether this bijection may be extended from quadratic to arbitrary Dirichlet characters. The answer is yes, and the primitive Dirichlet characters correspond to cyclotomic number fields.

Proposition 10.9 *Let $N = |\Delta|$ be a natural number. If χ is a primitive quadratic Dirichlet character defined modulo N, then*

$$\chi(-1) = \begin{cases} +1 & \text{for } \Delta > 0, \\ -1 & \text{for } \Delta < 0. \end{cases}$$

In particular, $\Delta = \chi(-1) \cdot N$.

In fact we have $\chi_\Delta(-1) = \text{sgn}(\Delta)$ for each primitive quadratic Dirichlet character with prime conductor $N = |\Delta|$; this follows from the observation that for odd prime conductors N we have

$$\chi_\Delta(-1) = \left(\frac{-1}{p}\right) = \begin{cases} -1 & \text{for } \Delta = -p, \ p \equiv 3 \bmod 4, \\ +1 & \text{for } \Delta = +p, \ p \equiv 1 \bmod 4, \end{cases}$$

and since the claim is also true for the three primitive quadratic Dirichlet characters defined modulo 4 and 8, the proposition is now completely proved.

10.1.4 Modularity and Reciprocity

Quite often in mathematics there is a deep conceptual reason why bijections such as the one in Theorem 10.8 exist. In our case, the existence of the bijection would be explained by the fact that for quadratic number fields with discriminant Δ there exists a Dirichlet character χ with conductor $N = |\Delta|$. This is indeed true: The Kronecker symbol $\left(\frac{\Delta}{p}\right)$ introduced in Sect. 3.2, which describes the splitting of primes p in the quadratic number field with discriminant Δ (see Thm. 6.14), defines a "Kronecker character" $\kappa_\Delta(a) = \left(\frac{\Delta}{a}\right)$ for all natural numbers $a \geq 1$, which assigns the value $+1$ or -1 to all integers a coprime to Δ. It is, however, not at all obvious that κ is a Dirichlet character, i.e., that there exists a modulus m with

$$\left(\frac{\Delta}{a}\right) = \left(\frac{\Delta}{a + km}\right) \quad \text{for all} \quad k \geq 0.$$

It is the Modularity Theorem for Kronecker characters that guarantees the existence of such a modulus m:

Theorem 10.10 (Modularity Theorem) *Every Kronecker character is modular. More exactly,* $\kappa(a) = \left(\frac{\Delta}{a}\right)$ *defines a primitive quadratic Dirichlet character with conductor* $N = |\Delta|$.

We have already proved this Theorem in Chap. 3 using elementary means, namely Gauss's Lemma. Here we will present an approach using generating functions.

Dirichlet used this bijection between Dirichlet and Kronecker characters in his proof of the theorem on primes in arithmetic progression in order to turn quadratic Dirichlet characters into Kronecker characters: This allowed him to reduce the non-vanishing of his L-series (we will say a few things about this below) to the arithmetic of quadratic number fields. Harvey Cohn [22, 23] called this bijection *Dirichlet's Lemma*.

10.2 Pell Forms

For proving the modularity of Kronecker characters we will proceed as in our derivation of Binet's formula (2.5) and study the generating function for a Kronecker character κ, namely $f_\kappa(q) = \sum_{n \geq 1} \kappa(n) q^n$. Without modularity, however, we know next to nothing about f_κ, and we are not in a position to derive essential properties of f_κ.

For this reason we will investigate the generating function

$$f_\chi(q) = \sum_{n=1}^{\infty} \chi(n) q^n$$

of a Dirichlet character χ defined modulo N. To this end we set $\chi(a) = 0$ for all integers a that are not coprime to N. Clearly the geometric series majorizes $f_\chi(q)$, hence this series converges absolutely for all complex numbers q with $|q| < 1$. Let us now compute $f_\chi(q)$ for the two discriminants $\Delta = -4$ and $\Delta = 8$:

- $\Delta = -4$: For $\kappa(p) = \left(\frac{-4}{p}\right)$ we obtain

$$f_\chi(q) = q - q^3 + q^5 - q^7 + \ldots = q(1 - q^2 + q^4 - q^6 + \ldots) = \frac{q}{1 + q^2}.$$

This is a rational function with poles at the primitive 4-th roots of unity. In addition, we find

$$f_\chi\left(\frac{1}{q}\right) = \frac{\frac{1}{q}}{1 + \frac{1}{q^2}} = \frac{q}{q^2 + 1} = f_\chi(q),$$

hence f_χ satisfies the functional equation $f_\chi\left(\frac{1}{q}\right) = f_\chi(q)$, which connects the values of f_χ inside the unit circle, where f_χ converges, with values of f_χ outside the domain of convergence.

- $\Delta = 8$: For $\kappa(p) = \left(\frac{-4}{p}\right)$ we find in a similar way

$$f_\chi(q) = q - q^3 - q^5 + q^7 + q^9 - \ldots = (q - q^3 - q^5 + q^7)(1 + q^8 + q^{16} + \ldots)$$

$$= \frac{q - q^3 - q^5 + q^7}{1 - q^8} = \frac{q - q^3}{q^4 + 1}$$

because of

$$q - q^3 - q^5 + q^7 = q(q-1)^2(q+1)^2(q^2+1) \quad \text{and}$$

$$q^8 - 1 = (q-1)(q+1)(q^2+1)(q^4+1).$$

Here we obtain

$$f_\chi\left(\frac{1}{q}\right) = \frac{\frac{1}{q} - \frac{1}{q^3}}{\frac{1}{q^4} + 1} = \frac{q^3 - q}{q^4 + 1} = -f_\chi(q).$$

For general Dirichlet characters we obtain in a similar way

$$f_\chi(q) = \sum_{n=1}^{\infty} \chi(n)q^n = \sum_{k=0}^{\infty}\left(\sum_{n=1}^{N} \chi(n)q^n\right)q^{kN}$$

$$= \left(\sum_{n=1}^{N} \chi(n)q^n\right)(1 + q^N + q^{2N} + \ldots) = \frac{\mathrm{Fek}_\chi(q)}{1 - q^N},$$

Table 10.1 Fekete polynomials with small conductor

Δ	N	$\text{Fek}_\chi(q)$	Δ	N	$\text{Fek}_\chi(q)$
-3	3	$q - q^2$	-7	7	$q + q^2 - q^3 + q^4 - q^5 - q^6$
-4	4	$q - q^3$	8	8	$q - q^3 - q^5 + q^7$
5	5	$q - q^2 - q^3 + q^4$	-8	8	$q + q^3 - q^5 - q^7$

where

$$\text{Fek}_\chi(q) = \sum_{n=1}^{N-1} \chi(n)q^n$$

denotes the Fekete polynomial for the Dirichlet character χ with conductor N (see Table 10.1).

Proposition 10.11 *The Pell form f_χ of a Dirichlet character χ with conductor N represents, for all $q \in \mathbb{C}$ with $|q| < 1$, a rational function*

$$f_\chi(q) = \frac{\text{Fek}_\chi(q)}{1 - q^N}$$

that can be extended, except for possible poles at the N-th roots of unity, to the whole complex plane.

Fekete polynomials first occurred explicitly in Dirichlet's proof of the theorem on primes in arithmetic progression, according to which there exist infinitely many primes in each coprime residue class modulo some integer N. Implicitly, Fekete polynomials already showed up in Gauss's sixth proof [44] of the quadratic reciprocity law; later Cauchy, Jacobi and Eisenstein published variants of this proof in which they replaced x by a p-th root of unity.

Yet Fekete polynomials have remained mathematical wallflowers; one of the few articles that underline the importance of Fekete polynomials for the arithmetic of quadratic number fields is Ayoub [6].

The periodicity of χ allowed us to write the generating function f_χ as a rational function; but rational functions can be extended to meromorphic functions on the whole complex plane, and the only possible poles are at the N-th roots of unity.

Our first task is the determination of the poles of Pell forms f_χ, which we know can only occur at the N-th roots of unity. A few calculations for Pell forms with small conductor show that f_χ does not have poles at each N-th root of unity. If we factor numerator and denominator of the rational function f_χ and cancel as many factors as possible, then we find, for small values of N:

$$f_{-3}(q) = \frac{q - q^2}{1 - q^3} = \frac{q(1 - q)}{(1 - q)(1 + q + q^2)} = \frac{q}{1 + q + q^2},$$

$$f_{-4}(q) = \frac{q - q^3}{1 - q^4} = \frac{q(1 - q^2)}{(1 - q^2)(1 + q^2)} = \frac{q}{1 + q^2},$$

$$f_5(q) = \frac{q - q^2 - q^3 + q^4}{1 - q^5} = \frac{q - q^3}{1 + q + q^2 + q^3 + q^4},$$

Already these few examples suggest that the poles of the function f_χ are exactly at the *primitive* N-th roots of unity. Here an N-th root of unity ζ is called primitive if the equation $\zeta^m = 1$ holds for $m = N$, but not for any smaller value $1 \le m < N$.

For proving this claim we proceed as in our derivation of Binet's formulas: We determine the partial fraction decomposition of f_χ. To this end we set

$$f_\chi(q) = \sum_{k=0}^{N-1} \frac{a_k}{\zeta^k - q};$$

then a simple application of Euler's formulas (2.4) shows that the coefficients a_k are given by

$$a_k = \frac{\mathrm{Fek}_\chi(\zeta^k)}{-N\zeta^{k(N-1)}} = -\frac{\zeta^k \, \mathrm{Fek}_\chi(\zeta^k)}{N}.$$

The expression

$$\mathrm{Fek}_\chi(\zeta^k) = \sum_{n=1}^{N-1} \chi(n)\zeta^{kn} =: \tau_k(\chi)$$

is called a *Gauss sum*. If χ is a quadratic Dirichlet character, then τ is called a *quadratic Gauss sum*. Gauss sums are important tools in number theory; in our approach, these objects show up naturally.

Thus we have

$$f_\chi(q) = -\frac{1}{N} \sum_{k=1}^{N-1} \frac{\zeta^k \tau_k(\chi)}{q - \zeta^k} = \frac{1}{N} \sum_{k=1}^{N-1} \frac{\tau_k(\chi)}{1 - q\zeta^{-k}}. \qquad (10.2)$$

It is clear that $f_\chi(q)$ has a pole in $q = \zeta^k$ if and only if $\tau_k(\chi) \ne 0$. The question of the location of the poles of Pell forms thus boils down to determining the values of k for which the quadratic Gauss sums $\tau_k(\chi)$ vanish.

It turns out that the quadratic Gauss sums τ_k are, up to a root of unity, equal to $\tau = \tau_1(\chi)$. In fact we have:

Proposition 10.12 *For primitive Dirichlet characters defined modulo N and all natural numbers k we have*

$$\tau_k(\chi) = \overline{\chi}(k) \cdot \tau, \qquad (10.3)$$

where $\overline{\chi}$ is the conjugate character of χ, which is defined by $\overline{\chi}(a) = \overline{\chi(a)}$.
In particular, we have $\tau_k(\chi) = 0$ if $\gcd(k, N) \neq 1$.

Proof Assume first that $\gcd(k, N) = 1$. Then

$$\tau_k(\chi) = \sum_{a=1}^{N-1} \chi(a)\zeta^{ka} = \overline{\chi}(k) \sum_{a=1}^{N-1} \chi(a)\zeta^{ka} = \overline{\chi}(k) \sum_{b=1}^{N-1} \chi(b)\zeta^{b} = \overline{\chi}(k)\,\tau,$$

where we have used that $b = ka$ runs through all coprime residue classes of $(\mathbb{Z}/N\mathbb{Z})^\times$ when a does.

If $\gcd(k, N) = d > 1$, on the other hand, then we write $N = dn$ and $k = ds$ for coprime integers n and s. We first claim that there exists an integer $b \equiv 1 \bmod n$ with $\chi(b) \neq 1$. Since χ is primitive, χ is not trivial on the kernel of the projection map $(\mathbb{Z}/N\mathbb{Z})^\times \longrightarrow (\mathbb{Z}/n\mathbb{Z})^\times$, and this is exactly what we have claimed.

Next we have $k \equiv bk \bmod N$ since $bk - k = k(b-1) \equiv 0 \bmod dn$; in particular, we have $\zeta^k = \zeta^{bk}$. Now we get

$$\chi(b)\tau_k(\chi) = \sum_{a=1}^{N-1} \chi(ab)\zeta^{ka} = \sum_{a=1}^{N-1} \chi(ab)\zeta^{kab} = \tau_k(\chi),$$

and since $\chi(b) \neq 1$ we obtain $\tau_k(\chi) = 0$ as claimed. $\qquad\square$

This simple result implies

Theorem 10.13 *The partial fraction decomposition of $f_\chi(q)$ is given by*

$$f_\chi(q) = -q \cdot \frac{\tau}{N} \sum_{k=1}^{N-1} \frac{\chi(k)}{q - \zeta^k} = \frac{\tau}{N} \sum_{k=1}^{N-1} \frac{\overline{\chi}(k)}{1 - q\zeta^{-k}}. \qquad (10.4)$$

Observe that this implies $\tau \neq 0$ since we already know that $f_\chi(q)$ is a nontrivial rational function.

Before we continue, let us give two simple examples of Gauss sums.

- Consider the Dirichlet character $\chi(a) = \left(\frac{2}{a}\right)$ defined modulo 8. If ζ denotes a primitive 8th root of unity, then

$$\tau_1(\chi) = \zeta - \zeta^3 - \zeta^5 + \zeta^7 = \zeta(1 - i + 1 - i) = \zeta(2 - 2i),$$

which implies that $\tau_1^2 = i(2 - 2i)^2 = 8$.

- Now let $\chi(a) = (\frac{a}{5})$ denote the quadratic Dirichlet character defined modulo 5, and let ζ denote a primitive 5th root of unity. Then

$$\tau_1 = \zeta - \zeta^2 - \zeta^3 + \zeta^4$$

and thus

$$\tau_1^2 = \zeta^2 + \zeta^4 + \zeta^6 + \zeta^8 - 2\zeta^3 - 2\zeta^4 + 2\zeta^5 + 2\zeta^5 - 2\zeta^6 - 2\zeta^7$$

$$= \zeta + \zeta^2 + \zeta^3 + \zeta^4 4 - 2\zeta - 2\zeta^2 - 2\zeta^3 - 2\zeta^4$$

$$= -1 + 4 + 2 = 5,$$

where we have used $1 + \zeta + \zeta^2 + \zeta^3 + \zeta^4 = 0$ several times.

These calculations suggest that quadratic Gauss sums for primitive quadratic Dirichlet characters with conductor N have absolute value $|\tau| = \sqrt{N}$. From the many possible proofs we choose one that uses the partial fraction decomposition of f_χ.

Theorem 10.14 *For primitive Dirichlet characters with conductor N we have $|\tau| = \sqrt{N}$. If χ is a quadratic character, then we even have*

$$\tau^2 = \Delta \tag{10.5}$$

for a discriminant Δ with $|\Delta| = N$.

Among the many possibilities of proving this theorem we choose the one based on Pell forms. According to (10.2) the partial fraction decomposition of the Pell form $f_\chi(q)$ is given by

$$f_\chi(q) = \frac{\tau}{N} \sum_{k=1}^{N-1} \frac{\overline{\chi}(k)}{1 - \zeta^{-k}q}. \tag{10.6}$$

Expanding the left side into a power series we get $\sum_{n\geq1} \chi(n)q^n$, and on the right side we obtain, when we develop the fractions into geometric series,

$$\frac{\overline{\chi}(k)}{1 - \zeta^{-k}q} = \overline{\chi}(k)(1 + \zeta^{-k}q + \zeta^{-2k}q^2 + \ldots) = \overline{\chi}(k) + \overline{\chi}(k)\zeta^{-k}q + \ldots.$$

Comparing the coefficients of q on both sides of (10.6) we find

$$1 = \frac{\tau}{N} \sum_{k=1}^{N} \overline{\chi}(k)\zeta^{-k} = \frac{\tau}{N} \overline{\sum_{k=1}^{N} \chi(k)\zeta^k} = \frac{\tau}{N} \cdot \overline{\tau},$$

hence $\tau\overline{\tau} = N$ and $|\tau| = \sqrt{N}$.

For proving the second claim we observe that, in the case of quadratic characters, we have $\chi = \overline{\chi}$, hence

$$\overline{\tau} = \sum_{k=1}^{N} \overline{\chi}(k)\zeta^{-k} = \sum_{k=1}^{N} \chi(k)\zeta^{-k} = \chi(-1)\tau.$$

Thus it follows from the proof above that $1 = \frac{\tau}{N} \cdot \chi(-1)\tau$, and so, taking Proposition 10.9 into account,

$$\tau^2 = \chi(-1)N = \Delta.$$

Since τ by definition is an element of $\mathbb{Q}(\zeta_N)$, this implies that each quadratic number field is a subfield of some cyclotomic number field $\mathbb{Q}(\zeta_N)$, and in fact that we can choose $N = |\Delta|$.

10.3 Fekete Polynomials

Gauss's sixth proof of the quadratic reciprocity law is today usually presented in the form given by Jacobi and Cauchy, who used the basic arithmetic of cyclotomic number fields. These proofs have the advantage of being very slick and short. Here we will present Gauss's original sixth proof of the quadratic reciprocity law in such a way that the role of the Fekete polynomials becomes clearly visible. The necessary changes are mainly of a cosmetic nature. Apart from Fekete polynomials, Gauss also uses the cyclotomic polynomial

$$\Phi_p(x) = 1 + x + x^2 + \ldots + x^{p-1} = \frac{x^p - 1}{x - 1}.$$

This polynomial is known to be irreducible over the rationals, as can be seen most easily using a method due[1] to Schönemann and Eisenstein: one shows that $\Phi_p(x+1)$ is an "Eisenstein polynomial," i.e., that the it has the form

$$\Phi_p(x + 1) = x^{p-1} + a_{p-2}x^{p-2} + \ldots + a_1 x + a_0,$$

where all coefficients a_j are divisible by p, and a_0 is not divisible by p^2. We now claim

Lemma 10.15 *Let p denote an odd prime number and n a natural number. Then*

$$\Phi_p(x^n) \equiv \begin{cases} 0 \mod \Phi_p(x) & \text{if } p \nmid n, \\ p \mod \Phi_p(x) & \text{if } p \mid n. \end{cases} \tag{10.7}$$

---

[1]See [26].

In fact we have

$$\frac{\Phi_p(x^n)}{\Phi_p(x)} = \frac{x^{np} - 1}{x^n - 1} \cdot \frac{x - 1}{x^p - 1}.$$

If $p \nmid n$, let m denote a natural number such that $mn \equiv 1 \bmod p$. With $mn = hp + 1$ it follows that

$$\frac{\Phi_p(x^n)}{\Phi_p(x)} = \frac{x^{np} - 1}{x^n - 1} \cdot \frac{x^{mn} - 1 + x - x^{hp+1}}{x^p - 1}$$

$$= \frac{x^{np} - 1}{x^p - 1} \cdot \frac{x^{mn} - 1}{x^n - 1} - \frac{x(x^{np} - 1)}{x^n - 1} \cdot \frac{x^{hp} - 1}{x^p - 1},$$

and this implies the claim.

If $p \mid n$, on the other hand, then $n = mp$ and

$$\Phi_p(x^n) - p = 1 + x^n + x^{2n} + \ldots + x^{n(p-1)} - (1 + 1 + \ldots + 1)$$

$$= x^n - 1 + x^{2n} - 1 + \ldots + x^{n(p-1)} - 1.$$

Clearly each term $x^{kn} - 1$ is divisible by $x^n - 1$, and from

$$\frac{x^n - 1}{x - 1} = \frac{x^{mp} - 1}{x - 1} = \frac{x^{mp} - 1}{x^p - 1} \cdot \frac{x^p - 1}{x - 1}$$

we deduce that it is divisible by $\Phi_p(x) = \frac{x^p - 1}{x - 1}$.

In the following, let $\mathrm{Fek}_p(x)$ be the Fekete polynomial for the primitive quadratic Dirichlet character with odd prime conductor p.

Lemma 10.16 *For every natural number $1 \le q < p$, the polynomial*

$$\mathrm{Fek}_p(x^q) - \left(\frac{q}{p}\right) \mathrm{Fek}_p(x)$$

is divisible by $x^p - 1$, that is, we have the congruence

$$\mathrm{Fek}_p(x^q) \equiv \left(\frac{q}{p}\right) \mathrm{Fek}_p(x) \bmod (x^p - 1)$$

in the polynomial ring $\mathbb{Z}[x]$.

Let ζ denote a primitive p-th root of unity. Then $\mathrm{Fek}_p(\zeta) = \tau$ and $\mathrm{Fek}_p(\zeta^q) = \tau_k(\chi)$ are quadratic Gauss sums. Thus plugging $x = \zeta$ into the identity

$$\mathrm{Fek}_p(x^q) - \left(\frac{q}{p}\right) \mathrm{Fek}_p(x) = g(x)(x^p - 1)$$

implies that

$$\tau_q(\chi) = \left(\frac{q}{p}\right)\tau.$$

This is just Eq. (10.3) in the special case where $\chi = \overline{\chi} = \left(\frac{\cdot}{p}\right)$.

Proof of Lemma 10.16 We have

$$\text{Fek}_p(x^q) = \sum_{a=1}^{p-1} \chi_p(a)x^{aq} = \chi_p(q) \sum_{a=1}^{p-1} \chi_p(aq)x^{aq}.$$

Thus if a runs through a coprime system of residue classes modulo p, then so does aq. Each exponent aq is thus congruent modulo p to exactly one number c with $1 \le c < p$, i.e., we have $aq = c + k_a p$ for an integer k_a depending on a. This implies

$$x^{aq} = x^{c+k_a p} = x^c x^{k_a p} = x^c + x^c(x^{k_a p} - 1) \equiv x^c \mod (x^p - 1),$$

hence

$$\text{Fek}_p(x^q) = \chi_p(q) \sum_{a=1}^{p-1} \chi_p(aq)x^{aq} = \chi_p(q) \sum_{c=1}^{p-1} \chi_p(c)x^c \mod (x^p - 1)$$

as claimed.   □

Next Gauss turns his attention to the polynomial $\text{Fek}_p(x)^2$. Clearly

$$\text{Fek}_p(x)^2 = \sum_{k=1}^{p-1} \left(\frac{k}{p}\right)x^k \text{Fek}_p(x).$$

According to Lemma 10.16 we have

$$\text{Fek}_p(x)^2 \equiv \sum_{k=1}^{p-1} x^k \text{Fek}_p(x^k) \mod (x^p - 1).$$

Now we develop the second Fekete polynomial and find

$$\text{Fek}_p(x)^2 \equiv \sum_{k=1}^{p-1} x^k \sum_{h=1}^{p-1} \left(\frac{h}{p}\right)x^{kh} = \sum_{h=1}^{p-1} \left(\frac{h}{p}\right) \sum_{k=1}^{p-1} x^{kh+k}$$

$$= \sum_{h=1}^{p-1} \left(\frac{h}{p}\right)(\Phi_p(x^{h+1}) - 1) = \sum_{h=1}^{p-1} \left(\frac{h}{p}\right)\Phi_p(x^{h+1}) \mod \Phi_p(x),$$

where we have used that $\sum(\frac{h}{p}) = 0$. Using Lemma 10.15 we now obtain

$$\mathrm{Fek}_p(x)^2 \equiv \left(\frac{p-1}{p}\right)\Phi_p(x^p) \equiv \left(\frac{-1}{p}\right)p \bmod \Phi_p(x).$$

We have proved

Proposition 10.17 *Fekete polynomials satisfy the congruence*

$$\mathrm{Fek}_p(x)^2 \equiv \left(\frac{-1}{p}\right)p \bmod \Phi_p(x).$$

If we set $x = \zeta$ for a primitive p-th root of unity ζ, then the congruence above turns into the equation

$$\tau^2 = \left(\tfrac{-1}{p}\right)\tau,$$

which is a special case of (10.5).

10.3.1   Gauss's Sixth Proof

The heart of the proof is simple: We combine the congruences

$$\mathrm{Fek}_p(x)^2 \equiv p^* \bmod \Phi_p(x), \tag{10.8}$$

$$\mathrm{Fek}_p(x)^q \equiv \mathrm{Fek}_p(x^q) \bmod q, \tag{10.9}$$

$$\mathrm{Fek}_p(x^q) \equiv \left(\frac{q}{p}\right)\mathrm{Fek}_p(x) \bmod \Phi_p(x), \tag{10.10}$$

that we have proved above. Instead of working with double congruences modulo q and modulo $\Phi_p(X)$ we write the congruences as equations—another possibility would be working modulo q in cyclotomic number fields. The congruences above then become the following equations:

$$\mathrm{Fek}_p(x)^2 = p^* + \Phi_p(x)A(x),$$

$$\mathrm{Fek}_p(x)^q = \mathrm{Fek}_p(x^q) + qB(x),$$

$$\mathrm{Fek}_p(x^q) = \left(\frac{q}{p}\right)\mathrm{Fek}_p(x) + \Phi_p(x)C(x).$$

Here $A, B, C \in \mathbb{Z}[x]$ are suitably chosen polynomials. Now

$$\mathrm{Fek}_p(x)^q = \left(\mathrm{Fek}_p(x)^2\right)^{\frac{q-1}{2}} \mathrm{Fek}_p(x) = \left(p^* + \Phi_p(x)A(x)\right)^{\frac{q-1}{2}} \mathrm{Fek}_p(x)$$

$$= (p^*)^{\frac{q-1}{2}} \mathrm{Fek}_p(x) + \Phi_p(x)A_1(x)\,\mathrm{Fek}_p(x)$$

$$= \left(\frac{p^*}{q}\right)\mathrm{Fek}_p(x) + qh\,\mathrm{Fek}_p(x) + \Phi_p(x)A_1(x)\,\mathrm{Fek}_p(x),$$

as well as

$$\mathrm{Fek}_p(x)^q = \mathrm{Fek}_p(x^q) + qB(x) = \left(\frac{q}{p}\right)\mathrm{Fek}_p(x) + \Phi_p(x)C(x) + qB(x).$$

Thus

$$\left(\frac{q}{p}\right)\mathrm{Fek}_p(x) - \left(\frac{p^*}{q}\right)\mathrm{Fek}_p(x) = qR(x) + \Phi_p(x)S(x)$$

for polynomials $R, S \in \mathbb{Z}[x]$. Our goal is showing that the polynomial on the left hand side is 0. To this end we first write

$$\mathrm{Fek}_p(x) = \varepsilon\Phi_p(x) + F(x)$$

with $\varepsilon = \left(\frac{-1}{p}\right)$ and some polynomial $F(x)$ of degree $\leq p - 2$.
    Next $F(0) = \mathrm{Fek}_p(0) - \varepsilon\Phi_p(0) = -\varepsilon$. Thus we have

$$\left[\left(\frac{q}{p}\right) - \left(\frac{p^*}{q}\right)\right]F(x) = qR(x) + \Phi_p(x)T(x).$$

Now we write $R(x) = \Phi_p(x)q(x) + r(x)$ for some polynomial r of degree $\leq p-2$, and we find

$$\left[\left(\frac{q}{p}\right) - \left(\frac{p^*}{q}\right)\right]F(x) - qr(x) = \Phi_p(x)U(x).$$

The polynomial on the left side has degree $\leq p - 2$ and is divisible by $\Phi_p(x)$. Since Φ_p is irreducible, this is only possible if the polynomial vanishes:

$$\left[\left(\frac{q}{p}\right) - \left(\frac{p^*}{q}\right)\right]F(x) - qr(x) = 0.$$

Plugging in $x = 0$ yields

$$\left[\left(\frac{q}{p}\right) - \left(\frac{p^*}{q}\right)\right] F(0) - qr(0) = -\varepsilon\left[\left(\frac{q}{p}\right) - \left(\frac{p^*}{q}\right)\right] - qr(0) = 0,$$

and since $\varepsilon = \pm 1$ is not divisible by q, the expression in the bracket must be a multiple of q But since $q > 2$ this is only possible if the two Legendre symbols coincide.

10.4   The Analytic Class Number Formula

In this last section we will sketch possible extensions of our investigations. We have already seen that the factor $1 - q$ of the Fekete polynomial may be canceled with the corresponding factor in $1 - q^N$. This fact allows us to determine the value $f_\chi(1)$ (see Table 10.2).

The fact that $f_\chi(1) = 0$ for $\Delta > 0$ follows immediately from the functional equation of f_χ (see Exercise 10.9). The values for negative discriminants are mysterious; if we extend the table far enough, then it turns out that, for negative discriminants $\Delta < -3$, the value $f_\chi(1)$ is related to the class number of $\mathbb{Q}(\sqrt{\Delta})$ in a very simple and striking way:

Theorem 10.18 *We have $f_\chi(1) = 0$ if and only if the unit group of the quadratic number field with discriminant Δ has rank 1, i.e., if and only if the Pell equation $T^2 - \Delta U^2 = 4$ has a nontrivial solution.*

Table 10.2 The values $f_\chi(1)$

Δ	N	$\chi(-1)$	$f_\chi(q)$	$f_\chi(1)$
-3	3	-1	$\dfrac{q}{1+q+q^2}$	$\frac{1}{3}$
-4	4	-1	$\dfrac{q}{1+q^2}$	$\frac{1}{2}$
5	5	$+1$	$\dfrac{q-q^3}{1+q+q^2+q^3+q^4}$	0
-7	7	-1	$\dfrac{q+2q^2+q^3+2q^4+q^5}{1+q+q^2+q^3+q^4+q^5+q^6}$	1
8	8	$+1$	$\dfrac{q-q^3}{1+q^4}$	0
-8	8	-1	$\dfrac{q+q^3}{1+q^4}$	1
12	12	$+1$	$\dfrac{q-q^3}{1-q^2+q^4}$	0
-15	15	$+1$	$\dfrac{q-q^3+2q^4-q^5+q^7}{1-q+q^3-q^4+q^5-q^7+q^8}$	2

If $\Delta < 0$, on the other hand, then

$$f_\chi(1) = \frac{2h}{w} = \frac{h}{w/2} = \frac{(\#Cl(K) : \#Cl(\mathbb{Q}))}{(\#W_K : \#W_\mathbb{Q})}, \tag{10.11}$$

where $h = \#Cl(K)$ denotes the class number, w the number of roots of unity in $K = \mathbb{Q}(\sqrt{\Delta}\,)$, and W_K the group of roots of unity in K.

Observe that $Cl(\mathbb{Q}) = 1$ since $\mathbb{Z}$ has unique factorization, and that $W_\mathbb{Q} = \{\pm1\}$, hence $\#W_\mathbb{Q} = 2$.

The expression on the right shows that the formula $f_\chi(1) = \frac{2h}{w}$ is actually a relative class number for the quadratic extension $K/\mathbb{Q}$ and beautifully explains the occurrence of the factor 2 in the numerator.

The investigation of the generating functions of Kronecker and Dirichlet characters has led us into rather deep waters. Although the terms in (10.11) all are closely related to the arithmetic of number fields, the natural proof of this equation uses analytic methods.

In this proof, a central role is played by Dirichlet L-series, which Dirichlet had also used for proving his theorem on primes in arithmetic progression. L-series provide a second possibility of writing down a generating function for Dirichlet characters χ, which is different from the Pell form, which is a power series. We set

$$L(s, \chi) = \sum_{n=1}^{\infty} \chi(n)n^{-s}$$

and then show that this series converges absolutely for all $s > 1$.

By manipulating divergent series without fear and evaluating the L-series $L(s, \chi)$ at places where it is not defined we find

$$f_\chi(1) = \lim_{q \to 1} \sum_{n \geq 1} \chi(n)q^n = \sum_{n \geq 1} \chi(n) = \lim_{s \to 0} \sum_{n \geq 1} \chi(n)n^{-s} = L(0, \chi).$$

We can assign a value to the meaningless expression $L(0, \chi)$ by extending the function analytically to the whole complex plane. This function then satisfies a functional equation relating the values of the L-series at s and $1 - s$; in particular, it allows to compute $L(0, \chi)$ from $L(1, \chi)$. It is rather easy to see that the series $L(1, \chi)$ converges conditionally for all quadratic Dirichlet characters with conductor $N > 1$. For $\chi = \chi_{-4}$, for example, we have, according to Leibniz,

$$L(1, \chi) = 1 - \frac{1}{3} + \frac{1}{5} - \frac{1}{7} + \ldots = \frac{\pi}{4}.$$

There is a connection between this Leibniz series and Pell forms: Clearly

$$F_\chi(q) = q - \frac{q^3}{3} + \frac{q^5}{5} - \frac{q^7}{7} + \dots$$

is a primitive of

$$1 - q^2 + q^4 - q^6 + \dots = \frac{1}{1 + q^2} = \frac{f_\chi(q)}{q},$$

and this can be done for arbitrary Dirichlet characters since we have

$$\int_0^1 \frac{f_\chi(q)}{q} \, dq = \int_0^1 \sum_{n \geq 1} \chi(n) q^{n-1} \, dq = \sum_{n \geq 1} \chi(n) \int_0^1 q^{n-1} \, dq = \sum_{n \geq 1} \frac{\chi(n)}{n} = L(1, \chi),$$

where we once more point out that interchanging the order of taking limits requires a proof.

For $\chi = \chi_8$, for example, we obtain

$$L(1, \chi) = 1 - \frac{1}{3} - \frac{1}{5} + \frac{1}{7} + \dots = \int_0^1 \frac{1 - q^2}{1 + q^4} \, dq.$$

A numerical integration shows that

$$L(1, \chi) \approx 0.31161262007011525669701004 \approx \frac{\log(1 + \sqrt{2})}{\sqrt{8}}.$$

The fact that $L(1, \chi) = \frac{\log(1 + \sqrt{2})}{\sqrt{8}}$ can be proved using the partial fraction decomposition of $\frac{f_\chi(q)}{q}$.

For arbitrary quadratic Dirichlet character we have the following important and deep

Theorem 10.19 *Let K be a quadratic number field with discriminant Δ and class number h, and let χ be the quadratic Dirichlet character defined modulo $N = |\Delta|$ which is attached to K. Moreover, let $\varepsilon > 1$ denote the fundamental unit of the real quadratic number field K if $\Delta > 0$. Then*

$$L(1, \chi) = \begin{cases} \frac{2\pi h}{w\sqrt{N}}, & \text{if } \Delta < 0, \\ \frac{h \log \varepsilon}{\sqrt{N}}, & \text{if } \Delta > 0, \end{cases}$$

where the value of the L-series on the left is given by

$$L(1, \chi) = \sum_{n=1}^{\infty} \frac{\chi(n)}{n} = \int_0^1 \frac{f_\chi(q)}{q} \, dq.$$

These class number formulas, which underline again the central importance of the Pell forms f_χ and the Fekete polynomials for the arithmetic of quadratic number fields, are due to Dirichlet, who proved them for quadratic forms rather than for quadratic number fields. The integral representation of $L(1, \chi) = \int_0^1 \frac{f_\chi(q)}{q} \, dq$ may be transformed via the partial fraction decomposition of f_χ into a finite sum, which has a certain charm, but is not very well suited for the computation of class numbers except for small discriminants.

Dirichlet's main motivation for working out the class number formula was the obvious corollary that $L(1, \chi) \neq 0$ for all quadratic Dirichlet characters. Since the corresponding claim for Dirichlet characters that attain nonreal values may be proved rather easily, Dirichlet obtained that $L(1, \chi) \neq 0$ for all Dirichlet characters modulo N. This in turn quickly implies (by an idea going back to Euler) that for any pair of coprime integers a and N there exist infinitely many prime numbers p with $p \equiv a \bmod N$. This is Dirichlet's famous theorem on primes in arithmetic progression.

For a proof of these results we refer the reader to the wonderful books by Scharlau and Opolka [111] and by Zagier [134].

10.5 Modularity

Euler's Modularity Theorem is not a reciprocity law in the sense of Legendre because it does not connect the solvability of the congruence $x^2 \equiv p \bmod q$ with that of $x^2 \equiv q \bmod p$. But, as Kronecker has made clear, the Modularity Theorem is more fundamental than the reciprocity law because higher reciprocity laws are governed by modularity.

Already in the rational integers, the Modularity Theorem has certain advantages. For example, the formulation of Legendre's reciprocity law requires two supplementary laws for computing the symbols $\left(\frac{-1}{p}\right)$ and $\left(\frac{2}{p}\right)$; the Modularity Theorem, on the other hand, also holds for $a = -1$ and $a = \pm 2$.

The generalization of Legendre's quadratic reciprocity law from rational integers to algebraic integers in general number fields turns out to be very difficult, since there is no simple formula for the inversion factor $\left(\frac{\alpha}{\beta}\right)\left(\frac{\beta}{\alpha}\right)$. More seriously, Legendre's reciprocity law only makes sense for principal ideals, which restricts the applicability of the reciprocity law considerably; in particular, the reciprocity law cannot be used directly for computing power residue symbols of the form $\left(\frac{\alpha}{\mathfrak{p}}\right)$ for nonprincipal ideals $\mathfrak{p}$.

The generalization of the Modularity Theorem to arbitrary number fields is rather straightforward. Essentially we have $(\frac{\alpha}{\beta}) = (\frac{\alpha}{\gamma})$ if $\beta \equiv \gamma$ mod 4α and if certain sign conditions are satisfied. The modulus 4α may often be replaced by a smaller one, for example, the relative discriminant of the quadratic extension $K(\sqrt{\alpha})/K$.

In addition, the Modularity Theorem may be extended to arbitrary ideals coprime to 2α. In fact we have $(\frac{\alpha}{\mathfrak{b}}) = (\frac{\alpha}{\mathfrak{c}})$ if $\mathfrak{b} \equiv \mathfrak{c}$ mod 4α, by which we mean that $\mathfrak{a}$ and $\mathfrak{b}$ are coprime to (2α), and that the ideal $\mathfrak{b}\mathfrak{c}^{-1} = (\delta)$ is principal and generated by an element $\delta \equiv 1$ mod 4α. Again there are a couple of sign conditions that must be observed.

We also remark that the most general reciprocity law for abelian extensions, namely Artin's reciprocity law, is by its nature a modularity theorem. Similar remarks apply to generalizations to non-abelian extensions inside Langlands' program, and for corresponding results in the theory of elliptic curves.

10.5.1   Modularity of Polynomials

Let $f \in \mathbb{Z}[x]$ be a monic polynomial with integral coefficients. We denote by $\mathrm{Spl}(f)$ the set of all prime numbers p not dividing the discriminant disc f such that f is a product of linear factors when considered over the ring $\mathbb{F}_p[x]$. For example, the polynomial $f(x) = x^2 - 2$ splits into two linear factors modulo all primes $p \equiv \pm 1$ mod 8.

We say that such a polynomial f is modular[2] if there exists a natural number N such that the set of primes $\mathrm{Spl}(f)$ can be described (up to at most finitely many exceptions) by congruence relations modulo N. As an example consider $f(x) = x^3 - 3x + 1$. This polynomial is a cube modulo 3 since $f(x) \equiv (x+1)^3$ mod 3, and it is easily checked that f does not have roots modulo primes $5 \le p \le 13$ and therefore is irreducible modulo these primes. For $p = 17$, however, we have $f(x) \equiv (x+3)(x+4)(x-1)$ mod 17. If we continue these calculations we are led to suspect that f splits into three distinct linear factors if and only if $p \equiv \pm 1$ mod 9.

The reason for this behavior of f modulo p has to do with the fact that the roots of f are elements of the field of 9th roots of unity. In fact, if ζ is a primitive 9th root of unity, then $\zeta^6 + \zeta^3 + 1 = 0$; setting $\alpha = \zeta + \zeta^{-1}$ we then find

$$\alpha^3 = \zeta^3 + 3\zeta + 3\zeta^{-1} + \zeta^{-3} = \zeta^3 + \zeta^6 + 3\alpha = 3\alpha - 1,$$

hence α is a root of f. Thus the roots of f generate a cubic subfield of $\mathbb{Q}(\zeta)$, and this implies that f is a polynomial with an abelian Galois group.

---

[2]In the theory of complex multiplication there exists something called "the modular polynomial."

The ring homomorphism σ sending ζ to ζ^2 is an automorphism of $\mathbb{Q}(\zeta)$, and we find

$$\sigma(\alpha) = \sigma(\zeta + \zeta^{-1}) = \zeta^2 + \zeta^{-2} = \alpha^2 - 2.$$

This map permutes the roots of f and makes the fact that f has an abelian Galois group explicit.

The classification of modular polynomials is achieved by class field theory; the result is

Theorem 10.20 *A polynomial is modular if and only if its Galois group is abelian.*

The Galois group of a polynomial is abelian if and only if its splitting field (up to isomorphism, the smallest extension of $\mathbb{Q}$ containing all the roots of f) is abelian.

The "finitely many exceptions" have to do with the choice of f. Clearly $f(x) = x^2 - x - 1$ and $g(x) = x^2 - 5$ have the same splitting behavior for every odd prime since $4f(x) = (2x - 1)^2 - 5 = g(2x - 1)$. But f is irreducible modulo 2 (which corresponds to the fact that 2 is inert in $\mathbb{Q}(\sqrt{5})$), yet $x^2 - 5 \equiv (x + 1)^2 \bmod 5$. Note that disc $f = 5$ and disc $g = 20$.

10.5.2 Modularity of Number Fields

Observe, for example, that $\mathbb{Q}(\sqrt{-3}) = \mathbb{Q}(\zeta_3)$ and $\mathbb{Q}(i) = \mathbb{Q}(\zeta_4)$ are fields of roots of unity. The quadratic number fields $\mathbb{Q}(\sqrt{2})$ and $\mathbb{Q}(\sqrt{-2})$ are contained in the field of eighth roots of unity since $\zeta_8 = \frac{\sqrt{2}+\sqrt{-2}}{2}$, which implies that $\sqrt{-2} = \zeta_8 + \zeta_8^3$ and $\sqrt{2} = \zeta_8 + \zeta_8^{-1}$. In all these examples, the quadratic number field with discriminant Δ is contained in the field of N-th roots of unity with $N = |\Delta|$. This is no coincidence: Let us call a quadratic number field with discriminant Δ modular if it is contained in some field of N-th roots of unity. We may (and will) assume in addition that $N \not\equiv 2 \bmod 4$ since $\mathbb{Q}(\zeta_{2m+1}) = \mathbb{Q}(\zeta_{2(2m+1)})$.

Proposition 10.21 *If the quadratic number field k is contained in $\mathbb{Q}(\zeta_m)$ and $\mathbb{Q}(\zeta_n)$, then it is also contained in $\mathbb{Q}(\zeta_{\gcd(m,n)})$.*

If k is contained in both fields, then it is contained in their intersection. Thus the claim follows from the observation

$$\mathbb{Q}(\zeta_m) \cap \mathbb{Q}(\zeta_n) = \mathbb{Q}(\zeta_{\gcd(m,n)}),$$

which we do not prove here.

The smallest positive integer N for which the quadratic number field k is contained in $\mathbb{Q}(\zeta_N)$ is called the conductor of k.

Theorem 10.22 *Each quadratic number field is contained in some cyclotomic field. In fact, if $\Delta = \mathrm{disc}\, k$ and $N = |\Delta|$, then N is the conductor of k.*

   This follows by writing the discriminant of k as a product of prime discriminants. It is a very special case of a more general result. Let us call a number field K *modular* if there is a natural integer N such that $K \subset \mathbb{Q}(\zeta_N)$. Then we have

Theorem 10.23 (Theorem of Kronecker-Weber) *A number field K is modular if and only if it is a Galois extension of $\mathbb{Q}$ with abelian Galois group.*

This theorem was conjectured by Kronecker, who claimed to have a partial proof. The first published proofs are due to Weber and Hilbert, and nowadays there are many different proofs.

10.5.3 Pell Forms

Let us now call the Pell form f_κ of a Kronecker symbol $\kappa = (\frac{\Delta}{\cdot})$ *modular* if the following conditions are satisfied:

- There exist polynomials A, $B \in \mathbb{Z}[q]$, with B monic, such that $f_\kappa(q) = \pm\frac{A(q)}{B(q)}$;
- f_κ satisfies a functional equation of the form $f_\kappa(\frac{1}{q}) = \pm f_\kappa(q)$ for some choice of the sign.

We say that f_κ is *strongly modular* if we can choose $B(q) = q^N - 1$ for $N = |\Delta|$. The following theorem tells us that the modularity of the Kronecker symbol is a consequence of analytic properties of the associated Pell forms:

Theorem 10.24 *The modularity of f_κ implies the modularity of the Kronecker symbol $\kappa(p) = (\frac{\Delta}{p})$.*

   Assume that $f_\kappa(q) = \frac{A(q)}{B(q)}$ is rational and satisfies the functional equation $f_\kappa(\frac{1}{q}) = \pm f_\kappa(q)$. Since f_κ converges absolutely inside the unit circle, f_κ does not have any poles there. By the functional equation, it cannot have any poles outside the unit circle. Thus the rationality and the functional equation imply that f_κ has all its poles on the unit circle.
   Since $B(q)$ is monic, the poles of f_κ must be algebraic integers. Thus if f_κ has a pole in $q = \zeta$, then ζ and all of its conjugates lie on the unit circle. Now we invoke the following result due to Kronecker:

Proposition 10.25 (Kronecker) *If η is an algebraic integer with the property that all of its conjugates lie on the unit circle, then η is a root of unity.*

   Let η be a root of a monic polynomial with degree n. By Dirichlet's pigeonhole principle, there exist natural numbers $r < s$ such that $|\eta^s - \eta^r| < 2^{-n}$. The conjugates η_j^k of η^k all lie on the unit circle, hence $|\eta_j^s - \eta_j^r| \le 2$. Since the norm of an algebraic number is the product of its conjugates (see Exercise 2.46), we have $|N(\eta^s - \eta^r)| < 2^{-n}2^{n-1} = \frac{1}{2}$. Since η is an algebraic integer, its norm is a rational integer, and we conclude that its norm is 0. But then $\eta^s = \eta^r$, hence $\eta^{s-r} = 1$, and this implies that η is a root of unity.

Here is a second proof based on a similar idea: Let η be an algebraic integer, that is, a root of a monic polynomial of degree n. Then η^k is an algebraic integer of degree $m \leq n$, and its minimal polynomial is

$$(x - \eta_1^k)(x - \eta_2^k) \cdots (x - \eta_m^k) = x^m + a_{m-1}x^{m-1} + a_1x + a_0 \in \mathbb{Z}[x],$$

where $\eta_1^k, \ldots, \eta_m^k$ are the conjugates of η^k. Clearly

$$a_{m-1} = \eta_1^k + \eta_2^k + \ldots + \eta_m^k,$$
$$a_{m-1} = \eta_1^k\eta_2^k + \eta_1^k\eta_3^k + \ldots + \eta_{m-1}^k\eta_m^k,$$
$$\ldots = \ldots,$$
$$a_1 = \eta_1^k \cdots \eta_{m-1}^k + \ldots + \eta_2^k \cdots \eta_m^k,$$
$$a_0 = \eta_1^k\eta_2^k \cdots \eta_m^k.$$

Since the absolute values of the η_j^k are $= 1$, this implies that

$$|a_{m-1}| \leq m, \ |a_{m-2}| \leq \binom{m}{2}, \ \ldots, \ |a_k| \leq \binom{m}{k}, \ \ldots, \ |a_1| \leq m, \ |a_0| = 1.$$

These bounds show that there are only finitely many such polynomials, hence there must exist natural numbers $r < s$ with $\eta^r = \eta^s$, and this implies as above that η is a root of unity.

Now we can finish the proof of Theorem 10.24. Since we have just shown that the poles of f_κ are roots of unity, we can choose an integer N such that the poles are roots of $x^N - 1$. Then we can write

$$f_\kappa(q) = \frac{C(q)}{1 - q^N},$$

where the rational function is not necessarily in written in lowest terms.

The functional equation tells us that $f(\frac{1}{q}) = \pm f(q)$. Since $f(0) = 0$ and f is continuous at $q = 0$, this implies that $f(x) \longrightarrow 0$ for $x \to \infty$, and this implies that $\deg A < \deg B$ and therefore $\deg C < N$.

If we now compare the power series expansion

$$f_\kappa(q) = \frac{C(q)}{1 - q^N} = C(q) + C(q)q^N + C(q)q^{2N} + \ldots$$

with the definition of $f_\kappa(q)$ we find that $\kappa(m)$ only depends on the value of m modulo N, and this finally shows that κ is modular.

10.6   Modularity of Elliptic Curves

The notion of modularity is essential for understanding quadratic (and higher) reciprocity, but it was developed in a different area of mathematics (if we disregard the early insights of Euler and Kronecker). The idea that all elliptic curves $y^2 = f(x)$ with a cubic polynomial $f \in \mathbb{Z}[x]$ should be modular goes back to the Japanese mathematicians Yutaka Taniyama and Goro Shimura.

   We only present the material in this section in order to emphasize the importance of the notion of modularity in modern number theory. We also remark in passing that the modularity of elliptic curves was used in an essential way by Andrew Wiles in his proof of Fermat's Last Theorem for arbitrary exponents $p \geq 5$.

10.6.1   Group Law

The affine points on an elliptic curve $E : y^2 = x^3 + ax + b$, together with the point at infinity, carry a group law that has a geometric interpretation: Given two affine points P and Q, the line through P and Q (or the tangent in P if $P = Q$) intersects the elliptic curve in a third point R; the reflection of R at the x-axis is, by definition, the sum $P \oplus Q$. The neutral element is the point at infinity. Checking the axioms of the group structure is easy except for associativity (Fig. 10.1).

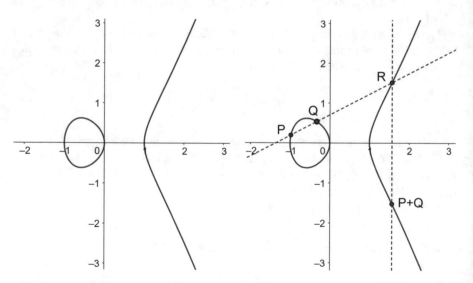

Fig. 10.1   The elliptic curve $y^2 = x^3 - x$ over the reals and the group law

10.6.2 Curves with Complex Multiplication

Let us consider the elliptic curve $E : y^2 = x^3 - x$. This curve is said to have complex multiplication by $\mathbb{Z}[i]$; we will be content with mentioning that this has to do with the fact that the substitutions $y = iy_1$ and $x = -x_1$ leave the equation of E invariant.

We now count the number of $\mathbb{F}_p$-rational points on this elliptic curve. For $p = 5$, for example, there are the following $\mathbb{F}_5$-rational points on E besides the point at infinity:

$$E(\mathbb{F}_5) = \{\infty, (0, 0), (1, 0), (2, \pm 1), (3, \pm 2), (-1, 0)\}.$$

Thus $N_5 = \#E(\mathbb{F}_5) = 8$.

Similar calculations for other odd primes yield the following table:

p	3	5	7	11	13	17	19	23	29	31	37	41
N_p	4	8	8	12	8	16	20	24	40	32	40	32

This table suggests that $N_p = p + 1$ for primes $p \equiv 3 \bmod 4$. For understanding what is going on for primes $p \equiv 1 \bmod 4$ we write $N_p = p + 1 - a_p$. Then $a_p = 0$ for primes $p \equiv 3 \bmod 4$; and for primes $p \equiv 1 \bmod 4$ we obtain

p	5	13	17	29	37	41
a_p	-2	6	2	-10	-2	10

The pattern becomes visible if we write these primes p as sums of two squares: In fact, if $p = a^2 + b^2$, where $a \equiv 1 \bmod 4$, then $a_p = -2a$. We have already proved this result in Theorem 3.31.

10.6.3 Hasse's Theorem

It follows from the formula $N_p = p + 1 - 2a$, where $E : y^2 = x^3 - x$ and p is an odd prime $p = a^2 + b^2$ with $a \equiv 1 \bmod 4$, that $|N_p - (p + 1)| = 2a \leq 2\sqrt{p}$. The fact that this bound holds for all elliptic curves $y^2 = x^3 + ax + b$ defined over $\mathbb{Q}$ was proved by Helmut Hasse in the 1930s. Hasse's theorem can be interpreted as a Riemann conjecture for the zeta function attached to the elliptic curve over $\mathbb{F}_p$. For the history of this result, see [110].

10.6.4   Modularity of Elliptic Curves

It is rather difficult to explain the modularity of elliptic curves from scratch.[3] As for Pell conics, the main content is the existence of a modulus N for each elliptic curve E such that the values $a_p = p + 1 - N_p$, where $N_p = \#E(\mathbb{F}_p)$ is the number of $\mathbb{F}_p$-rational points on the elliptic curve E, is determined by what is called a *modular form* on $\Gamma_0(N)$.

Consider the following example, which is essentially already contained in Shimura's work. Let $E : y^2 - y = x^3 - x^2$ be an elliptic curve; its discriminant is $\Delta = -11$. Consider the function

$$f(q) = q \prod_{n=1}^{\infty} (1 - q^n)^2 (1 - q^{11n})^2 = \sum_{n=1}^{\infty} a_n q^n.$$

This is a cusp form of weight 2 living on $\Gamma_0(11)$. What this means is that this function satisfies a functional equation of the following form: Set $q = e^{2\pi i z}$ and interpret f as a function of the complex variable z. Then

$$f\left(\frac{az + b}{cz + d}\right) = (cz + d)^2 f(z)$$

for all matrices $\begin{pmatrix} a & b \\ c & d \end{pmatrix}$ with $a, b, c, d \in \mathbb{Z}$, determinant $ad - bc = 1$ and $c \equiv 0 \bmod 11$. Since $\begin{pmatrix} 1 & 1 \\ 0 & 1 \end{pmatrix} \in \Gamma_0(N)$ for every N, we have in particular $f(z+1) = f(z)$; but this was clear from the fact that f is a function of $q = e^{2\pi i z}$.

A simple calculation yields

$$f(q) = q - 2q^2 - q^3 + 2q^4 + q^5 + 2q^6 - 2q^7 - 2q^9 - 2q^{10} + q^{11}$$
$$- 2q^{12} + 4q^{13} + 4q^{14} - q^{15} - 4q^{16} - 2q^{17} + 4q^{18} + 2q^{20} + \cdots$$

If we compute the numbers $a_p = p + 1 - N_p$ for E and some small primes p, we obtain

p	2	3	5	7	11	13	17	19
a_p	−2	−1	1	−2	−1	4	−2	0

The pattern now is clear: For all prime numbers p not dividing the modulus $N = 11$, the coefficient a_p in the power series expansion of $f(q)$ coincides with the number $a_p = p + 1 - N_p$ that determines the number of $\mathbb{F}_p$-rational points on E.

---

[3]I highly recommend the books [4, 5] to everyone interested in learning more about the big picture.

And there is more: Define the Tribonacci numbers T_n by $T_0 = 0$, $T_1 = T_2 = 1$ and $T_{n+3} = T_{n+2} + T_{n+1} + T_n$ for $n \geq 3$. Then the following theorem[4] holds:

Theorem 10.26 *For all primes $p \neq 2, 11, 19$, the following assertions are equivalent:*

(1) $T_{p-1} \equiv 0$ mod p;
(2) $p = x^2 + 11y^2$;
(3) a_p is even and $(\frac{p}{11}) = +1$.

The first few primes p for which $p \mid T_{p-1}$ are $p = 19, 47, 53, 103$ and 163. The prime $p = 19$ is exceptional, and the others are represented by $x^2 + 11y^2$:

$$47 = 6^2 + 11, \quad 53 = 3^2 + 11 \cdot 2^2, \quad 103 = 2^2 + 11 \cdot 3^2, \quad 163 = 8^2 + 11 \cdot 3^2.$$

The Tribonacci numbers can be expressed explicitly using the roots of the cubic polynomial $f(x) = x^3 - x^2 - x - 1$; the corresponding elliptic curve $E_2 : y^2 = x^3 - x^2 - x - 1$ has discriminant $\Delta_2 = -2^6 \cdot 11$ and is a "quadratic twist" of the elliptic curve E above. In fact, consider the elliptic curve $E : y^2 - y = x^3 - x^2$. Multiplying through by 4 and completing the square shows that $(2y - 1)^2 = 4x^3 - 4x^2 + 1$, hence

$$[\sqrt{-2}(2y - 1)]^2 = -8x^3 + 8x^2 - 2 = (-2x)^3 + 2(-2x)^2 - 2.$$

Setting $Y = \sqrt{-2}(2y - 1)$ and $\xi = -2x$ we obtain the equation $Y^2 = \xi^3 + 2\xi^2 - 2$. Finally, setting $\xi = X - 1$ gives $E_2 : Y^2 = X^3 - X^2 - X - 1$.

This implies that both curves have the same number $\mathbb{F}_p$-rational points if $(\frac{-2}{p}) = +1$; if $(\frac{-2}{p}) = -1$, on the other hand, then the number of points is $N_p = p+1-a_p$ on one and $N_p = p+1+a_p$ on the other curve. In all cases, $a_p(E) = (\frac{-2}{p}) \cdot a_p(E_2)$. This in turn implies that the condition $a_p \equiv 0$ mod 2 in Theorem 10.26 is equivalent to the character sum

$$S = \sum_{x=0}^{p-1} \left(\frac{x^3 - x^2 - x - 1}{p} \right)$$

being odd. I do not know whether there is a link between the value of S and the representations of p in the form $x^2 + 11y^2$.

These examples are only the tip of a massive iceberg. Following the breakthrough by Wiles it was shown that every elliptic curve defined over $\mathbb{Q}$ is modular. The modularity theorems for quadratic number fields (and actually for all abelian number fields) and for elliptic curves are pieces of a large area of conjectures due to Robert Langlands.

---

[4]See Evink and Helminck [40].

10.7 Exercises

10.1. Let A and B be finite abelian groups. Show that the set $A \oplus B$ of all pairs (a, b) with $a \in A$ and $b \in B$ becomes a group by setting $(a_1, b_1) \cdot (a_2, b_2) = (a_1 a_2, b_1 b_2)$.
This is a purely formal exercise. If 1_A and 1_B denote the neutral elements of A and B, then $(1_A, 1_B)$ is the neutral element of $A \oplus B$. The inverse element of (a, b) is (a^{-1}, b^{-1}), and associativity is directly inherited from A and B.

10.2. Let A and B be finite abelian groups. Show that each subgroup of $A \oplus B$ has the form $A_1 \oplus B_1$, where A_1 and B_1 are subgroups of A and B, respectively.

10.3. Let χ_1 and χ_2 denote two Dirichlet characters defined modulo N_1 and modulo N_2, respectively, and assume that N_1 and N_2 are coprime. Then

$$\chi(a + N\mathbb{Z}) = \chi_1(a + N_1\mathbb{Z})\chi_2(a + N_2\mathbb{Z})$$

defines a Dirichlet character defined modulo $N = N_1 N_2$.

10.4. If χ is a Dirichlet character defined modulo N, and if χ is defined modulo N_1 and modulo N_2, where $N_1 \mid N$ and $N_2 \mid N$, then χ is also defined modulo $\gcd(N_1, N_2)$.

10.5. The set $X((\mathbb{Z}/N\mathbb{Z})^\times)$ of all Dirichlet characters defined modulo N becomes a multiplicatively written abelian group by setting

$$(\chi_1 \chi_2)(a + N\mathbb{Z}) = \chi_1(a + N\mathbb{Z}) \cdot \chi_2(a + N\mathbb{Z}).$$

Show that, for coprime moduli N_1 and N_2 with $N = N_1 N_2$ we have

$$X((\mathbb{Z}/N\mathbb{Z})^\times) \simeq X((\mathbb{Z}/N_1\mathbb{Z})^\times) \oplus X((\mathbb{Z}/N_2\mathbb{Z})^\times).$$

10.6. Let χ be a Dirichlet character defined modulo 5. Show that χ is determined by the value $\chi(2 + 5\mathbb{Z})$, and that $\chi(2 + 5\mathbb{Z})^4 = 1$. Conclude that there are exactly four nontrivial Dirichlet character defined modulo 5.

10.7. Show that a Dirichlet character χ defined modulo N is primitive if and only if χ is nontrivial on the kernel of the projection map $\pi : (\mathbb{Z}/N\mathbb{Z})^\times \longrightarrow (\mathbb{Z}/n\mathbb{Z})^\times$ for each proper divisor $n \mid N$.

10.8. Write the Pell forms for the Kronecker characters with conductor $N \leq 12$ as rational functions.

10.9. Show that the Pell form $f_\chi(q)$ satisfies the functional equation

$$f_\chi(\tfrac{1}{q}) = \begin{cases} -f_\chi(q), & \text{for } \Delta > 0, \\ f_\chi(q), & \text{for } \Delta < 0. \end{cases}$$

10.10. Prove without using Euler's formulas (2.4) that the partial fraction decomposition of the Pell form for the primitive Dirichlet character χ defined modulo 4 is given by

$$f_\chi(q) = \frac{q}{1+q^2} = \frac{1}{2i}\left(\frac{1}{1-qi} - \frac{1}{1+qi}\right).$$

10.11. Compute the partial fraction decomposition of the Pell form for the primitive Dirichlet character modulo 8 defined by $\chi(n) = (\frac{2}{n})$.

10.12. Since $\mathrm{Fek}_\chi(q)$ is divisible by q, we can determine the partial fraction decomposition of

$$\frac{\mathrm{Fek}_\chi(q)}{q(1-q^N)}.$$

Show that an application of Euler's formulas yields

$$\frac{\mathrm{Fek}_\chi(q)}{q(1-q^N)} = -\frac{1}{N}\sum_{k=1}^{N-1}\frac{\mathrm{Fek}_\chi(\zeta^k)}{q-\zeta^k}.$$

10.13. Prove the congruence

$$\mathrm{Fek}_p(x) \equiv \sum_{n=1}^{p-1} n^m x^n \bmod p. \tag{10.12}$$

Deduce that $\mathrm{Fek}_p(1) \equiv 0 \bmod p$.

10.14. Show that $\mathrm{Fek}_p^{(k)}(1) \equiv 0 \bmod p$ for $0 \le k < m$, and that $\mathrm{Fek}_p^{(m)}(1) \equiv -1 \bmod p$.

10.15. Show that the cyclotomic polynomial $\Phi_p(x)$ is irreducible for prime values of p.
Hint: Consider $\Phi_p(x+1)$.

10.16. Count the number N_p of solutions of the congruence $y^2 = x^3 + 1 \bmod p$ for various prime numbers p. Observe that multiplying x by a cube root of unity ρ does not change the equation $y^2 = x^3 + 1$, so this elliptic curve has complex multiplication by ρ. Use this information for writing down a conjecture for N_p.

10.17. Let x be a p-th root of unity, i.e., assume that $x^p = 1$. Show that

$$\mathrm{Fek}_p(x)^n = \sum_{a=0}^{p-1} J_n(a)x^a,$$

where $J_n(a)$ is the character sum we have studied at the end of Sect. 3.4 defined by

$$J_n(a) = \sum \left(\frac{t_1 t_2 \cdots t_n}{p} \right),$$

where the sum is over all t_j mod p with $t_1 + t_2 + \ldots + t_n \equiv a$ mod p.

Appendix A
Computing with Pari and Sage

A.1 Pari

Pari[1] is very easy to use, and the basic version is installed within seconds.

A.1.1 Arithmetic in Integers

For computing $3^{N-1} \bmod N$ for $N = 2^{67} - 1$, simply type

```
N = 2^67-1; Mod(3,N)^(N-1)
```

(Observe that

```
Mod(3^(N-1),N)
```

is doing something completely different, even if the result is the same); the answer

```
Mod(95591506202441271281, 147573952589676412927)
```

tells us that $3^{N-1} \equiv 95591506202441271281 \bmod N$; in particular, N is not a prime. The command

```
factor(2^67-1)
```

immediately yields the prime factorization of N in the form

```
[193707721 1]

[761838257287 1].
```

---

[1]pari was developed at the University of Bordeaux by Henri Cohen and his colleagues.

The meaning of these "vectors" becomes clear by factoring, e.g., $N = 48$.

It is easy to realize small programs using `pari`. In order to implement the Lucas–Lehmer-test, for example, we type the program between two braces into a text file and then copy it by right-clicking the black `pari` window:

```
{p=61; N=2^p-1; S=Mod(4,N);
 for(n=1,p-2,S=S^2-2);
 print(lift(S)) }
```

The command `lift` transforms the residue class `Mod(2,127)` into the integer 2; its only purpose is to produce a nicer output. Other methods for programming loops may be found by typing `?11`; the command `?while` produces an explanation of how to program a while-loop.

The result 0 in the above computation shows that $2^{61} - 1$ is prime. Putting a loop around the commands we obtain a program that finds the small Mersenne primes:

```
{forstep(p=3,2000,2,if(isprime(p),
  N=2^p-1; S=Mod(4,N);
  for(n=1,p-2,S=S^2-2);
  if(S,,print(p)))) }
```

Within a few seconds, `pari` gives the following exponents:

$$p = 3, 5, 7, 13, 17, 19, 31, 61, 89, 107, 127, 521, 607, 1279.$$

A.1.2   Arithmetic in Quadratic Number Fields

You can obtain a generator of the ring of integers of the quadratic number field with discriminant $d = 12$ by typing

```
w = quadgen(12)
```

Squaring this element you can convince yourself that $w^2 = 3$. Similarly

```
w = quadgen(13)
```

generates an element with $w^2 = 3+w$, thus $w = \frac{1+\sqrt{13}}{2}$. Using these generators, the basic arithmetic operations $+$, $-$, $\cdot$ and $:$ can easily performed in quadratic number fields.

The solutions of the Pell equation are given with respect to this integral basis:

```
quadunit(4*67)
```

yields `48842 + 5967*w`, that is, the fundamental unit $\varepsilon = 48842 + 5967\sqrt{67}$, whereas

```
quadunit(21)
```

produces the result $\varepsilon = 2 + w = 2 + \frac{1+\sqrt{21}}{2} = \frac{5+\sqrt{21}}{2}$. The norm and the trace of the fundamental unit are computed via

```
eps = quadunit(21); print(norm(eps)," ",trace(cps))
```

Class numbers and the class groups of quadratic number fields are obtained easily:

```
quadclassunit(-84)[1]
```

yields the class number 4 of $\mathbb{Q}(\sqrt{-21}\,)$, and

```
quadclassunit(-84)[2]
```

gives the structure of the class group: $[2, 2]$ denotes the abelian group $\mathbb{Z}/2\mathbb{Z}\oplus\mathbb{Z}/2\mathbb{Z}$.

Residue classes modulo rational primes may be realized via

```
w = quadgen(21)*Mod(1,7)
```

using

```
(3-2*w)^7
```

then shows that $(3 - 2\frac{1+\sqrt{21}}{2})^7 \equiv 2 \bmod 7$.

For other calculations, e.g., with ideals we have to define a number field. This is accomplished by

```
nf = bnfinit(x^2-79);
```

This command computes the basic invariants of the quadratic number field $\mathbb{Q}(\sqrt{79}\,)$, namely an integral basis, the discriminant, the fundamental unit, and the class group. The semicolon at the end tells pari not to print the results of these calculations. We have access to the individual results by commands such as

```
nf.zk
```

Here $[1, x]$ denotes the integral basis $\{1, x\}$, where x is the root of the polynomial $x^2 - 79$.

We get the ideal class group of this number field with

```
nf.clgp
```

the expression

```
[3, [3], [[3, 2; 0, 1]]]
```

gives the class number 3, the structure of the class group ($[3]$ denotes the cyclic group of order 3), and an ideal (here it is a prime ideal q above 3) that generates the class group. Using

```
idealfactor(nf,5)
```

we obtain the prime ideal factorization of 5 in K. With

```
p = %[1,1]
```

we choose the first prime ideal, which we will denote by $\mathfrak{p}$ in the following. The command

```
bnfisprincipal(nf,p)
```

then yields

```
[[2]~, [19/9, -2/9]~]
```

which means that the prime ideal $\mathfrak{p}$ lies in the ideal class $\mathfrak{q}^2$, where $\mathfrak{q}$ is the prime ideal above (3) found above, and that the principal ideal $\mathfrak{p}\mathfrak{q}^{-2}$ is generated by $\frac{19}{9} - \frac{2}{9}\sqrt{79}$, i.e., that we have $9\mathfrak{p} = (19 - 2\sqrt{79})\mathfrak{q}^2$ ist.

The prime ideal decomposition of $19 - 2\sqrt{79}$ may be controlled by

```
idealfactor(nf,19-2*x)
```

A.2   Sage

In pari, only a few very basic functions for doing arithmetic with elliptic functions are implemented. For computing on elliptic curves it is a good idea to familiarize yourself with sage.[2] As a matter of fact, sage is also more comfortable for doing arithmetic in number fields, and you can access pari from within sage.

A.2.1   Number Fields

In sage we define number fields by

```
K.<a> = NumberField(x^2-79)
```

Now a is a root of the polynomial $x^2 - 79$ in the number field K, i.e., $a^2 = 79$. With

```
K.class_group()
```

you obtain the information

```
Class group of order 3 with structure C3 of Number Field,
```

and

```
K.units()
```

---

[2]This program was developed under William Stein and uses work done for other computer algebra systems such as pari, Cremona's mwrank or GAP, to mention but a few.

yields the fundamental unit $9\sqrt{79} - 80$. The command

```
K.integral_basis()
```

explains itself. The ideal $I = [5, 2 + \sqrt{79}\,]$ is defined by

```
I = K.ideal([5,2+a])
```

and the order of the ideal class generated by I is found by typing

```
C = K.class_group()
order(C(I))
```

For everything else we refer the readers to several introductions to sage that can be found quickly on the world wide web.

A.2.2   Elliptic Curves

In sage , the elliptic curve $y^2 = x^3 + ax + b$ is defined by the command

```
E = EllipticCurve([a,b])
```

In order to find the integral solutions of the equation $y^2 = x^3 - 26$, we first define the elliptic curve

```
E = EllipticCurve([0,-26])
```

The command

```
E.rank()
```

then shows that the group of rational points on this elliptic curve has rank 2; with

```
E.gens()
```

we find the generating points

```
(3 : 1 : 1), (35 : 207 : 1),
```

corresponding to the affine points $(x, y) = (3, 1)$ and $(35 : 207)$. Finally,

```
E.integral_points()    ,
```

shows that these are the only integral solutions of the equation $y^2 = x^3 - 26$. By copying

```
for a in [1..30]:
E = EllipticCurve([0,-a])
print(a, E.integral_points())
```

into the sage window and pressing Enter you obtain the following table with all
integral solutions of the equations $y^2 = x^3 - d$ for $1 \le d \le 30$:

d	$y^2 = x^3 - d$	d	$y^2 = x^3 - d$	d	$y^2 = x^3 - d$
1	$(1, 0)$	11	$(3, 4), (15, 58)$	21	
2	$(3, 5)$	12		22	
3		13	$(17, 70)$	23	$(3, 2)$
4	$(2, 2), (5, 11)$	14		24	
5		15	$(4, 7)$	25	$(5, 10)$
6		16		26	$(3, 1), (35, 207)$
7	$(2, 1), (32, 181)$	17		27	$(3, 0)$
8	$(2, 0)$	18	$(3, 3)$	28	$(4, 6), (8, 22), (37, 225)$
9		19	$(7, 18)$	29	
10		20	$(6, 14)$	30	

In order to test the truth of Theorem 6.20 we run the following program in sage

```
for t in [2,4,..20]:
d = 3*t^2+1
E = EllipticCurve([0,-d])
K.<a> = QuadraticField(-d)
print(d, K.class_number(), E.integral_points())
```

and obtain

d	h	Solutions	d	h	Solutions
13	2	$(17, 70)$	433	12	$(13, 42), (577, 13860)$
49	1	$(65, 524)$	589	16	$(785, 21994)$
109	6	$(5, 4), (145, 1746)$	769	20	$(1025, 32816)$
193	4	$(257, 4120)$	973	12	$(1297, 46710)$
301	8	$(401, 8030)$	1201	16	$(1601, 64060)$

The limits of sage become visible by extending the loop until $t = 20$: For
$t = 14$ and $d = 2353$, sage does not produce an answer. The reason for this
behavior is that either the generators of the group of rational points on E are huge,
or that the elliptic curve has a nontrivial Tate–Shafarevich group.

By looking at the table one is led to the conjecture that the class numbers of
$\mathbb{Q}(\sqrt{m})$ for all $m = 12t^2 + 1$ are divisible by 3 whenever t is a multiple of 3. This is
indeed true, but quite likely this is very difficult to prove without class field theory.

Appendix B
Solutions

Chapter 1

1.1. Let $(m^2 - n^2, 2mn, m^2 + n^2)$ be the first triple. If we choose m even and n odd, then $m^2 + n^2$ is odd, and we can find integers r and s such that $m^2 + n^2 = r^2 - s^2$; for example, we can set $r - s = 1$ and $r + s = m^2 + n^2$, that is $r = \frac{1}{2}(m^2 + n^2 + 1)$ and $s = \frac{1}{2}(m^2 + n^2 - 1)$. With these values, we have

$$(r^2 + s^2)^2 = (r^2 - s^2)^2 + (2rs)^2 = (m^2 + n^2)^2 + (2rs)^2$$
$$= (m^2 - n^2)^2 + (2mn)^2 + (2rs)^2$$

as desired.

1.2. Write $a = m^2 - n^2$, $b = 2mn$ and $c = m^2 + n^2$. For making $b = 2mn$ a square it is only necessary to set $m = a^2$ and $n = 2b^2$. For making $c = m^2 + n^2$ a square, write $m^2 + n^2 = p^2$ and set $m = r^2 - s^2$, $n = 2rs$ and $p = r^2 + s^2$.

1.3. For solving $2a^2 = c^2 - b^2 = (c - b)(c + b)$ it is sufficient to set $c - b = 4s^2$ and $c + b = 2r^2$. Then $2a^2 = 8r^2 s^2$ shows that $a = 2rs$. Moreover, $2c = (c + b) + (c - b) = 2r^2 + 4s^2$, which gives us $c = r^2 + 2s^2$. Similarly, $b = r^2 - 2s^2$.

For a geometric parametrization of the ellipse $x^2 + 2y^2 = 1$, consider the lines through $P(-1, 0)$. These have the equation $y = m(x + 1)$; intersecting these lines with the ellipse we get $x^2 + 2m^2(x + 1)^2 = 1$, which is equivalent to

$$0 = x^2 - 1 + 2m^2(x + 1)^2 = (x + 1)(x - 1 - 2m^2(x + 1)).$$

© The Author(s), under exclusive license to Springer Nature Switzerland AG 2021
F. Lemmermeyer, *Quadratic Number Fields*, Springer Undergraduate
Mathematics Series, https://doi.org/10.1007/978-3-030-78652-6

The equation $x + 1 = 0$ gives us P; from $x - 1 - 2m^2(x+1) = 0$ we obtain

$$x = \frac{1 + 2m^2}{1 - 2m^2} \quad \text{and} \quad y = m(x+1) = \frac{2m}{1 - 2m^2}.$$

Writing $m = \frac{s}{r}$ and simplifying the resulting expressions we get

$$x = \frac{r^2 + 2s^2}{r^2 - 2s^2} \quad \text{and} \quad y = \frac{2rs}{r^2 - 2s^2},$$

which leads to the same formulas as above.

1.4. Consider the lines through the point $(-1, 0, 0)$ on the unit sphere; these are given by the equations

$$x = -1 + at, \quad y = bt, \quad z = ct.$$

Intersecting this line with the unit sphere results in a quadratic equation with the solutions $t_1 = 0$ and $t_2 = \frac{2a}{a^2 + b^2 + c^2}$, which then provides us with the points

$$x = \frac{a^2 - b^2 - c^2}{a^2 + b^2 + c^2}, \quad y = \frac{2ab}{a^2 + b^2 + c^2}, \quad z = \frac{2ac}{a^2 + b^2 + c^2}.$$

1.5. This is clear:

$$(a^2 - mb^2)(c^2 - md^2) = (ac + mbd)^2 - m(bc + ad)^2.$$

1.6. Implicit differentiation yields $2yy' = 3x^2$, hence $y' = \frac{3x^2}{2y}$. Thus the tangent in the point (u, v) on Bachet's curve $y^2 = x^3 - k$ has the equation $y = \frac{3u^2}{2v}(x - u) + v$. Intersecting this line with Bachet's curve gives rise to the equation

$$x^3 - \left(\frac{3u^2}{2v}(x - u) + v\right)^2 - k = 0.$$

This equation has a double root $x = u$; if we denote the third root by x_3, then the sum of the roots $2u + x_3$ is the negative coefficient of x^2 in this equation:

$$2u + x_3 = -\frac{9u^4}{4v^2}.$$

This gives us (1.5):

$$x_3 = \frac{9u^4}{4v^2} - 2u = \frac{u^4 + 8ku}{4v^2}.$$

Plugging this value of $x = x_3$ into the line equation $y = \frac{3u^2}{2v}(x - u) + v$ we also get (1.6).

1.7. Applying the duplication formulas to $(3, 5)$ on the elliptic curve $y^2 + 2 = x^3$ we find

$$x_1 = \frac{3^4 + 8 \cdot 2 \cdot 3}{4 \cdot 5^2} = \frac{129}{10^2} \qquad y_1 = \frac{383}{10^3},$$

$$x_2 = \frac{2340922881}{7660^2} \qquad y_2 = \frac{113259286337292}{7660^3}$$

1.8. We find

$$x_1 = \frac{2^4 + 8 \cdot 4 \cdot 2}{4 \cdot 2^2} = 5, \quad y_1 = \frac{-2^6 + 20 \cdot 4 \cdot 2^3 + 8 \cdot 4^2}{8 \cdot 2^3} = 11.$$

Two more applications of the duplication formula yield

$$x_2 = \frac{785}{22^2}, \qquad y_2 = -\frac{5497}{22^3},$$

$$x_3 = \frac{3227836439105}{241868^2}, \qquad y_3 = -\frac{5799120182710629023}{241868^3}.$$

As in the case of Bachet's equation $y^2 + 2 = x^3$ it can be shown easily that the power of 2 dividing the denominator is strictly increasing. This shows that $(5, 11)$ is the only integral point obtained through duplication.

1.9. Assume that $p^2 + q^2 = r^2$, where p and q are prime. Then $p = m^2 - n^2$ and $q = 2mn$. Since q is prime, we must have $mn = \pm 1$, hence $p = m^2 - n^2 = 0$: Contradiction.

Now assume that (p, b, q) is a Pythagorean triple in which one leg p and the hypotenuse q are primes. Then $p = m^2 - n^2 = (m + n)(m - n)$ and $q = m^2 + n^2$ are prime, which is only possible if $m - n = 1$. In this case, $p = m^2 - n^2 = 2n + 1$ and $q = m^2 + n^2 = 2n^2 + 2n + 1$ and $b = 2m(m+1) = 2m^2 + 2m = q - 1$.

1.10. If (a, b, c) is a Pythagorean triple, then the equation

$$(t - a)^2 + (t - b)^2 = (t + c)^2$$

is a quadratic equation in t with the solution $t = 0$. The other solution is given by $t = 2a + 2b + 2c$, which produces the new Pythagorean triple

$(a+2b+2c, 2a+b+2c, 2a+2b+3c)$. This construction is due to B. Berggren (1934) and F.J.M. Barning (1963).
The equation

$$(t + a)^2 + (t + b)^2 = (2t + c)^2$$

leads to the same result. The same approach works for equations of the form $x_1^2 + x_2^2 + \ldots + x_n^2 = y^2$.

1.11. We have $24 \cdot 13 = 17^2 + 23 \cdot 1^2$ and $24 = 1^2 + 23 \cdot 1^2$, yet 13 cannot be represented by the form $x^2 + 23y^2$.

1.12. Let $p_1, \ldots, p_t$ be distinct primes of the form $4n+1$, and set $N = 4p_1^2 \cdots p_t^2 + 1$. Clearly N is not divisible by any p_j. If N is prime, then we have found a new prime of the form $4n + 1$. If N is composite, any of its prime divisors q divides N, which is a sum of coprime squares; but then q has the form $4N+1$. If $q_1, \ldots, q_t$ are primes of the form $4N-1$, set $N_3 = 4q_1 \cdots q_t - 1$. Clearly N is not divisible by any q_j. If N is prime, then we are done; if not, then at least one of the prime factors q of N has the form $4n + 3$: in fact, if all of them had the form $4n+1$, then the product N would also have this form. Since $q \neq p_j$, this completes the proof.

1.13. The discriminant of the quadratic equation $x^4 - 2ty^2 = t^2$ in t is $D = 4(x^4 + y^4)$. If $x^4 - 2y^2 = 1$ has a rational solution with y, then the quadratic equation in t must have a rational solution, so its discriminant is a square; this leads to $y^4 + x^4 = w^2$, which implies that $x = 0$ or $y = 0$. But then $xy = 0$ in $x^4 - 2y^2 = 1$, which is only possible if $y = 0$. This proves the claim.

1.14. Assume that $y^2 = x^3 - dx$ has a nontrivial rational solution. Then the Diophantine equation $dxt^2 + y^2t - x^3 = 0$ has a rational solution with $t = 1$, hence its discriminant $D = y^4 + 4dx^4$ must be a square, hence $y^4 + 4dx^4 = w^2$ must have a rational solution.
Observe that the trivial solution $x = 0$ of the second equation corresponds to the trivial solution $x = 0$ of $y^2 = x^3 - dx$.

1.15. Assume first that $p = 8n + 1$. Then

$$a^{p-1} - 1 = a^{8n} - 1 = (a^{4n} - 1)(a^{4n} + 1).$$

Choose a in such a way that p divides $a^{4n} + 1$. The identity $a^{4n} + 1 = (a^{2n} - 1)^2 + 2(a^n)^2$ then shows that -2 is a quadratic residue modulo p.
If $p \equiv 3 \bmod 8$, then $p \equiv 3 \bmod 4$ implies that -1 is a quadratic nonresidue modulo p. We also claim that 2 is a quadratic nonresidue modulo primes $p \equiv \pm 3 \bmod 8$. Assume to the contrary that p divides $c^2 - 2d^2$ for integers c and d not divisible by p. Reducing c and d modulo p we may assume that $|c|, |d| < \frac{p}{2}$. This implies that $c^2 - 2d^2 < p^2$, hence we can write $c^2 + 2d^2 = kp$ for some nonzero integer k with $|k| < p$. Cancelling common divisors of c and d from our equation we may assume that c and d are coprime. In particular, c is odd, and this implies $c^2 - 2d^2 \equiv \pm 1 \bmod 8$. Since $p \equiv \pm 3 \bmod 8$ we

must have $k \equiv \pm 3 \bmod 8$. This implies that k must be divisible by some prime number $q \equiv \pm 3 \bmod 8$. This prime number is smaller than p, and it satisfies $c^2 \equiv 2d^2 \bmod q$. Thus for every prime $p \equiv \pm 3 \bmod 8$ for which 2 is a quadratic residue there is a smaller prime q with the same properties. Infinite descent now yields a contradiction.

1.16. We claim that all primes $p \equiv 1, 3 \bmod 8$ can be represented in the form $p = c^2 + 2d^2$. This is true for $p = 3$ since $3 = 1^2 + 2 \cdot 1^2$. Assume that the result is true for all primes less than some prime number $p \equiv 1, 3 \bmod 8$.

We know from the last exercise that -2 is a quadratic residue modulo p; thus there is an integer x with $x^2 + 2 = kp$. Reducing x modulo p we may assume that $|x| < \frac{p}{2}$, and this implies that $|k| < p$. If $x = 2x_1$ is even, then $4x_1^2 + 2 = kp$ implies $k = 2k_1$ and $1 + 2x_1^2 = k_1 p$; thus we may assume that $c^2 + 2d^2 = kp$ for coprime integers c and d, where c is odd.

For each prime factor q of k we have $c^2 \equiv -2d^2 \bmod q$, which implies that -2 is a quadratic residue modulo q. By induction assumption, q has the form $q = e^2 + 2f^2$. We now claim that by suitably choosing the sign of f, we can make sure that the left hand side in the equation

$$\frac{c^2 + 2d^2}{e^2 + 2f^2} = \frac{k}{q} \cdot p$$

is an integer of the form $c_1^2 + 2d_1^2$. In this way we can eliminate all prime factors of k and end up with a representation of p in the form $p = c^2 + 2d^2$. Note that

$$e^2(c^2 + 2d^2) = c^2 e^2 + 2d^2 e^2 \quad \text{and} \quad c^2(e^2 + 2f^2) = c^2 e^2 + 2c^2 f^2$$

are both divisible by q, hence so is their difference

$$2(d^2 e^2 - c^2 f^2) = 2(de - cf)(de + cf).$$

Thus we can choose the sign of f in such a way that $q \mid (de - cf)$. Now consider the identity

$$qkp = (c^2 + 2d^2)(e^2 + 2f^2) = (ce + 2df)^2 + 2(cf - de)^2.$$

Since q divides $cf - de$, it must also divide $ce + 2df$. Setting $c_1 = (ce + 2df)q$ and $d_1 = (cf - de)q$ we obtain

$$\frac{k}{q} \cdot p = c_1^2 + 2d_1^2$$

as desired.

1.17. From $21 = 1^2 + 5 \cdot 4^2 = 4^2 + 5 \cdot 1^2$ we read off the elements $1 + 2\sqrt{-5}$ and $4 + \sqrt{-5}$ with norm 21. Squaring these elements we obtain

$$(1 + 2\sqrt{-5})^2 = -19 + 4\sqrt{5} \quad \text{and} \quad (4 + \sqrt{-5})^2 = 11 + 8\sqrt{-5}.$$

The representations $21^2 = 6^2 + 5 \cdot 9^2 = 14^2 + 5 \cdot 7^2$ come from $7^2 = 2^2 + 5 \cdot 3^2$ and $3^2 = 2^2 + 5 \cdot 1^2$.

1.18. The problem whether $y^2 + 2 = x^3$ can be solved by elementary means, for example, by writing $(y-5)(y+5) = y^2 - 25 = x^3 - 27 = (x-3)(x^2+3x+9)$, remains open.

1.19. The equation $y\sqrt{2} + \sqrt{5} = (a\sqrt{2} + b\sqrt{5})^3$ leads to $y = 2a^3 + 15ab^2$ and $1 = 6a^2b + 5b^3$; the second equation implies $b = \pm 1$ since it is equivalent to $1 = b(6a^2 + 5b^2)$. But then $\pm 1 = 6a^2 + 5$ does not have a solution in the reals let alone in integers.

On the other hand we have

$$\frac{4\sqrt{2} + \sqrt{5}}{\sqrt{2} + \sqrt{5}} = \frac{(4\sqrt{2} + \sqrt{5})(-\sqrt{2} + \sqrt{5})}{(\sqrt{2} + \sqrt{5})(-\sqrt{2} + \sqrt{5})} = \frac{-3 + 3\sqrt{10}}{3} = -1 + \sqrt{10},$$

as well as $(\sqrt{2} + \sqrt{5})^2 = 7 + 2\sqrt{10}$, hence

$$\frac{4\sqrt{2} + \sqrt{5}}{(\sqrt{2} + \sqrt{5})^3} = \frac{-1 + \sqrt{10}}{7 + 2\sqrt{10}} = -3 + \sqrt{10}.$$

This shows that the equation

$$4\sqrt{2} + \sqrt{5} = (a\sqrt{2} + b\sqrt{5})^3(3 + \sqrt{10})^n$$

has the solution $a = b = 1$ and $n = -1$.

1.20. If a is even, then b must be odd and $b^4 \equiv 1 \bmod 16$; thus $-q \equiv 1 \bmod 16$ contradicting our assumption. Thus a must be odd and we find $qb^4 = a^4 - 1 = (a^2 - 1)(a^2 + 1)$ with $\gcd(a^2 - 1, a^2 + 1) = 2$. Since $q \nmid a^2 + 1$ we have $a^2 - 1 = 8qc^4$, $a^2 + 1 = 2d^4$ and therefore $d^4 - 4qc^4 = 1$. Thus $4qc^4 = (d^2 - 1)(d^2 + 1)$, and since $q \nmid d^2 + 1$ we find $d^2 + 1 = 2e^4$ and $d^2 - 1 = 2qf^4$, which yields $e^4 - qf^4 = 1$. By infinite descent we conclude that the only integral solution is the one with $b = 0$.

1.21. If x is even, then $y^2 \equiv 7 \bmod 8$ is not solvable. Therefore x is odd. Now write $y^2 + 1 = x^3 + 8 = (x + 2)(x^2 - 2x + 4)$. If $x \equiv 1 \bmod 4$, then $x + 2 \equiv 3 \bmod 4$. Thus $x + 2$ must be divisible by a prime number $q \equiv 3 \bmod 4$, but such primes cannot divide a sum of two coprime squares. If $x \equiv 3 \bmod 4$, then $x^2 - 2x + 4 \equiv 3 \bmod 4$, and we get a contradiction in the same way.

1.22. Here $y^2 + 25 = x^3 + 8 = (x + 2)(x^2 - 2x + 4)$.

If x is even, then y is odd and $y^2 + 25 \equiv 2 \bmod 4$; this is impossible since $x^3 + 8 \equiv 0 \bmod 8$.

Thus x is odd and y is even, hence $x^3 \equiv 17 \bmod 4$ and therefore $x \equiv 1 \bmod 4$. But then $x + 2 \equiv 3 \bmod 4$, hence the first factor is divisible by a prime $q \equiv 3 \bmod 4$. Such primes divide the sum of squares $y^2 + 5^2$ only if $q \mid 5$, which is impossible.

1.23. If y is even, then $x^3 = y^2 + k \equiv k \equiv 3 \bmod 4$, hence $x \equiv 3 \bmod 4$. Write the equation in the form

$$y^2 + B^2 = x^3 - A^3 = (x - A)(x^2 + Ax + A^2).$$

Since $x^2 + Ax + A^2 \equiv 3 \bmod 4$, this number is divisible by an odd prime $q \equiv 3 \bmod 4$. But such a prime can divide a sum of squares $y^2 + B^2$ only if $q \mid B$ and $q \mid y$. Since B is not divisible by primes $q \equiv 3 \bmod 4$, this is a contradiction.

1.24. We write the equation in the form $y^2 + B^2 = x^3 - A^3 = (x - A)(x^2 + Ax + A^2)$. If y is odd, then x is even and $y^2 + B^2 \equiv 2 \bmod 4$, which contradicts the fact that $x^3 - A^3$ is divisible by 8. Thus y is even and x is odd. Now $x^2 + xA + A^2 \equiv 1 + A \equiv 3 \bmod 4$. Now $x^2 + xA + A^2 > 0$, hence there exists a prime number $q \equiv 3 \bmod 4$ dividing this number. But then $q \mid (y^2 + B^2)$ implies $q \mid B$, and we have a contradiction.

For $k = (-2)^3 + 21^2 = 433$, the equation $y^2 = x^3 - k$ has two integral points (13, 42) and (577, 13860). This example shows that the condition on B is necessary.

1.25. Clearly $(x, y) = (A, B)$ is an integral point.

For $k = 17$, sage finds the following integral points on $y^2 = x^3 + 17$:

$$(x, y) = (-2, 3), (-1, 4), (2, 5), (4, 9), (8, 23), (43, 282), (52, 375), (5234, 378661).$$

1.26. We have

$$\begin{vmatrix} a & b \\ c & d \end{vmatrix} = p^2 + q^2 + r^2 + s^2 \quad \text{and} \quad \begin{vmatrix} a' & b' \\ c' & d' \end{vmatrix} = p'^2 + q'^2 + r'^2 + s'^2.$$

Thus the left hand side is the product of two sums of four squares. Now

$$(aa' + bc')(cb' + dd') = [(p + qi)(p' + q'i) + (r + si)(-r' + s'i)]$$
$$\cdot [(-r + si)(r' + s'i) + (p - qi)(p' - q'i)]$$
$$= (pp' - qq' - rr' - ss')^2 + (pq' + qp' + rs' - r's)^2,$$

$$-(ca' + dc')(ab' + bd') = [(r - si)(p' + q'i) + (p - qi)(r' - s'i)]$$
$$\cdot [(p + qi)(r' + s'i) + (r + si)(p' - q'i)]$$
$$= (rp' + sq' + pr' - qs')^2 + (rq' - sp' - ps' - qr')^2.$$

Thus the determinant on the right hand side is a sum of four squares.

1.27. Multiplying through by a shows that it is sufficient to consider equations of the form $x^2 + ay^2 = bz^2$. Assume now that (ξ, η, ζ) is a nontrivial solution of this equation (such solutions exist by the Local–Global Principle if and only if the conic has nontrivial points in every completion of $\mathbb{Q}$). Then multiplying $bz^2 = x^2 + ay^2$ through by $b\zeta^2$ gives

$$(b\zeta z)^2 = b\zeta^2 x^2 + ab\zeta^2 y^2 = (\xi^2 + a\eta^2)x^2 + (a\xi^2 + a^2\eta^2)y^2$$
$$= (\xi x + a\eta y)^2 + a(\xi y - \eta x)^2.$$

Similarly,

$$(a\eta y)^2 = ab\eta^2 z^2 - a\eta^2 x^2 = b(b\zeta^2 - \xi^2)z^2 - (b\zeta^2 - \xi^2)x^2$$
$$= (\xi x + b\zeta z)^2 - b(\xi z + \zeta x)^2,$$

or

$$(\xi X)^2 = b\xi^2 z^2 - a\xi^2 y^2 = b(b\zeta^2 - a\eta^2)z^2 - a(b\zeta^2 - a\eta^2)y^2$$
$$= (b\zeta z + a\eta y)^2 - ab(\eta z + \zeta y)^2.$$

Thus "Euler's trick" provides us with three different factorizations of the form $AB = mC^2$, which we have collected in the following table:

	A	B	C	m
I	$b\zeta z + a\eta y + \xi x$	$b\zeta z - a\eta y - \xi x$	$\xi y - \eta x$	a
II	$b\zeta z + a\eta y + \xi x$	$b\zeta z - a\eta y + \xi x$	$\xi z + \zeta x$	b
III	$b\zeta z + a\eta y + \xi x$	$b\zeta z + a\eta y - \xi x$	$\eta z + \zeta y$	ab

Chapter 2

2.1. Clearly we can add, subtract, and multiply numbers of the form $a + b\sqrt{m}$, where $a, b \in \mathbb{Q}$, in the obvious way. For example,

$$(a + b\sqrt{m})(c + d\sqrt{m}) = ac + mbd + (ad + bc)\sqrt{m}.$$

The quotient of two elements is given by

$$\frac{a+b\sqrt{m}}{c+d\sqrt{m}} = \frac{(a+b\sqrt{m})(c-d\sqrt{m})}{c^2-md^2} = \frac{ac-mbd}{c^2-md^2} + \frac{bc-ad}{c^2-md^2}\sqrt{m}.$$
(2.1)

This formula works except when $c^2 - md^2 = 0$, which happens if and only if either $c = d = 0$ or if $m = (\frac{c}{d})^2$ is a square, which we have excluded.

2.2. Since $\alpha, \beta \in K$ and since $\{1, \sqrt{m}$ is a $\mathbb{Q}$-basis of K, there exist rational numbers a, b, c, d with $\alpha = a + b\sqrt{m}$ and $\beta = c + d\sqrt{m}$. Thus $M = \begin{pmatrix} a & b \\ c & d \end{pmatrix}$. Clearly $\{\alpha, \beta\}$ is also a $\mathbb{Q}$-basis of K if 1 and $\sqrt{m}$ can be expressed as $\mathbb{Q}$-linear combinations of α and β, i.e., if and only if M has an inverse. This is the case if and only if $\det M \neq 0$.

2.3. Since $\alpha, \beta \in \mathcal{O}_K$ and since $\{1, \sqrt{m}$ is an integral basis of K, there exist integers $a, b, c, d \in \mathbb{Z}$ with $\alpha = a + b\omega$ and $\beta = c + d\omega$. Thus $M = \begin{pmatrix} a & b \\ c & d \end{pmatrix}$. Clearly $\{\alpha, \beta\}$ is also an integral basis of K if 1 and ω can be expressed as $\mathbb{Z}$-linear combinations of α and β, i.e., if and only if M has an inverse. This is the case if and only if $\det M = \pm 1$.

It is now easily checked that the discriminant of the integral basis $\{\alpha, \beta\}$, namely $\left| \begin{pmatrix} \alpha & \beta \\ \alpha' & \beta' \end{pmatrix} \right|^2 = (\alpha\beta' - \alpha'\beta)^2$ is equal to $(\det M)^2 \operatorname{disc} k$, hence equal to $\operatorname{disc} k$ since $\det M = \pm 1$.

2.4. We have

$$\begin{pmatrix} U_0 & U_1 \\ U_1 & U_2 \end{pmatrix} = \begin{pmatrix} 0 & 1 \\ 1 & 1 \end{pmatrix} \quad \text{and} \quad \begin{pmatrix} U_1 & U_2 \\ U_2 & U_3 \end{pmatrix} = \begin{pmatrix} 1 & 1 \\ 1 & 2 \end{pmatrix} = \begin{pmatrix} 0 & 1 \\ 1 & 1 \end{pmatrix}^2.$$

The first claim is now proved by induction, the induction step being

$$\begin{pmatrix} U_n & U_{n+1} \\ U_{n+1} & U_{n+2} \end{pmatrix} \cdot \begin{pmatrix} 0 & 1 \\ 1 & 1 \end{pmatrix} = \begin{pmatrix} U_{n+1} & U_n + U_{n+1} \\ U_n + U_{n+1} & U_{n+1} + U_{n+2} \end{pmatrix} = \begin{pmatrix} U_{n+1} & U_{n+2} \\ U_{n+2} & U_{n+3} \end{pmatrix}.$$

For diagonalizing the matrix T we determine its eigenvalues. These are the roots of the characteristic polynomial

$$\det(T - \lambda I) = \lambda^2 - \lambda - 1 = 0,$$

which gives $\lambda_1 = \frac{1+\sqrt{5}}{2} = \omega$ and $\lambda_2 = \frac{1+\sqrt{5}}{2} = \omega'$.

The corresponding eigenvectors are $v_1 = \begin{pmatrix} 1 \\ \omega \end{pmatrix}$ and $v_2 = \begin{pmatrix} 1 \\ \omega' \end{pmatrix}$: In fact $\begin{pmatrix} 0 & 1 \\ 1 & 1 \end{pmatrix}\begin{pmatrix} 1 \\ \omega \end{pmatrix} = \begin{pmatrix} \omega \\ \omega+1 \end{pmatrix} = \begin{pmatrix} \omega \\ \omega^2 \end{pmatrix} = \omega\begin{pmatrix} 1 \\ \omega \end{pmatrix}$.

Therefore the diagonalizing matrix S is given by $S = \begin{pmatrix} 1 & 1 \\ \omega & \omega' \end{pmatrix}$. We now find $S^{-1} = \frac{1}{\omega-\omega'} \cdot \begin{pmatrix} -\omega' & 1 \\ \omega & -1 \end{pmatrix}$, hence $D = S^{-1}TS = \frac{1}{\omega-\omega'} \cdot \begin{pmatrix} -\omega' & 1 \\ \omega & -1 \end{pmatrix}\begin{pmatrix} 0 & 1 \\ 1 & 1 \end{pmatrix}\begin{pmatrix} 1 & 1 \\ \omega & \omega' \end{pmatrix} = \frac{1}{\omega-\omega'} \cdot \begin{pmatrix} 2+\omega & 0 \\ 0 & -2-\omega' \end{pmatrix} = \begin{pmatrix} \omega & 0 \\ 0 & \omega' \end{pmatrix}$. Since $T^n = (S^{-1}DS)^n = S^{-1}D^nS$ we now find

$$T^n = \begin{pmatrix} U_n & U_{n+1} \\ U_{n+1} & U_{n+2} \end{pmatrix} = S^{-1} \begin{pmatrix} \omega^n & 0 \\ 0 & \omega'^n \end{pmatrix} S,$$

and this implies Binet's formula

$$U_n = \frac{\omega^n - \omega'^n}{\omega - \omega'}.$$

2.5. Let $\alpha = \frac{a+b\sqrt{m}}{2}$ be an algebraic integer in some quadratic number field $\mathbb{Q}(\sqrt{m})$. Since the binomial coefficients $\binom{p}{k}$ are divisible by p for $1 \le k \le p-1$ (since p divides the numerator $p!$, but not the denominator $k!(p-k)!$), we find

$$\alpha^p \equiv \frac{a^p + b^p \sqrt{m}^p}{2^p} \mod p.$$

By Fermat's Little Theorem we have $a^p \equiv a$, $b^p \equiv b$ and $2^p \equiv 2 \mod p$. Moreover, $\sqrt{m}^p = m^{\frac{p-1}{2}} \sqrt{m} \equiv (\frac{m}{p})\sqrt{m} \mod p$ by Euler's criterion. This shows that

$$\left(\frac{a+b\sqrt{m}}{2} \right)^p \equiv \begin{cases} \frac{a+b\sqrt{m}}{2} \mod p & \text{if } (\frac{m}{p}) = +1, \\ \frac{a-b\sqrt{m}}{2} \mod p & \text{if } (\frac{m}{p}) = -1. \end{cases}$$

In the special case $\alpha = \omega = \frac{1+\sqrt{5}}{2}$ we have

$$\omega^p \equiv \omega \mod p \qquad\qquad\qquad \text{if } (\tfrac{5}{p}) = +1,$$

$$\omega^p \equiv \omega' \mod p \qquad\qquad\qquad \text{if } (\tfrac{5}{p}) = -1,$$

where $\omega' = \frac{1-\sqrt{5}}{2}$. In the first congruence we may cancel ω, and we find $\omega^{p-1} \equiv 1 \mod p$. If we multiply the second congruence by ω we obtain $\omega^{p+1} \equiv \omega\omega' \equiv -1 \mod p$.

Applying these congruences to Binet's formula we find, if $(\frac{5}{p}) = +1$,

$$U_{p-1} = \frac{\omega^{p-1} - \omega'^{p-1}}{\omega - \omega'} \equiv 0 \mod p, \qquad U_p = \frac{\omega^p - \omega'^p}{\omega - \omega'} \equiv \frac{\omega - \omega'}{\omega - \omega'} \equiv 1 \mod p.$$

In the case $(\frac{5}{p}) = -1$, we find similarly

$$U_p \equiv \frac{\omega' - \omega}{\omega - \omega'} \equiv -1 \mod p \quad \text{and} \quad U_{p+1} \equiv 0 \mod p.$$

2.6. By definition, $N(\alpha) = \alpha\alpha'$, hence $N(\alpha\beta) = (\alpha\beta)(\alpha\beta)' = \alpha\beta\alpha'\beta' = \alpha\alpha'\beta\beta' = N(\alpha) \cdot N(\beta)$. Similarly, $\mathrm{Tr}(\alpha + \beta) = \alpha + \beta + \alpha' + \beta' = \mathrm{Tr}(\alpha) + \mathrm{Tr}(\beta)$.

For $\alpha = a + b\sqrt{m}$, $\mathrm{Tr}(\alpha) = 2a$, so clearly $\mathrm{Tr}(\alpha) = 2a$ for $\alpha = a + b\sqrt{m}$, hence $\mathrm{Tr}(\alpha) = 0$ if and only if $a = 0$.

Next $\mathrm{disc}\,\alpha = (\alpha - \alpha')^2 = (2b\sqrt{m})^2 = 4mb^2$, hence $\mathrm{disc}\,\alpha = 0$ if and only if $b = 0$.

Finally $0 = N\alpha = \alpha\alpha'$ if and only if $\alpha = 0$ or $\alpha' = 0$; but $\alpha' = 0$ implies $\alpha = 0$ and thus $a = b = 0$.

2.7. If $\alpha \mid \beta$, then $\beta = \alpha\gamma$; taking norms this implies $N\beta = N\alpha N\gamma$, and this means that $N\beta \mid N\alpha$ as claimed.

2.8. We have $\omega + \omega' = -p$ and $\omega\omega' = q$, hence

$$\mathrm{disc}\,\omega = (\omega - \omega')^2 = (\omega + \omega')^2 - 4\omega\omega' = q^2 - 4p.$$

If $ax^2 + bx + c = 0$, then a straightforward calculation yields $\mathrm{disc}\,\omega = \frac{b^2 - 4ac}{a^2}$.

2.9. Clearly $\mathbb{Q}(\sqrt{m})$ is a field if we can divide by any nonzero element $a + b\sqrt{m}$. By (2.1) this holds if and only if $a^2 - mb^2 = N(a + b\sqrt{m}) = 0$, which in turn is true if and only if $a = b = 0$ or m is a square. Thus division by nonzero elements is possible if and only if m is not a square in $\mathbb{Q}$, which is equivalent to $x^2 - m$ being an irreducible polynomial in $\mathbb{Q}[x]$.

2.10. Clearly $\sqrt{b} \in K$ if and only if $\sqrt{b} = r + s\sqrt{m}$. Squaring this equation we get $b = r^2 + ms^2 + 2rs\sqrt{m}$. Since $\sqrt{m}$ is irrational, we deduce that $rs = 0$. If $s = 0$, then $b = r^2$ is a square; if $r = 0$, then $b = ms^2$.

2.11. We already have shown that σ respects the ring operations since $\sigma(\alpha) = \alpha'$. It remains to prove the last claim. But for $\alpha = a + b\sqrt{m}$ we find that $\alpha = \sigma(\alpha)$ is equivalent to $a + b\sqrt{m} = a - b\sqrt{m}$, i.e., to $b = 0$. Thus $\alpha = \sigma(\alpha)$ if and only if $\alpha \in \mathbb{Q}$.

2.12. The elements of K form an abelian additive group, and multiplication by elements of $\mathbb{Q}$ is the scalar multiplication. Vector space axioms such as $(ab)v = a(bv)$ or $1v = v$ for $a, b \in \mathbb{Q}$ and $v \in K$ follow immediately from the field axioms for K.

For $v \in K$ let $\mu : K \longrightarrow K$ denote the multiplication of elements of K by $\alpha = a + b\sqrt{m}$, i.e., set $\mu(v) = \alpha v$. This map μ is $\mathbb{Q}$-linear, that is, we have $\mu(u + v) = \mu(u) + \mu(v)$ and $\mu(rv) = r\mu(v)$.

Fix the $\mathbb{Q}$-basis $\{1, \sqrt{m}\}$ of K. Then

$$\mu(1) = (a + b\sqrt{m}) \cdot 1 = a + b\sqrt{m},$$

$$\mu(\sqrt{m}) = (a + b\sqrt{m}) \cdot \sqrt{m} = bm + a\sqrt{m},$$

which shows that the matrix describing multiplication by α has columns $\binom{a}{b}$ and $\binom{mb}{a}$ and thus is given by $A = \left(\begin{smallmatrix} a & mb \\ b & a \end{smallmatrix}\right)$.

Clearly $\det A = a^2 - mb^2 = N(\alpha)$ and $\mathrm{Tr}(A) = 2a = \mathrm{Tr}(\alpha)$.

Changing the basis corresponds to replacing A by a matrix of the form $B^{-1}AB$ for some nonsingular matrix B. Clearly $\det(B^{-1}AB) = \det(B)^{-1}\det(A)\det(B) = \det(A)$.

For proving the invariance of the trace we first show that $\mathrm{Tr}(AB) = \mathrm{Tr}(BA)$ for arbitrary matrices; then clearly $\mathrm{Tr}(B^{-1}AB) = \mathrm{Tr}(AB^{-1}B) = \mathrm{Tr}(A)$.

2.13. By definition, an element $\alpha \in k$ is integral if and only if $\alpha + \alpha'$ and $\alpha \cdot \alpha'$ are integers. Since these expressions are invariant under switching α and α', the claim follows.

2.14. Assume that a, b, c, d are integers with $ad - bc = 0$. Given an integral basis $\{\omega_1, \omega_2\}$ of $\mathcal{O}_k$, set $\left(\begin{smallmatrix} \omega_1' \\ \omega_2' \end{smallmatrix}\right) = \left(\begin{smallmatrix} \omega_1 \\ \omega_2 \end{smallmatrix}\right)\left(\begin{smallmatrix} a & c \\ b & d \end{smallmatrix}\right)$. We have to show that ω_1 and ω_2 are $\mathbb{Z}$-linear combinations of ω_1' and ω_2'. But this follows from $\left(\begin{smallmatrix} a & c \\ b & d \end{smallmatrix}\right) \cdot \left(\begin{smallmatrix} d & -c \\ -b & a \end{smallmatrix}\right) = \left(\begin{smallmatrix} 1 & 0 \\ 0 & 1 \end{smallmatrix}\right)$ and $\left(\begin{smallmatrix} \omega_1 \\ \omega_2 \end{smallmatrix}\right) = \left(\begin{smallmatrix} \omega_1' \\ \omega_2' \end{smallmatrix}\right)\left(\begin{smallmatrix} d & -c \\ -b & a \end{smallmatrix}\right)$.

2.15. Assume first that $m \equiv 2, 3 \bmod 4$. Then elements of norm 2 or 3 exist if and only if the Diophantine equations $x^2 + |m|y^2 = 2$ and $x^2 + |m|y^2 = 3$ have solutions in integers. This is the case for $m = -1$ and $m = -2$, where the elements $1 + i$ and $\sqrt{-2}$ have norm 2, and where $1 + \sqrt{-2}$ has norm 3.

If $m \equiv 1 \bmod 4$, then elements of norm 2 and 3 exist if the norm equations $x^2 + |m|y^2 = 8$ and $x^2 + |m|y^2 = 12$ are solvable in integers. Clearly $m \leq 11$, and a case by case analysis shows that the elements $\frac{1 \pm \sqrt{-7}}{2}$ have norm 2, and that $\sqrt{-3}$ and $\frac{1 \pm \sqrt{-11}}{2}$ have norm 3.

Thus the only such m are $m = -1, -2, -3, -7$, and -11.

2.16. These are formal verifications of the axioms. In the case of $k^\times$, for example, we have to verify that

a. $\sigma(\alpha\beta) = \sigma(\alpha)\sigma(\beta)$,
b. $(\sigma\tau)\alpha = \sigma(\tau\alpha)$,
c. $\mathrm{id}\,\alpha = \alpha$

for all $\sigma, \tau \in \mathrm{Gal}(k/\mathbb{Q})$ and all elements $\alpha, \beta \in k^\times$.

2.17. Multiplying $x^2 + y^2 = 2z^2$ through by 2 we obtain. $(x + y)^2 + (x - y)^2 = (2z)^2$. Thus $(x+y, x-y, 2z)$ is a Pythagorean triple, hence there exist integers m and n with $x + y = m^2 - n^2$, $x - y = 2mn$ and $2z = m^2 + n^2$, where m and n have the same parity. Solving for the unknowns we find

$$x = \frac{m^2 + 2mn - n^2}{2}, \quad y = \frac{m^2 - 2mn - n^2}{2}, \quad z = m^2 + n^2.$$

2.18. If $m \equiv 1 \bmod 4$, then clearly $\omega = \frac{1+\sqrt{m}}{2}$ and $\sigma(\omega) = \frac{1-\sqrt{m}}{2}$ form an integral basis since $1 = \omega + \sigma(\omega)$.

If $m \equiv 2, 3 \bmod 4$, then $\{1, \sqrt{m}\}$ is an integral basis. Assume that $\omega = a + b\sqrt{m}$ generates a normal integral basis. Then there exist integers $r, s \in \mathbb{Z}$ such that

$$1 = r(a + b\sqrt{m}) + s(a - b\sqrt{m}).$$

Comparing the coefficients of $\sqrt{m}$ we find that $r = s$, hence $1 = 2ar$, which is a contradiction.

2.19. It is clear that $K = \mathbb{Q}(\sqrt[3]{2})$ is closed with respect to addition and multiplication.

For proving that it is always possible to divide by nonzero elements we observe that it is sufficient to show that $\frac{1}{\alpha}$ is an element of K. If we write $\frac{1}{\alpha} = \beta$, then $\alpha\beta = 1$. With $\alpha(x) = a + bx + cx^2$ we have $\alpha(\sqrt[3]{2}) = \alpha$. The polynomials $\alpha(x)$ and $x^3 - 2$ are coprime, hence there exist polynomials β and f with $\alpha(x)\beta(x) - (x^3 - 2)f(x) = 1$. Setting $x = \sqrt[3]{2}$ we obtain $\alpha\beta = 1$.

Clearly the elements of the form $a + b\sqrt[3]{2}$ are not multiplicatively closed since $\sqrt[3]{2} \cdot \sqrt[3]{2} = \sqrt[3]{4}$ does not have this form: If $\sqrt[3]{4} = a + b\sqrt[3]{2}$, then $2 = a\sqrt[3]{2} + b\sqrt[3]{4} = a\sqrt[3]{2} + b(a + b\sqrt[3]{2})$ implies that $\sqrt[3]{2}$ is rational, which is a contradiction.

2.20. The identities are easy to check. For explaining how to find them we start with the observation that

$$-2(2 + \sqrt{-5}) = (1 - \sqrt{-5})^2.$$

Now the identities follow from $-2 = i(1 + i)^2 = \sqrt{2}^2$.
Similarly, $2(31 + 6\sqrt{26}) = (6 + \sqrt{26})^2$, hence

$$31 + 6\sqrt{26} = \left(\frac{6 + \sqrt{26}}{\sqrt{2}}\right)^2 = (3\sqrt{2} + \sqrt{13})^2.$$

2.21. We have $17 + 4\sqrt{15} = (2\sqrt{3} + \sqrt{5})^2$. Finding more examples is trivial: $(\sqrt{2} + \sqrt{5})^2 = 7 + 2\sqrt{10}$.

2.22. If $a + b\sqrt{m} = (r + s\sqrt{m})^2$, then $a = r^2 + ms^2$. Since $1 \leq a \leq m$ we must have $s = 0$, contradicting our assumptions.

2.23. We have $u^2 = 2^2 + m = (2 + \sqrt{-m})(2 - \sqrt{-m})$; since the only units are ± 1 and since the elements $\pm(2 \pm \sqrt{-m})$ are not squares (see the preceding exercise), this means that the Square Product Theorem does not hold in $\mathbb{Z}[\sqrt{-m}]$.

2.24. We have to show that the factors in the equation

$$169 = 13 \cdot 13 = (4 + 3\sqrt{-17})(4 - 3\sqrt{-17})$$

are irreducible in $\mathbb{Z}[\sqrt{-17}]$. Clearly $13 = a^2 + 17b^2$ is not solvable in integers, which implies that 13 is irreducible.

If $4 + 3\sqrt{-17}) = \alpha\beta$, then taking norms we find $13^2 = N\alpha \cdot N\beta$, and unless α or β is a unit, this implies $N\alpha = N\beta = 13$, which is impossible.

2.25. The verification of

$$\frac{5 + \sqrt{-7}}{2} = \left(\frac{-1 - \sqrt{-7}}{2}\right)^3$$

is straightforward. Similarly,

$$\frac{181 + \sqrt{-7}}{2} = -\left(\frac{1+\sqrt{-7}}{2}\right)^3.$$

2.26. Consider the points $P(x_1, y_1)$ and $P_2(x_2, y_2)$ on the unit circle. Then

$$P \oplus Q = (x_1 x_2 - y_1 y_2, x_1 y_2 + x_2 y_1).$$

Since

$$x_1 = \cos(\alpha),\ y_1 = \sin(\alpha),\ x_2 = \cos(\beta),\ y_2 = \sin(\beta)$$

and

$$x_3 = \cos(\alpha + \beta), \quad y_3 = \sin(\alpha + \beta)$$

we can deduce that

$$\cos(\alpha + \beta) = \cos(\alpha)\cos(\beta) - \sin(\alpha)\sin(\beta),$$
$$\sin(\alpha + \beta) = \sin(\alpha)\cos(\beta) + \cos(\alpha)\sin(\beta).$$

2.27. Consider the point $P(r, s)$ on the unit circle, where $r = \cos\alpha$ and $s = \sin\alpha$. By similarity, the projection $P'(1|t)$ of P onto the line $x = 1$ satisfies $\frac{t}{2} = \frac{s}{1+r}$, so $t = \frac{2s}{1+r}$.
By elementary geometry, addition of points P' and Q' on the tangent line corresponds to adding the angles $\sphericalangle P'ZN$ and $\sphericalangle Q'ZN$; for $P'(1|t)$, the angle is given $\tan \sphericalangle P'ZN = \frac{t}{2}$, hence the sum of two points $P_1'(1|t_1)$ and $P_2'(1|t_2)$ is given by $P_1' \oplus P_2' = P_3'(1|t_3)$, where

$$t_3 = \frac{t_1 + t_1}{1 - t_1 t_2}.$$

2.28. Let $P = (x_1, y_1)$, where $x_1 = \sqrt{\frac{1+x}{2}}$ and $y_1 = \sqrt{\frac{1-x}{2}}$ (thus the sign of xy is positive). The addition formula gives $2P = (x_3, y_3)$ for

$$x_3 = x_1^2 - y_1^2 = \frac{1+x}{2} - \frac{1-x}{2} = x, \quad y_3 = 2x_1 y_1 = 2\sqrt{\frac{1-x^2}{4}} = y.$$

Applying this procedure to the point $P(\frac{\sqrt{2}}{2}, \frac{\sqrt{2}}{2})$ corresponding to the angle $\frac{\pi}{4}$ we obtain $\frac{1}{2}P = (\frac{1}{2}\sqrt{2 + \sqrt{2}}, \frac{1}{2}\sqrt{2 - \sqrt{2}})$ as claimed.

2.29. We have to verify two properties:

- If $x_1 y_1 = x_2 y_2 = 1$, then $x_1 x_2 \cdot y_1 y_2 = 1$. This is trivially true.
- The slope $m_1 = \frac{y_2 - y_1}{x_2 - x_1}$ is equal to $m = \frac{y_3 - 1}{x_3 - 1}$. But after clearing denominators in the equation $m_1 = m$ we get, using $x_1 y_1 = x_2 y_2 = 1$,

$$(y_1 y_2 - 1)(x_2 - x_1) = (x_1 x_2 - 1)(y_2 - y_1)$$

$$y_1 x_2 y_2 - x_2 - x_1 y_1 y_2 + x_1 = x_1 x_2 y_2 - y_2 - x_1 x_2 y_1 + y_1$$

$$y_1 - x_2 - y_2 + x_1 = x_1 - y_2 - x_2 + y_1,$$

and this is clearly true.

2.30. We have to verify two properties:

- If $y_1 = x_1^2$ and $y_2 = x_2^2$, then $y_1 + y_2 + 2x_1 x_2 = (x_1 + x_2)^2$. This is obviously true.
- The slopes $m_1 = \frac{y_2 - y_1}{x_2 - x_1}$ and $m = \frac{y_3}{x_3}$ are equal. But clearly

$$m = \frac{y_3}{x_3} = \frac{y_1 + y_2 + 2x_1 x_2}{x_1 + x_2} = \frac{(x_1 + x_2)^2}{x_1 + x_2} = x_1 + x_2, \quad \text{and}$$

$$m_1 = \frac{y_2 - y_1}{x_2 - x_1} = \frac{x_2^2 - x_1^2}{x_1 + x_2} = x_1 + x_2.$$

2.31. We have $f(q) = \frac{q}{1 - q - q^2}$, hence

$$f\left(\frac{1}{q}\right) = \frac{\frac{1}{q}}{1 - \frac{1}{q} - \frac{1}{q^2}} = \frac{q}{q^2 - q - 1} = \frac{-q}{1 + q - q^2} = f(-q).$$

2.32. This is an immediate consequence of Binet's formula:

$$\frac{U_{n+1}}{U_n} = \frac{\omega^{n+1} - \omega'^{n+1}}{\omega^n - \omega'^n} = \omega \cdot \frac{1 - \alpha^{n+1}}{1 - \alpha^n},$$

where $\alpha = \frac{\omega'}{\omega} = -\frac{1}{\omega^2}$. Since $|\alpha| < 1$, the fraction tends to 1, and this implies the claim.

2.33. We have to show:

1. If $x_1^2 - x_1 y_1 - y_1^2 = x_2^2 - x_2 y_2 - y_2^2 = 1$, then $x_3^2 - x_3 y_3 - y_3^2 = 1$ for

$$x_3 = x_1 x_2 + y_1 y_2, \qquad y_3 = x_1 y_2 + x_2 y_1 - y_1 y_2.$$

This can be done by brute force:

$$x_3^2 - x_3y_3 - y_3^2 = (x_1x_2 + y_1y_2)^2 - (x_1x_2 + y_1y_2)(x_1y_2 + x_2y_1 - y_1y_2)$$
$$- (x_1y_2 + x_2y_1 - y_1y_2)^2$$
$$= (x_1^2 - x_1y_1 - y_1^2)(x_2^2 - x_2y_2 - y_2^2).$$

2. The slopes $m_1 = \frac{y_2-y_1}{x_2-x_1}$ and $m = \frac{y_3}{x_3-1}$ are equal. Clearing denominators yields the equation

$$y_2(x_1^2 - x_1y_1 - y_1^2 - 1) = y_1(x_2^2 - x_2y_2 - y_2^2 - 1),$$

which is clearly correct.

2.34. Starting with $(1, 0)$ we obtain, using Vieta jumping, the sequence

$$(1,0), (1, -1), (-2, -1), (-2, 3), (5, 3), (5, -8), \ldots$$

of integral points on the Fibonacci hyperbola $x^2 - xy - y^2 = 1$. The second sequence is given by

$$(-1,0), (-1, 1), (2, 1), (2, -3), (-5, -3), (-5, 8), \ldots.$$

Observe that the coordinate with the largest absolute value is conserved. If, for example, (x, y) is an integral point with $x > y > 0$, then the second solution of the quadratic equation $Y^2 + xY - x^2 + 1 = 0$ is $Y' = -y - x$, and we obtain the second integral point $(x, -x - y)$. In all examples, $|x| + |y|$ is increasing.

We claim that every integral point on the Fibonacci hyperbola belongs to one of these two sequences. Assume therefore that (x, y) is an integral point on the Fibonacci hyperbola with $|y| > 0$. Vieta jumping gives us a new point (x_1, y_1) with $|y_1| \le |y|$ and $|x_1| + |y_1| < |x| + |y|$ until we find one with y-coordinate 0. But then $x = \pm 1$, and (x, y) belongs to one of the two sequences above.

2.35. We will show more generally that the group law on the Pell conic $x^2 - my^2 = 1$ is given by $P_1 \oplus P_2 = P_3$, where $P_j = (x_j, y_j)$ and

$$x_3 = x_1x_2 + my_1y_2, \quad y_3 = x_1y_2 + x_2y_1.$$

We show that P_3 is on the conic and that the slopes of the lines P_1P_2 and NP_3 coincide:

$$x_3^2 - my_3^2 = (x_1x_2 + my_1y_2)^2 - m(x_1y_2 + x_2y_1)^2 = (x_2^2 - my_2^2)x_1^2 - m(x_2^2 - my_2^2)y_1^2$$
$$= x_1^2 - my_1^2 = 1,$$

hence P_3 lies on the conic.

Next we compare the slopes $m_1 = \frac{y_2-y_1}{x_2-x_1}$ and $m_1 = \frac{y_3}{x_3-1}$. We will show that $m_1 = m_2$; this equation is equivalent to

$$\frac{y_2 - y_1}{x_2 - x_1} = \frac{y_3}{x_3 - 1}$$

$$(x_3 - 1)(y_2 - y_1) = y_3(x_2 - x_1)$$

$$(x_1 x_2 + m y_1 y_2 - 1)(y_2 - y_1) = (x_1 y_2 + x_2 y_1)(x_2 - x_1)$$

$$y_2(x_1^2 - m y_1^2) - y_2 = y_1(x_2^2 - m y_2^2) - y_1$$

$$y_2 - y_1 = y_2 - y_1,$$

and the last equation is true.

Clearly the integral multiples of $(2, 1)$ are integral points on the hyperbola $x^2 - 3y^2 = 1$. If $Q = (x, y)$ is any integral point not of this form, then it must lie between two consecutive multiples kP and $(k + 1)P$. But then $Q - kP$ is an integral point between N and P: Contradiction.

Finally we have $2(x, y) = (x^2 + 3y^2, 2xy)$; since $x^2 - 3y^2 = 1$, we can write this equation as $2(x, y) = 2x^2 - 1, 2xy)$.

2.36. Assume that n is prime. Then $a^{n-1} \equiv 1 \bmod n$ by Fermat's Little Theorem. If we choose a as a primitive root modulo n, then $a^k \not\equiv 1 \bmod n$ for each proper divisor k of $n - 1$.

Now assume conversely that $a^{n-1} \equiv 1 \bmod n$ for some integer a and $a^k \not\equiv 1 \bmod n$ for each proper divisor k of $n - 1$. Then a is coprime to n, and so are all the powers a^k for $1 \leq k \leq n - 1$. If we can show that these powers are distinct modulo n, then there are $n - 1$ coprime residue classes, and then n must be prime.

But if $a^r \equiv a^s \bmod n$ for $1 \leq s < r \leq n - 1$, then $a^{r-s} \equiv 1 \bmod n$, and this contradicts our assumptions.

2.37. Let E be the field with p^2 elements, and let $\mathbb{F}_p$ denote its subfield with p elements. We represent the elements of E in the form $a + b\sqrt{3}$, where $a, b \in \mathbb{Z}/p\mathbb{Z}$ and where $a^2 - 3b^2 = 1$.

Observe that for primes p with $(\frac{3}{p}) = -1$, we have $(a + b\sqrt{p})^p \equiv a - b\sqrt{p}$, hence the norm map $E^\times \longrightarrow \mathbb{F}_p^\times$ is given by $a^2 - 3b^2 = N(\alpha) = \alpha \cdot \alpha^p$. Since its kernel has at most $p + 1$ elements, it must be onto. Thus the kernel $\mathcal{C}(\mathbb{Z}/p\mathbb{Z})$, which consists of the points on the conic $x^2 - 3y^2 = 1$ over $\mathbb{Z}/p\mathbb{Z}$, has exactly $p + 1$ elements. Since the multiplicative groups of finite groups are cyclic, and since subgroups of cyclic groups are cyclic, it follows that the points modulo p on the conic $x^2 - 3y^2 = 1$ form a cyclic group of order $p + 1$.

If $P \in \mathcal{C}(\mathbb{Z}/p\mathbb{Z})$ is any point on the conic $x^2 - 3y^2 = 1$ modulo p, where $(\frac{3}{p}) = -1$, then $(p + 1)P = N = (1, 0)$. Since the group $\mathcal{C}(\mathbb{Z}/p\mathbb{Z})$ is cyclic, $\frac{p+1}{2}P = N$ or $\frac{p+1}{2}P = T$, where $T(-1, 0)$ is the unique point of order 2.

Clearly $\frac{p+1}{2}P = N$ for all points P of the form $P = 2Q$, and $\frac{p+1}{2}P = T$ otherwise.

We now claim that $P = (2, 1)$ does not have the form $P = 2Q$ for some $Q \in C(\mathbb{Z}/p\mathbb{Z})$. In fact, if $Q = (a, b)$, then $2Q = (a^2 + 3b^2, 2ab)$; but the system of congruences $a^2 + 3b^2 \equiv 2 \bmod p$ and $2ab \equiv 1 \bmod p$ is equivalent to $a^2 + 3(\frac{1}{2a})^2 \equiv 2 \bmod p$, i.e., to $0 \equiv 4a^4 - 8a^2 + 3 = (2a^2 - 1)(2a^2 - 3) \bmod p$. The congruence $2a^2 \equiv 3 \bmod p$ is not solvable for primes $p = 2^q - 1 \equiv 7 \bmod 8$ with $(\frac{3}{p}) = -1$. Let $a \equiv \frac{\sqrt{2}}{2} \bmod p$ be a solution of the congruence $2a^2 \equiv 1 \bmod p$. Then $b \equiv a \bmod p$, but the point (a, b) is not in $C(\mathbb{Z}/p\mathbb{Z})$.

Thus $\frac{p+1}{2}P = (-1, 0)$ for $P = (2, 1)$ if $p = 2^q - 1$ is prime. Assume conversely that $\frac{p+1}{2}P = (-1, 0)$. Since $\frac{p+1}{2} = 2^{q-1}$ is a power of 2, this implies that P has order $p+1$. This in turn implies that $C(\mathbb{Z}/p\mathbb{Z})$ is cyclic, and that p is prime: If $p = qr$ is a product of coprime integers, then $C(\mathbb{Z}/p\mathbb{Z}) \simeq C(\mathbb{Z}/q\mathbb{Z}) \times C(\mathbb{Z}/r\mathbb{Z})$, which contradicts the fact that $C(\mathbb{Z}/p\mathbb{Z})$ is cyclic; if p is divisible by the square q^2 of an odd prime q, then the order of $C(\mathbb{Z}/p\mathbb{Z})$ is a multiple of q.

Finally let us compute the point Q with $2Q = \frac{p+1}{2}P = (-1, 0)$. Setting $Q = (a, b)$ we obtain $2Q = a^2 + 3b^2, 2ab)$. If $b = 0$, then $a^2 \equiv -1 \bmod p$, which is impossible since $p \equiv 3 \bmod 4$. Thus $a = 0$ and $3b^2 \equiv -1 \bmod p$. Since $(\frac{3}{p}) = (\frac{-1}{p}) = -1$, this congruence is solvable, and Q has the form $(0, \pm b)$ as claimed.

2.38. Jumping upwards we construct the sequence of integral points $(2, 0)$, $(8, 2)$, $(30, 8)$, $(112, 30)$, ...; the remaining integral points are obtained from these by switching x and y and by replacing (x, y) with $(-x, -y)$. Inverting the process we easily see that there are no other integral points.

There are explicit formulas of Binet type for these integral points. Using $\varepsilon = 2 + \sqrt{3}$ we define integers U_n, V_n via $U_n + V_n\sqrt{3} = \varepsilon^n$; then induction shows that the integral points in the first quadrant are given by $(2V_{n+1}, 2V_n)$.

2.39. The conic $C : x^2 + y^2 - 3xy + 1 = 0$ has the integral point $(x, y) = (1, 1)$. Applying Vieta jumping we obtain the sequence of integral points

$$(1, 1), (1, 2), (2, 5), (5, 13), (13, 34), \ldots$$

and one in which x and y are interchanged. We now prove that there are no other integral points on the conic lying in the first quadrant (we obtain similar sequences in the third quadrant by switching the signs of x and y) (Fig. B.1). In fact, assume that (x, y) is an integral point on C lying in the first quadrant. If $y > x$, then (x, y') with $y' = 3x - y$ is another integral point on C in the first quadrant with $y' \leq x$. In fact, the equation $yy' = x^2 - 1$ immediately implies $y' = \frac{x^2-1}{y} < \frac{x^2}{y} < x$.

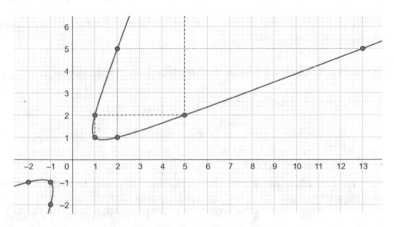

Fig. B.1 Vieta Jumping on the conic $x^2 + y^2 - 3xy + 1 = 0$

Continuing in this way we eventually must find an integral point (ξ, η) in the first quadrant with $\xi = \eta$, and this implies $(\xi, \eta) = (1, 1)$. Thus (x, y) must arise by Vieta jumping from $(1, 1)$, and this is what we wanted to prove.

2.40. We first study the conic $x^2 + y^2 - 3xy + x = 0$. It has the integral point $(x, y) = (1, 1)$. Applying Vieta jumping we obtain the sequence of integral points

$$(1, 1), (1, 2), (4, 2), (4, 10), (10, 25), \ldots.$$

In fact, assume that (a, b) is an integral point. Then the equation $x^2 + b^2 + (1 - 3b)x = 0$ has the solution $x_1 = a$; and the second solution x_2 satisfies $x_1 x_2 = b^2$ and $x_1 + x_2 = 3b - 1$. The second equation tells us that $x_2 = 3b - 1 - a$. Observe that exactly one of x_1 and x_2 is $< b$, and the other is $> b$.

Now consider the equation $\frac{x+1}{y} + \frac{y}{x} = k$, i.e., the conic $C_k : x^2 - kxy + y^2 + x = 0$. Assume that we have an integral point (a, b) on C_k; we have to show that $k = 3$.

We will show that if $b \neq a$, then there exists an integral point (a', b') with $0 < b' < b$ or $0 < a' < a$. Applying this step sufficiently often we will arrive at an integral point (a, a) with $a > 0$. But then $(2 - k)a^2 + a = 0$ implies $a = \frac{1}{k-2}$, and this is a positive integer if and only if $k = 3$.

It remains to construct (a', b'). If $b > a$, then (a, b') with $b' = ka - b$ is another integral point, and $bb' = a^2 + a$ implies that $b' < b$. Similarly, if $a > b$, then $a' = a - (1 - kb)$ is an integer with $aa' = b^2$, hence $a' < a$.

2.41. The equation of Platon's hyperbola $\mathcal{H} : x^2 - 2y^2 = 1$ in the new coordinates X and Y determined by the substitution $x = Y + Y$, $y = Y$ is $\mathcal{H}' : X^2 - 2XY - Y^2 = 1$. Given a point $P(x, y)$ on $C'(\mathbb{Z})$, Vieta jumping gives rise to $P_*(-x - 2y, y)$ and $P^*(x, -2x - y)$; starting from $(1, 0)$ we obtain the sequence of integral points

$$(1, 0), \ (1, 2), \ (-5, 2), \ (-5, -12), \ (29, -12), \ (29, 70), \ \dots,$$

which correspond to the points

$$(1, 0), \ (3, 2), \ (-3, 2), \ (-17, -12), \ (17, -12), \ (99, 70), \ \dots$$

on Platon's hyperbola $\mathcal{H}$.

The standard argument shows that every integral point on $\mathcal{H}$ comes from the sequence beginning with $(1, 0)$ or the one with $(-1, 0)$.

2.42. The substitution $x = X - 2Y$ and $y = Y$ transforms the hyperbola $\mathcal{C}$: $x^2 - 3y^2 = 1$ into $\mathcal{C}' : X^2 - 4XY + Y^2 = 1$. Given a point $P(x, y)$ on $\mathcal{C}'(\mathbb{Z})$, Vieta jumping gives rise to $P_*(4y - x, y)$ and $P^*(x, 4x - y)$; the sequence of integral points starting from $(1, 0)$ is

$$(1, 0), \ (1, 4), \ (15, 4), \ (15, 56), \ \dots;$$

it corresponds to the sequence of integral points

$$(1, 0), \ (-7, 4), \ (7, 4), \ (-97, 56), \ \dots$$

on $\mathcal{C}$. The other sequence of integral points on $\mathcal{C}'$ is

$$(0, 1), \ (4, 1), \ (4, 15), \ (56, 15), \ \dots$$

corresponding to the points

$$(-2, 1), \ (2, 1), \ (-26, 15), \ (26, 15), \ \dots$$

on $\mathcal{C}$ (Fig. B.2).

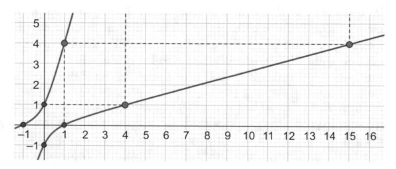

Fig. B.2 Vieta jumping on $\mathcal{C}' X^2 - 4XY + Y^2 = 1$

2.43. The substitution $x = X - nY$ and $y = Y$ transforms the conic $C : x^2 - (n^2 - 1)y^2 = 1$ into $C' : X^2 - 2nXY + Y^2 = 1$. Applying Vieta jumping to $P(x, y)$ we find the points $P^*(x, 2nx - y)$ and $P_*(2ny - x, y)$, hence to the sequences

$$(1, 0), \ (1, 2n), \ (4n^2 - 1, 2n), \ (4n^2 - 1, 8n^3 - 4n), \ \ldots$$

and

$$(0, 1), \ (2n, 1), \ (2n, 4n^2 - 1), \ (8n^3 - 4n, 4n^2 - 1), \ \ldots$$

of integral points on C'.

2.44. The field $\mathbb{Q}(\alpha, \beta)$ as basis $\{1, \beta, \beta^2\alpha, \alpha\beta, \alpha\beta^2\}$. We now multiply each basis element with $\alpha + \beta$ and find

$$
\begin{aligned}
(\alpha + \beta) \cdot 1 &= 0 \cdot 1 + 1 \cdot \beta + 0 \cdot \beta^2 + 1 \cdot \alpha + 0 \cdot \alpha\beta + 0 \cdot \alpha\beta^2 \\
(\alpha + \beta) \cdot \beta &= 0 \cdot 1 + 0 \cdot \beta + 1 \cdot \beta^2 + 0 \cdot \alpha + 1 \cdot \alpha\beta + 0 \cdot \alpha\beta^2 \\
(\alpha + \beta) \cdot \beta^2 &= 2 \cdot 1 + 0 \cdot \beta + 0 \cdot \beta^2 + 0 \cdot \alpha + 0 \cdot \alpha\beta + 1 \cdot \alpha\beta^2 \\
(\alpha + \beta) \cdot \alpha &= 3 \cdot 1 + 0 \cdot \beta + 0 \cdot \beta^2 + 0 \cdot \alpha + 1 \cdot \alpha\beta + 0 \cdot \alpha\beta^2 \\
(\alpha + \beta) \cdot \alpha\beta &= 0 \cdot 1 + 3 \cdot \beta + 0 \cdot \beta^2 + 0 \cdot \alpha + 0 \cdot \alpha\beta + 1 \cdot \alpha\beta^2 \\
(\alpha + \beta) \cdot \alpha\beta^2 &= 0 \cdot 1 + 0 \cdot \beta + 3 \cdot \beta^2 + 2 \cdot \alpha + 0 \cdot \alpha\beta + 0 \cdot \alpha\beta^2
\end{aligned}
$$

Using the vector $B = (1, \beta, \beta^2, \alpha, \alpha\beta, \alpha\beta^2)^t$ we write this set of equation in matrix form:

$$
(\alpha + \beta) \cdot B =
\begin{pmatrix}
0 & 1 & 0 & 1 & 0 & 0 \\
0 & 0 & 1 & 0 & 1 & 0 \\
2 & 0 & 0 & 0 & 0 & 1 \\
3 & 0 & 0 & 0 & 1 & 0 \\
0 & 3 & 0 & 0 & 0 & 1 \\
0 & 0 & 3 & 2 & 0 & 0
\end{pmatrix}
\cdot B.
$$

Call the matrix in this equation M; then $(\alpha + \beta)B = MB$ tells us that $\alpha + \beta$ is an eigenvalue of the eigenvector B, hence $\alpha + \beta$ must be a root of the equation $\det(xI - M) = 0$. Working out the determinant produces a monic polynomial f of degree 6 with $f(\alpha + \beta) = 0$, namely

$$f(x) = x^6 - 9x^4 - 4x^3 + 27x^2 - 36x - 23.$$

This procedure shows that if α and β are algebraic integers (if you cannot see where we are using this fact, go through the calculation with $\alpha = \frac{1+\sqrt{3}}{2}$), then so is $\alpha + \beta$.

The same procedure works for products of algebraic integers; here we multiply B with $\alpha \cdot \beta$.

2.45. If $f(a) = a_0 + a_1 x + \ldots + a_{n-1} x^{n-1}$ and $g(x) = b_0 + b_1 x + \ldots + b_{n-1} x^{n-1}$, then $\alpha = f(\zeta)$ and $\beta = g(\zeta)$ are added and multiplied like polynomials, i.e., we set $s(x) = f(x) + g(x)$, $m(x) = f(x)g(x)$ and $\alpha + \beta = s(zeta)$ and $\alpha\beta = m(\zeta)$; in the case of the product, we replace ζ^n (and all higher powers of ζ) by smaller powers using $\zeta^n = -1 - \zeta - \ldots - \zeta^{n-1}$. This gives $\mathbb{Z}[\zeta]$ a ring structure (distributivity and associativity follow from the corresponding properties of polynomial rings).

2.46. The claims are simple results of Galois theory and the fact that algebraic integers form a ring. Since the product of all conjugates is fixed by the Galois group of the normal closure of K, it is a rational number, and since it is a product of algebraic integers, it must be an integer.

If $N(\omega) = \pm 1$, then ω and $\omega' = \pm\omega_2 \cdots \omega_n$ are elements of K whose product is 1, hence ω is a unit. Conversely, if ω is a unit, then $\omega\omega' = 1$, and taking norms shows that $N(\omega)$ divides 1 in $\mathbb{Z}$. This implies that $N\omega = \pm 1$.

2.47. If $x^2 - 8y^2 = -1$, then $x^2 + 1 = 8y^2$, and clearly both x^2 and $y^2 + 1 = 8y^2$ are powerful. Similarly, x^2 and $x^2 - 1$ are powerful if $x^2 - 8y^2 = 1$. Since both equations have infinitely many integral solutions, the claim follows.

Chapter 3

3.1. For fields with characteristic $\neq 2$, the equation of $\mathcal{P}$ is equivalent to $(x + \frac{y}{2})^2 - \Delta(\frac{y}{2})^2 = 1$. According to Theorem 3.1, we have

$$x + \frac{y}{2} = \frac{\Delta s^2 + r^2}{\Delta s^2 - r^2}, \quad \frac{y}{2} = \frac{2rs}{\Delta s^2 - r^2}.$$

The number of $\mathbb{F}_p$-rational points on $\mathcal{C}$ for odd primes p follows from the corresponding result for Pell conics of the form $X^2 - dY^2 = 1$: We have

$$4x^2 + 4xy - 4my^2 = (2x + y)^2 - \Delta y^2 = 4,$$

and this implies $\#\mathcal{P}(\mathbb{F}_p) = p - (\frac{\Delta}{p})$ for all odd primes.

The equation $\#\mathcal{P}(\mathbb{F}_p) = p - (\frac{\Delta}{p})$ also holds for $p = 2$ if we define $(\frac{\Delta}{2}) = +1$ or $= -1$ according as $\Delta \equiv 1$ or $\Delta \equiv 5 \bmod 8$. In fact, if $\Delta \equiv 1 \bmod 8$, then m is even, and the congruence $x^2 + xy \equiv 1 \bmod 2$ has the unique solution $(x, y) = (1, 1)$ modulo 2. If $\Delta \equiv 5 \bmod 8$, on the other hand, then m is odd, and $x^2 + xy + y^2 \equiv 1 \bmod 2$ has three solutions $(x, y) = (1, 0)$, $(0, 1)$, and $(1, 1)$ modulo 2.

3.2. Intersecting the pencil of lines $y = tx$ with the curve $\mathcal{C} : y^2 = x^3$ yields the equation $t^2 x^2 = x^3$. The lines intersect the curve twice in $x = 0$; the other solution is given by $x = t^2$ and $y = tx = t^3$. Thus we obtain the parametrization $(x, y) = (t^2, t^3)$, and this parametrization includes the singular point $(0, 0)$. The same formulas hold for $\mathbb{F}_p$-rational points, hence $\#\mathcal{C}(\mathbb{F}_p) = p$.

For $\mathcal{C} : y^2 = x^3 + x^2$ we find with the same pencil $y = tx$ that $t^2 x^2 = x^3 + x^2$; this is equivalent to $x^2(x + 1 - t^2) = 0$. This yields the parametrization $(x, y) = (t^2 - 1, t^3 - t)$. The singular point $(0, 0)$ is parametrized twice, namely for $t = \pm 1$. The last fact is responsible for $\#\mathcal{C}(\mathbb{F}_p) = p - 1$.

3.3. The map λ is well defined since $\lambda(s + \mathbb{Z}) = \lambda(s + 1 + \mathbb{Z})$. Now we find

$$\lambda(s + \mathbb{Z}) + \lambda(t + \mathbb{Z}) = (\cos 2\pi s, \sin 2\pi s) + (\cos 2\pi t, \sin 2\pi t)$$

$$= (\cos(2\pi s)\cos(2\pi t) - \sin(2\pi s)\sin(2\pi t),$$

$$\sin(2\pi s)\cos(2\pi t) + \sin(2\pi t)\cos(2\pi s))$$

$$= (\cos(2\pi(s + t)), \sin(2\pi(s + t)) = \lambda(s + t + \mathbb{Z}).$$

The kernel of λ consists of all cosets $t + \mathbb{Z}$ with $\cos 2\pi t = 1$ and $\sin 2\pi t = 0$; these equations imply that t is an integer, hence $t + \mathbb{Z} = \mathbb{Z}$, and λ is injective. Since λ is clearly surjective, it must be bijective.

Since λ is an isomorphism, the image of the cyclic subgroup of order n generated by $\frac{1}{n} + \mathbb{Z}$ generates a cyclic subgroup of order n on the unit circle. The claims now follow since $e^{\frac{2\pi i}{n}} = \cos\frac{2\pi}{n} + i\sin\frac{2\pi}{n}$ is a primitive n-th root of unity in the complex plane.

3.4. This is clear since $(\frac{a}{p}) = (\frac{bc^2}{p}) = (\frac{b}{p})(\frac{c}{p})^2 = (\frac{b}{p})$.

3.5. For computing $(\frac{2}{15})$ we work modulo 15:

$$
\begin{array}{ll}
2 \cdot 1 \equiv 2 & 2 \cdot 4 \equiv -7 \\
2 \cdot 2 \equiv 4 & 2 \cdot 5 \equiv -5 \\
2 \cdot 3 \equiv 6 & 2 \cdot 6 \equiv -3 \\
& 2 \cdot 7 \equiv -1
\end{array}
$$

which implies that $(\frac{2}{15}) = (-1)^4 = +1$.

For computing $(\frac{3}{35})$, we choose the half system $\{1, 2, \ldots, 17\}$ and compute the remainders of the products with 3; the only negative remainders occur for

$$
\begin{array}{ll}
3 \cdot 1 \equiv -17 & 3 \cdot 4 \equiv -8 \\
3 \cdot 2 \equiv -14 & 3 \cdot 5 \equiv -5 \\
3 \cdot 3 \equiv -11 & 3 \cdot 6 \equiv -2,
\end{array}
$$

which implies that $(\frac{3}{35}) = (-1)^6 = +1$.

3.6. Write $m = p_1 \cdots p_r q_1 \cdots q_s$ for primes $p_j \equiv 1 \bmod 4$ and $q_j \equiv 3 \bmod 4$. Then $m \equiv (-1)^s \bmod 4$ and

$$\frac{m-1}{2} \equiv \begin{cases} 0 \bmod 2 & \text{if } s \equiv 0 \bmod 2, \\ 1 \bmod 2 & \text{if } s \equiv 1 \bmod 2. \end{cases}$$

Thus $\frac{m-1}{2} \equiv s \bmod 2$ and $(-1)^{\frac{m-1}{2}} = (-1)^s$.
On the other hand,

$$\left(\frac{-1}{m}\right) = \prod_{i=1}^{r} \left(\frac{-1}{p_i}\right) \cdot \prod_{j=1}^{s} \left(\frac{-1}{q_i}\right) = (-1)^s.$$

This proves the claim.

3.7. The permutation $(a_1 a_2)(a_2 a_3) \cdots (a_{n-1} a_n)$ maps a_1 to a_2, a_2 to a_3, ..., and a_{n-1} to a_n (recall that these maps are applied from right to left). The element a_n is first mapped to a_{n-1}, then to a_{n-2}, ..., and finally to a_1. This proves the claim.

3.8. We will give a proof by example. Consider the sets $A = \{1, 2\}$ and $B = \{1, 2, 3\}$. Then

$$A \times B = \{(1, 1), (1, 2), (1, 3), (2, 1), (2, 2), (2, 3)\}.$$

The permutation π_A that swaps 1 and 2 has signature -1, and the induced permutation on $A \times B$ swaps $\#B = 3$ elements.

3.9. If we take the product over all $\phi(mn)/2$ congruences $a \cdot a_j \equiv (-1)^{s_j} a'_j$, then we obtain, after cancelling the products of all integers in the half system, the congruence

$$a^{\phi(mn)/2} \equiv (-1)^{\sum s_j} \bmod mn.$$

Now $\phi(mn) = \phi(m)\phi(n)$, so from $a^{\phi(m)/2} \equiv \pm 1 \bmod m$ it follows that $a^{\phi(mn)/2} \equiv (\pm 1)^{\phi(n)} \equiv +1 \bmod m$ since $\phi(n)$ is even. Similarly we can show that $a^{\phi(mn)/2} \equiv 1 \bmod n$, and since m and n are coprime, this implies that $a^{\phi(mn)/2} \equiv 1 \bmod mn$, and our claim follows.

3.10. If $n = 2k + 1$ is odd, then for each prime p dividing $2^m - 1$ we have $p \mid 3^{2k+1} - 1$, hence $(\frac{3}{p}) = +1$. By modularity, this condition is equivalent to $p \equiv \pm 1 \bmod 12$. But then $2^n - 1 \equiv \pm 1 \bmod 12$, hence $2^n \equiv 0, 2 \bmod 12$, which is impossible.

3.11. If we sum over $-t$ instead of t, then we get

$$S = \sum_{t=0}^{p-1} \left(\frac{-t}{p}\right)\left(\frac{(-t)^2 - 1}{p}\right) = \left(\frac{-1}{p}\right) \sum_{t=0}^{p-1} \left(\frac{t}{p}\right)\left(\frac{t^2 - 1}{p}\right) = -S,$$

hence $S = 0$.

3.12. Each residue class $t \bmod mn$ can be written uniquely in the form $t = rm + sn$, where r and s run through the residue classes modulo n and m, respectively. Thus

$$\phi_{mn}(1) = \sum_{t=0}^{pq-1} \left(\frac{t}{mn}\right)\left(\frac{t^2-1}{mn}\right) = \sum_{r=0}^{n-1}\sum_{s=0}^{m-1} \left(\frac{rm+sn}{mn}\right)\left(\frac{(rm+sn)^2-1}{mn}\right)$$

$$= \sum_{r=0}^{n-1}\sum_{s=0}^{m-1} \left(\frac{sn}{m}\right)\left(\frac{rm}{n}\right)\left(\frac{(sn)^2-1}{m}\right)\left(\frac{(rm)^2-1}{n}\right)$$

$$= \sum_{s=0}^{m-1} \left(\frac{sn}{m}\right)\left(\frac{(sn)^2-1}{m}\right) \sum_{r=0}^{n-1} \left(\frac{rm}{n}\right)\left(\frac{(rm)^2-1}{n}\right)$$

$$= \phi_m(1)\phi_n(1).$$

3.13. I do not know how to prove this claim.

3.14. The first equation is clear since $Q_2(m, n) = m + n$. Write $m + n = 2k$ for some odd integer k.

$$\left(\frac{m+n}{Q_q(m,n)}\right) = \left(\frac{2}{Q_q(m,n)}\right)\left(\frac{k}{Q_q(m,n)}\right).$$

Since $Q_q(m, n) \equiv q \bmod 8$ we have $\left(\frac{2}{Q_q(m,n)}\right) = \left(\frac{2}{q}\right)$.

For evaluating the second symbol we use quadratic reciprocity; to this end we observe that since $mn \equiv 1 \bmod 8$ we have $k \equiv 1 \bmod 4$:

$$\left(\frac{k}{Q_q(m,n)}\right) = \left(\frac{Q_q(m,n)}{k}\right) = 1$$

because $n \equiv -m \bmod k$ implies that

$$Q_q(m,n) = \frac{m^q - n^q}{m - n} \equiv \frac{2m^q}{2m} \equiv m^{q-1} = \left(m^{\frac{q-1}{2}}\right)^2 \bmod k$$

is a square modulo k.

3.15. We have

$$Q_r(m,n) = \frac{m^{2t} - n^{2t}}{m^2 - n^2} \frac{m^2 - n^2}{m - n} = Q_t(m^2, n^2) \cdot (m + n),$$

hence

$$\left(\frac{Q_r(m,n)}{Q_q(m,n)}\right) = \left(\frac{Q_2(m^2,n^2)}{Q_q(m,n)}\right)\left(\frac{Q_t(m^2,n^2)}{Q_q(m,n)}\right).$$

The claim now follows from (3.22).

3.16. The claim is true for $q = 1$ and all odd integers p since $Q_1(m, n) = 1$. Assume that the claim holds for all odd integers $q' < q$ and all $p \neq q$. If $p < q$, we have, by the quadratic reciprocity law,

$$\left(\frac{Q_p(m, n)}{Q_q(m, n)}\right) = (-1)^{\frac{p-1}{2} \cdot \frac{q-1}{2}} \left(\frac{Q_q(m, n)}{Q_p(m, n)}\right) = (-1)^{\frac{p-1}{2} \cdot \frac{q-1}{2}} \left(\frac{q}{p}\right) = \left(\frac{p}{q}\right).$$

If $p > q$ write $p \equiv r \bmod q$ with $0 < r < q$ and use the preceding exercise. Applying this result inductively to $r = 2^j t$ for some odd integer $t < q$ we find

$$\left(\frac{Q_r(m, n)}{Q_q(m, n)}\right) = \left(\frac{2}{q}\right)^j \left(\frac{Q_t(m^2, n^2)}{Q_q(m, n)}\right) = \left(\frac{2}{q}\right)^j \left(\frac{t}{q}\right) = \left(\frac{r}{q}\right).$$

This implies the claim if $p > q$, and thus finishes the proof.

3.17. 1. For $p = 13$, only the pairs $(3, 4)$ and $(9, 10)$ are consecutive quadratic residues, hence $RR = 2$. Similarly, the pairs $(5, 6)$, $(6, 7)$, and $(7, 8)$ are consecutive nonresidues, hence $NN = 3$. Similarly we find $RN = NR = 3$.
2. The equation $RR + RN + NR + NN = p - 2$ follows from the fact that there are exactly $p - 2$ pairs of consecutive nonzero residue classes modulo p.
3. Assume that $a(a + 1)$ is a quadratic residue modulo p, and write $a(a + 1) = y^2$ in $\mathbb{F}_p$. completing the square gives $(2a + 1)^2 = 4y^2 + 1$, hence $(2a + 1 + 2y)(2a + 1 - 2y) = 1$. Setting $2a + 1 + 2y = t$ and $2a + 1 - 2y = \frac{1}{t}$ for some $t \in \mathbb{F}_p^{\times}$ we obtain a parametrization of the conic $a^2 + a = y^2$. The elements $t = \pm 1$ give rise to the trivial solutions $a = 0$ and $a + 1 = 0$; moreover, the values t and $\frac{1}{t}$ give rise to the same value of a. Thus we obtain $\frac{p-3}{2}$ pairs $(a, a+1)$ for which $a(a+1)$ is a nonzero quadratic residue modulo p.
The second claim follows from 2. and $RR + NN = \frac{p-3}{2}$.
4. Let $a = x^2$ and $a + 1 = y^2$ be squares in $\mathbb{F}_p^{\times}$; then $y^2 - x^2 = 1$. Thus each pair $(a, a+1)$ of consecutive quadratic residues gives rise to four points $(\pm x, \pm y)$ on the conic $\mathcal{H} : Y^2 - X^2 = 1$. Conversely, each of these points produces the pair (x^2, y^2) of consecutive quadratic residues.
For $p = 13$, this correspondence is given by

a	3	9
P	$(\pm 4, \pm 2)$	$(\pm 3, \pm 6)$

Observe that the group law on $\mathcal{C}$ with neutral element $N(0, 1)$ is given by

$$(x_1, y_1) \oplus (x_2, y_2) = (x_1 y_2 + x_2 y_1, x_1 x_2 + y_1 y_2).$$

Thus $P = (5, 0)$ has order 4 since $2P = (0, -1)$ is the element of order 2. Moreover, $(x, y) \oplus (0, -1) = (-x, -y)$ and $(x, y) \oplus (5, 0) = (5y, 5x)$.

3.18. Since $(\frac{-1}{p}) = -1$, the quadratic nonresidues $\{n_1, \ldots, n_m\}$ form a half system modulo $p = 2m + 1$.

If $(\frac{a}{p}) = -1$, then $an_i \equiv -n_j \bmod p$; similarly $an_i \equiv n_j \bmod p$ if $(\frac{a}{p}) = +1$. In the first case, there are an odd number of sign changes (namely m), in the second case there is none. Thus Gauss's Lemma holds in both cases.

3.19. We have $N_2 = p-1$ and $N_4 = p^2(p-1)+pN_2 = p^3 - p^2 + p^2 - p = p^3 - p$. Assume that $N_{n-2} = p^{n-3} - p^{\frac{n-4}{2}}$; then

$$N_n = p^{n-2}(p-1) + p(p^{n-3} - p^{\frac{n-4}{2}}) = p^{n-1} - p^{n-2} + p^{n-2} - p^{\frac{n-2}{2}} = p^{n-1} - p^{\frac{n-2}{2}}$$

as claimed.

3.20. It is easy to produce infinitely many integral solutions of $Q_4(m, n) = x^4$: If $Q_4(m, n) = c$, then $Q_4(mc, nc) = c^4$. If $m \equiv n \equiv 1 \bmod 4$, then

$$Q_4(m, n) = \frac{m^4 - n^4}{m - n} = (m + n)(m^2 + n^2) = 4c$$

for some odd integer c, hence $Q_4(mc, nc) = 4c^4$ and $mc \equiv nc \bmod 4$.

3.21. The first claim is Gauss's Lemma.

Assume that $hq \equiv r \bmod p$ for some $0 < r < \frac{p}{2}$; then $hq - kp = r$ for some integer k with $0 < k < \frac{q}{2}$, and then $kp \equiv -r \bmod q$. Thus if r is positive in the first row, then $-r$ shows up in the second row. The other negative remainders in the second row are the numbers $-r$ with $\frac{p}{2} < r < \frac{q}{2}$, and they come in pairs. In fact, if $k < \frac{q-1}{2}$ and

$$kp \equiv -r \bmod q \qquad \text{and} \qquad \frac{1}{2}p < r < \frac{1}{2}q,$$

then we also have

$$k'p \equiv -r' \bmod q \qquad \text{and} \qquad \frac{1}{2}p < r' < \frac{1}{2}q,$$

where

$$k' = \frac{q-1}{2} - k \quad \text{and} \quad r' = \frac{p+q}{2} - r.$$

Clearly $k = k'$ if and only if $k = \frac{q-1}{4}$, which happens if and only $q \equiv 1$ mod 4. Moreover, the value $k = \frac{q-1}{2}$ yields the positive remainder $\frac{q-p}{2}$ mod q since

$$\frac{q-1}{2}p = \frac{p-1}{2}q + \frac{q-p}{2}.$$

Thus the negative remainders consist of

- the $\frac{p-1}{2}$ numbers $-1, -2, \ldots, -r$;
- the pairs of negative numbers $(-r, -r')$ with $\frac{p}{2} < r, r' < \frac{q}{2}$, which is an even number;
- the number $\frac{-p\pm q}{4} \equiv \frac{q-1}{4}p$ mod q, which exists if $q \equiv 1$ mod 4 and is negative if, in addition, $p \equiv 3$ mod 4.

This shows that the number of negative remainders is odd if and only if $\frac{p-1}{2}$ is odd and $q \equiv 3$ mod 4. But this implies the quadratic reciprocity law.

Chapter 4

4.1. Euclidean division shows that $f(x) = (x - a)q(x) + r$ for some constant r. Plugging in $x = a$ shows that $f(a) = 0$ if and only if $r = 0$, i.e., if and only if $x - a$ divides $f(x)$. This argument holds in all polynomial rings over fields because they are Euclidean.
These results are in fact valid over arbitrary polynomial rings: if $f(x) = a_n x^n + \ldots + a_1 x + a_0$, then $a_0 = 0$ implies that $f(x) = x g(x)$ for $g(x) = a_n x^{n-1} + \ldots + a_1$. Conversely, $f(x) = x g(x)$ implies $f(0) = 0$. The substitution $X = x - a$ allows us to prove this for general a.
Observe, however, that in $(\mathbb{Z}/8\mathbb{Z})[x]$, the polynomial $x^2 - 1$ is divisible by $x - a$ for $a = 1, 3, 5, 7$ since $x^2 - 1 = (x - 1)(x + 1) = (x - 3)(x + 3)$. In $(\mathbb{Z}/6\mathbb{Z})[x]$, the linear polynomial $f(x) = 3x + 2$ does not have any root.

4.2. The relation $2 \cdot 3 = (1 + \sqrt{-5})(1 - \sqrt{-5})$ is a counterexample to the Four Numbers Theorem in $\mathbb{Z}[\sqrt{-5}]$ since all factors are irreducible. The Four Numbers Theorem claims that all factorizations can be explained by a common refinement.
The equation $2 \cdot 3 = (2 + \sqrt{-2})(2 - \sqrt{-2})$ is compatible with the Four Numbers Theorem in $\mathbb{Z}[\sqrt{-2}]$ since

$$2 = -\sqrt{-2} \cdot \sqrt{-2}, \qquad\qquad 3 = (1 + \sqrt{-2})(1 - \sqrt{-2}),$$
$$2 + \sqrt{-2} = \sqrt{-2}(1 - \sqrt{-2}), \quad 2 - \sqrt{-2} = -\sqrt{-2}(1 + \sqrt{-2}).$$

4.3. Clearly $(p, q) \sim (p, q)$ since $pq = qp$, and if $(p, q) \sim (r, s)$, then $(r, s) \sim (p, q)$ since $ps = qr$ is equivalent to $rq = ps$.

For checking transitivity assume that $(p, q) \sim (r, s)$ and $(r, s) \sim (t, u)$. Then $ps = qr$ and $ru = st$. Multiplying these equations yields $prsu = qrst$; since R is a domain, we can cancel rs and obtain $pu = qt$, which implies $(p, q) \sim (t, u)$.

The cancellation rule in domains may be proved as follows: If $ac = bc$ for $c \neq 0$, then $(a - b)c = 0$. Since R is a domain, this implies $a = b$ since $c \neq 0$ by assumption.

When verifying the next claims keep in mind that we think of (p, q) as the "fraction" $\frac{p}{q}$. Clearly $(p, q) + (0, 1) = (p, q)$ and $(p, q) \cdot (1, 1) = (p, q)$, so $(0, 1)$ and $(1, 1)$ are the neutral elements with respect to addition and multiplication, respectively. The additive inverse of (p, q) is $(-p, q)$, and the multiplicative inverse of (p, q) is (q, p) if $p \neq 0$.

Finally, the map $\iota : R \longrightarrow K : r \to (r, 1)$ is an injective ring homomorphism (which allows us to interpret R as a subring of the field we just have constructed). In fact, $1 \in R$ is mapped to $(1, 1) \in K$, and $\lambda(rs) = (rs, 1) = (r, 1)(s, 1) = \lambda(r)\lambda(s)$. Finally, $r \in \ker \lambda$ if and only if $(r, 1) \sim (1, 1)$, which is equivalent to $r \cdot 1 = 1 \cdot 1 = 1$, i.e., to $r = 1$.

4.4. We have $a \equiv b \bmod m$ in R if and only if $a - b = mq$ for some $q \in R$. Since $R \subseteq S$, this implies $a \equiv b \bmod m$ in S.

The converse does not hold in general; in the ring $S = \mathbb{Z}[\frac{1}{2}]$, we have $1 \equiv 0 \bmod 2$ since $1 - 0 = 2 \cdot \frac{1}{2}$.

4.5. Another example is

$$\frac{2 + \sqrt{-5}}{3} = \frac{3}{2 - \sqrt{-5}}.$$

4.6. Since $N\alpha = \alpha\alpha'$ we clearly have $\alpha \mid N\alpha$. If $\alpha \mid \beta$, then $\alpha = \beta\gamma$, and taking the norm yields $N\alpha = N\beta \cdot N\gamma$, which implies that $N\alpha \mid N\beta$ in $\mathbb{Z}$.

4.7. If $\sqrt{-2} \mid y$ in $\mathbb{Z}[\sqrt{-2}]$ for some $y \in \mathbb{Z}$, then $2 = N(\sqrt{-2}) \mid N(y) = y^2$, and since 2 is prime in $\mathbb{Z}$ we conclude that $2 \mid y$.

If $\sqrt{m} \mid y$, where m us squarefree, then $m \mid N(y) = y^2$ as above, and this implies $m \mid y$ (prove this one prime at a time).

If $\alpha = y \in \mathbb{Z}$, then clearly $\alpha \mid y$ but $N\alpha \nmid y$.

4.8. This follows from $a + bi - (a + b) = b(i - 1) = bi(1 + i)$.

4.9. Assume that $a \equiv b \bmod m$ and $c \equiv d \bmod m$. Then $a - b = mq$ and $c - d = mr$, hence $(a + c) - (b + d) = m(q + r)$, which means $a + c \equiv b + d \bmod m$. Moreover, $ac - bd = ac - ad + ad - bd = (a(c - d) + d(a - b)) = amr + dmq = m(ar + dq)$, hence $ac \equiv bd \bmod m$.

4.10. If $m \mid (a + b\omega)$, then $a + b\omega = m(c + d\omega) = mc + md\omega$, hence $a - mc = -(b + md)\omega$. Since ω is irrational, we must have $a = mc$ and $b = md$, i.e., $m \mid a$ and $m \mid b$.

4.11. This is a special case of Exercise 4.4.

4.12. If $e \in R$ is a unit, then $ee' = 1$ for some $e' \in R$. This implies that e has an inverse, namely e'. If e_1 and e_2 are units, then $e_1 e_1' = e_2 e_2' = 1$, hence $(e_1 e_2)(e_1' e_2') = 1$, which implies that the product $e_1 e_2$ is a unit.

4.13. In a field, each nonzero element has a multiplicative inverse and thus is a unit.

4.14. If R is a domain, then $\deg(fg) = \deg(f) + \deg(g)$ for polynomials $f, g \in R[X]$. Thus $f(x)g(x) = 1$ implies $\deg(f) + \deg(g) = 0$, i.e., $\deg(f) = \deg(g) = 0$. Thus f and g are constants and therefore units of R.
The polynomial $2X + 1$ in $(\mathbb{Z}/4\mathbb{Z})[X]$ is a unit since $(2X + 1)(2X + 1) = X^2 + 4X + 1 \equiv 1 \bmod 4$.

4.15. If $\varepsilon = r + s\sqrt{m}$ is a unit in $R = \mathbb{Z}[\sqrt{m}]$ for $m < -1$, then $1 = r^2 - ms^2 = r^2 + |m|s^2$ implies that $s = 0$ and $\varepsilon = x = \pm 1$.

4.16. Let σ denote the nontrivial automorphism of the quadratic number field. If ε is a unit then so is ε^σ since $\varepsilon\varepsilon^\sigma = N(\varepsilon) = \pm 1$. In addition, we have $(\varepsilon_1 \varepsilon_2)^\sigma = \varepsilon_1^\sigma \varepsilon_2^\sigma$ and $(\varepsilon^\sigma)^\sigma = \varepsilon = \varepsilon^{\sigma \cdot \sigma}$. The other properties are trivially true.

4.17. If $\pi \mid n$ for some natural number n and if $n = p_1 \cdots p_t$, then π must divide some p_j since it is prime.

4.18. Clearly $\alpha = \frac{a+bi}{a-bi}$ has norm $N\alpha = 1$. Such an element is a unit only if $a + bi$ is a product of powers of i and $1 + i$ times a natural number.

4.19. Given any pair of nonzero integers a and m, we can repeatedly subtract m from a until the remainder is smaller than $|m|$; then $a - qm = r$ with $0 \le r < |m|$, and hence $\mathbb{Z}$ is Euclidean with respect to the absolute value.

4.20. Given any pair of nonzero polynomials $a, b \in K[x]$, long division provides us with polynomials $q, r \in K[x]$ such that $a = bq + r$ and $\deg r < \deg b$. The last equality is equivalent to $2^{\deg r} < 2^{\deg b}$.

4.21. In $\mathbb{Z}[\sqrt{-6}]$, the elements 2, 3, and $\sqrt{-6}$ are irreducible since there do not exist elements with norms 2 or 3 (the equation $x^2 + 6y^2 = 2$, for example, clearly is not solvable in integers). Thus $2 \cdot 3 = -\sqrt{-6} \cdot \sqrt{-6}$ in $\mathbb{Z}[\sqrt{-6}]$ is an example of nonunique factorization in $\mathbb{Z}[\sqrt{-6}]$.
Similarly, $2 \cdot 7 = (2 + \sqrt{-10})(2 - \sqrt{-10})$ is an example of nonunique factorization in $\mathbb{Z}[\sqrt{-10}]$.
The factorization $2 \cdot 3 = \sqrt{6} \cdot \sqrt{6}$ in $\mathbb{Z}[\sqrt{6}]$, on the other hand, is not an example of nonunique factorization since 2, 3, and $\sqrt{6}$ are not irreducible. Clearly $2 = -(2 + \sqrt{6})(2 - \sqrt{6})$, $3 = (3 + \sqrt{6})(3 - \sqrt{6})$, and $\sqrt{6} = (2 + \sqrt{6})(3 - \sqrt{6})$.

4.22. We find the following factorizations; if none is given, the element is irreducible.

p	2	3	5
-5			$-(\sqrt{-5})^2$
-3		$-(\sqrt{-3})^2$	
-2	$-(\sqrt{-2})^2$	$(1+\sqrt{-2})(1-\sqrt{-2})$	
-1	$i(1-i)^2$		$(2+i)(2-i)$
2	$\sqrt{2}^2$		
3	$(2-\sqrt{3})(1+\sqrt{3})^2$	$\sqrt{3}^2$	
5			$\sqrt{5}^2$

4.23. Assume that $\pi = \alpha\beta$. Taking norms shows that $N\pi = N\alpha N\beta$. Since $N\pi$ is a prime by assumption, it is irreducible, hence $N\alpha = \pm 1$ or $N\beta = \pm 1$; this implies that α or β is a unit, hence π is irreducible.

4.24. Most of these claims can be proved using the prime factorization.

1. Here we write $a = \prod p_i^{a_i}$ and $b = \prod p_i^{b_i}$. Then $\gcd(a, b) = \prod p_i^{\min(a_i, b_i)}$ and $\gcd(a^2, b^2) = \prod p_i^{\min(2a_i, 2b_i)} = (\gcd(a, b))^2$ as claimed.
2. Assume there is a prime $p \mid \gcd(a^2, b)$. Then $p \mid a^2$ and $p \mid b$. Since p is prime, $p \mid a^2 = a \cdot a$ implies that $p \mid a$, and then $p \mid \gcd(a, b)$: Contradiction.
3. Let $d = \gcd(a, b)$; then $d \mid a$, $d \mid b$, hence $d \mid (a + b)$ and therefore $d \mid \gcd(a + b, b)$.
   Conversely, if $d = \gcd(a + b, b)$, then $d \mid (a + b)$ and $d \mid b$, hence d divides $(a + b) - b = a$, and we have $d \mid \gcd(a, b)$. But then $\gcd(a, b)$ and $\gcd(a + b, b)$ divide each other, hence differ at most by a unit.
4. Write $a = \prod p_i^{a_i}$, $b = \prod p_i^{b_i}$ and $r = \prod p_i^{r_i}$. Then

$$\gcd(ra, rb) = \prod p_i^{\min(a_i + r_i, b_i + r_i)} = \prod p_i^{r_i + \min(a_i, b_i)} = r \gcd(a, b).$$

4.25. Any common divisor of $a = 1 + \sqrt{-5}$ and $b = 1 - \sqrt{-5}$ divides their sum 2; since 2 is irreducible, the greatest common divisor is either 1 or 2. But it cannot be 2 since $\frac{1+\sqrt{-5}}{2}$ is not an element of $\mathbb{Z}[\sqrt{-5}]$.
On the other hand, $a^2 = -4 + 2\sqrt{-5}$ and $b^2 = -4 - 2\sqrt{-5}$ have common divisor 2, which is easily seen to be their greatest common divisor.

4.26. Clearly $\omega \in R[\frac{1}{2}]$; it is therefore sufficient to show that $\frac{1}{2} \in S$. But $(\frac{1+\sqrt{-5}}{2})^2 = \frac{-2+\sqrt{-5}}{2}$, so S contains $\frac{-2+\sqrt{-5}}{2} - \frac{1+\sqrt{-5}}{2} + 2 = \frac{1}{2}$. This proves that $S = R[\frac{1}{2}]$.

Every element of S has the form $\frac{a+b\sqrt{-5}}{2^m}$, and such an element is in $\mathbb{Q}$ if and only if $b = 0$. Thus $S \cap \mathbb{Q}$ consists of all elements of the form $\frac{a}{2^m}$, which is $\mathbb{Z}[\frac{1}{2}]$.

The factorization $6 = 2 \cdot 3 = (1 + \sqrt{-5})(1 + \sqrt{-5})$ is not an example of nonunique factorization in S. In fact, 2 is a unit, and

$$3 = \frac{1}{2}(1 - \sqrt{-5})(1 + \sqrt{-5})$$

is a factorization of 3 into the unit $\frac{1}{2}$ and the two irreducible (and even prime) elements $1 \pm \sqrt{-5}$.

The factorization $3 \cdot 3 = (2 - \sqrt{-5})(2 + \sqrt{-5})$ can be refined:

$$3 \cdot 3 = \frac{1}{4}(1 - \sqrt{-5})^2(1 + \sqrt{-5})^2,$$

and we also have

$$(2 + \sqrt{-5}) = -2(1 - \sqrt{-5})^2,$$

where -2 is a unit in S.

4.27. Setting $x + y\sqrt{-5} = (r + s\sqrt{-5})^2$ gives $x = r^2 - 5s^2$ and $y = 2rs$. Clearly the solution $(x, y, z) = (2, 1, 3)$ of $x^2 + 5y^2 = z^2$ does not have this form. Now let us work in the domain $S = \mathbb{Z}[\sqrt{-5}, \frac{1}{2}]$. Here $x + y\sqrt{-5} = \pm 2^n(r + s\sqrt{-5})^2$ for some unit $\pm 2^n$. This implies $x = \pm 2^n(r^2 - 5s^2)$ and $y = \pm 2^{n+1}rs$. We are interested in coprime integral solutions, so we may assume $\gcd(r, s) = 1$. Now there are two cases:

- r and s have different parity; then $n = 0$ and $x = \pm(r^2 - 5s^2)$, $y = \pm 2rs$.
- r and s are both odd; then $n = -1$ and $x = \frac{r^2 - 5s^2}{2}$, $y = rs$.
  Choosing $r = 1$, $s = -1$ and the negative sign gives us the solution $(x, y, z) = (2, 1, 3)$.

4.28. Assume that $ab = ex^n$ for some unit e, and that $\gcd(a, b) = p$. Then both a and b must be divisible by p, but not both of them are divisible by p^2. Assume therefore that $p \parallel a$, i.e., that $p \mid a$ and $p^2 \nmid a$. Since $ab = ex^n$ we must have $p^{n-1} \mid b$. Thus we can write $a = pa_1$ and $b = p^{n-1}b_1$; Proposition 4.12 shows that there exist units e_1 and e_2 with $a_1 = e_1c^n$ and $b_1 = e_2d^n$. This implies the claim.

4.29. We have $x^3 = 4y^2 - 1 = (2y - 1)(2y + 1)$. Since the factors on the right are coprime, we must have $2y - 1 = a^3$ and $2y + 1 = b^3$. This implies $b^3 - a^3 = 2$, and since there are no cubes that differ by 2, the equation does not have any integral solutions.

4.30. The only ring homomorphism $\kappa_2 : \mathbb{Z}[\sqrt{-5}] \longrightarrow \mathbb{Z}/2\mathbb{Z}$. is given by $\kappa_2(a + b\sqrt{-5}) = a + b + 2\mathbb{Z}$. Its kernel consists of all elements $a + b\sqrt{-5}$ for which a and b have the same parity; equivalently, $a + b\sqrt{-5}$ has even norm.

There are two ring homomorphisms κ_3 and κ_3' to $\mathbb{Z}/3\mathbb{Z}$, and they are defined by $\kappa_3(a + b\sqrt{-5}) = a + b + 3\mathbb{Z}$ and $\kappa_3(a + b\sqrt{-5}) = a - b + 3\mathbb{Z}$. The kernel of κ_3 consists of all elements that are congruent to $1 - \sqrt{-5}$ modulo 3. The only ring homomorphism $\kappa_5 : \mathbb{Z}[\sqrt{-5}] \longrightarrow \mathbb{Z}/5\mathbb{Z}$ is $\kappa_5(a + b\sqrt{-5}) = a + 5\mathbb{Z}$. Its kernel consists of all elements of the form $5c + b\sqrt{-5}$.

4.31. The set $m\mathbb{Z}$ of multiples of m is an ideal in $\mathbb{Z}$: it is closed under addition and subtraction since $am \pm bm = (a \pm b)m$ is also a multiple of m; it is also closed with respect to multiplication by arbitrary elements $r \in \mathbb{Z}$ since $r \cdot (am) = ra \cdot m$ is also a multiple of m.

4.32. If $a \mid b$, then $b = ar$ for some $r \in R$, which implies $\in (a)$ and therefore $(b) \subseteq (a)$. The converse is also clear.

As for the remaining claims, we immediately deduce that (1) implies (2).

Now assume that a and b divide each other. Then $b = ad$ and $a = be$, hence $a = ade$ and thus $de = 1$. Thus d and e are units, and we have $a = be$ as claimed.

Finally, if $a = be$ for some unit e, then $(a) = (b)$ since clearly $a \in (b)$ and $b \in (a)$.

4.33. The sum and the product of upper triangular matrices is upper triangular, which shows that T is a subring of R with unit $\left(\begin{smallmatrix} 1 & 0 \\ 0 & 1 \end{smallmatrix}\right)$. But T is not an ideal in R since the product $\left(\begin{smallmatrix} 1 & 0 \\ 0 & 1 \end{smallmatrix}\right) \cdot \left(\begin{smallmatrix} 0 & 0 \\ 1 & 0 \end{smallmatrix}\right) = \left(\begin{smallmatrix} 0 & 0 \\ 1 & 0 \end{smallmatrix}\right)$ is not upper triangular.

4.34. Clearly $I \cap R$ is closed with respect to addition and subtraction. Moreover it is closed with respect to multiplication by elements $r \in R$ since if $a \in I \cap R$, then $r \cdot a \in I$ since I is an ideal in S and $r \cdot a \in R$ since $r, a \in R$ and R is a domain. Thus $r \cdot a \in I \cap R$, and $I \cap R$ is an ideal in R as claimed.

4.35. Since I is a nonzero ideal in $\mathcal{O}_K$, it contains an element $\alpha \neq 0$. Since I is closed with respect to multiplication by elements of $\mathcal{O}_k$, the element $\alpha \cdot \alpha' = N(\alpha)$ is also in I. Thus I contains a nonzero integer.

The ideal (X) in $\mathbb{Z}[X]$ or $\mathbb{Q}[X]$, on the other hand, consists of multiples of X, and so the only constant polynomial in (X) is the zero polynomial 0.

4.36. Reducing a polynomial $f \in \mathbb{Z}[x]$ modulo a prime number p yields a polynomial in $\mathbb{F}_p[x]$; clearly this map is a ring homomorphism. Reducing f modulo x is the same as evaluating f at $x = 0$ and therefore also is a ring homomorphism. Since both maps commute our claims follow.

4.37. Clearly $\mathbb{Z}$ is a subring of $\mathcal{O}_k$. It is not an ideal since $\mathbb{Z}$ is not closed with respect to multiplication by ring elements. For example, $\sqrt{m} \cdot 1$ is not in $\mathbb{Z}$.

4.38. For showing that $I = (2a + \sqrt{2}b : a, b \in \mathbb{Z}\}$ is an ideal in $\mathbb{Z}[\sqrt{2}]$ we show that $I = (\sqrt{2})$. For proving that $I \subseteq (\sqrt{2})$ take an arbitrary element $2a + \sqrt{2}b \in I$; the claim follows from $2a + \sqrt{2}b = \sqrt{2} \cdot (a\sqrt{2} + b)$. On the other hand, every element in $(\sqrt{2})$ has the form $\sqrt{2}(b + a\sqrt{2}) = 2a + b\sqrt{2}$, and these are elements in I.

The order $\mathcal{O} = \mathbb{Z} + 2\sqrt{2}\,\mathbb{Z}$ is clearly a subring of $\mathbb{Z}[\sqrt{2}]$; But although $1 + 2\sqrt{2} \in \mathcal{O}$, the element $\sqrt{2} \cdot (1 + 2\sqrt{2}) = 4 + \sqrt{2}$ is not in $\mathcal{O}$.

4.39. If $f_1\omega \in \mathcal{O}_k$ and $f_2\omega \in \mathcal{O}$, then clearly $(f_1 \pm f_2)\omega \in \mathcal{O}$. Thus $\mathcal{F}$ is an additive group. For showing that it is an ideal observe that if $f \in \mathcal{F}$, then

$f\omega \in \mathcal{O}$ for all $\omega \in \mathcal{O}_k$, hence $rf\omega \in \mathcal{O}$ for all $\omega \in \mathcal{O}_k$ and all $r \in \mathcal{O}_k$, hence $\mathcal{F}$ is an ideal in $\mathcal{O}_k$.

If $\mathcal{O}$ is the maximal order, then $1 \cdot \omega \in \mathcal{O}$ for all $\omega \in \mathcal{O}$, hence the conductor is (1).

The order $\mathbb{Z}[\sqrt{m}\,]$ for squarefree integers $m \equiv 1 \bmod 4$ has conductor (2) since $2\omega \in \mathbb{Z}[\sqrt{m}\,]$ for all $\omega \in \mathcal{O}_k$.

4.40. The only divisors of 2 in $\mathbb{Z}[x]$ are ± 1 and ± 2; clearly $2 \nmid x$, and this shows that $\gcd(2, x) = 1$.

Yet there do not exist Bézout elements. In fact, if $1 = 2p(x) + xq(x)$ for polynomials p and q, then plugging in $x = 0$ yields $1 = 2f(0)$, which is a contradiction since $2 \nmid 1$ in $\mathbb{Z}$.

If $(2, x) = (f)$ for some polynomial f then $f(x) \mid 2$ implies that f is constant, which is impossible. Thus $(2, x)$ is not principal in $\mathbb{Z}[x]$. We do have, however, $(2, x) = (1)$ in $\mathbb{Q}[x]$ since 2 is a unit in $\mathbb{Q}[x]$.

4.41. If $(2, \sqrt{-6}) = (\alpha)$, then α must divide both 2 and $\sqrt{-6}$. But both elements are irreducible in $\mathbb{Z}[\sqrt{-6}\,]$. The proof in the second case is similar.

   In the last case, showing that 2 is irreducible is a little bit more challenging: If $\pm 2 = \alpha\alpha'$, then setting $\alpha = a + b\sqrt{10}$ gives $\pm 2 = a^2 - 10b^2$. Reducing this equation modulo 5 we find $\pm 2 = a^2 \bmod 5$, and this congruence is not solvable.

4.42. If π is a factor of 2, then it cannot be irreducible since $\sqrt{\pi}$ is also an algebraic integer and $\pi = \sqrt{\pi} \cdot \sqrt{\pi}$.

   We now claim that the ideal $I = (2, \sqrt{2}, \sqrt[4]{2}, \sqrt[8]{2}, \ldots)$ is not principal in R; it suffices to show that it is not finitely generated, i.e., that there do not exist elements $a_i \in R$ with $I = (a_1, \ldots, a_n)$. Assume therefore that $I = (a_1, \ldots, a_n)$. Let K denote the finite extension $K = \mathbb{Q}(a_1, \ldots, a_n)$. Since this extension is finite, there must be some integer $k = 2^m$ such that $\sqrt[k]{2}$ is not in K. This implies that $\sqrt[k]{2}$ is not in I.

4.43. Assume that a and b are coprime in $\mathbb{Z}$. Then there exist integers $x, y \in \mathbb{Z}$ with $ax + by = 1$. This relation also holds in R, hence any common divisor of a and b in R divides 1 and thus is a unit.

4.44. We find

$$21 = 15 + 6,$$
$$15 = 2 \cdot 6 + 3,$$
$$6 = 2 \cdot 3,$$

which shows that $\gcd(21, 15) = 3$. Working backwards we find

$$3 = 15 - 2 \cdot 6 = 15 - 2(21 - 15) = 3 \cdot 15 - 2 \cdot 21.$$

4.45. We claim that

$$\gcd(x^n + x^2 - 2, x^2 - 1) = \begin{cases} x - 1 & \text{if } n \text{ is odd,} \\ x^2 - 1 & \text{if } n \text{ is even.} \end{cases}$$

Set $f(x) = x^n + x^2 - 2$. Since $f(1) = 0$, $x - 1$ divides f_n. Observe that $x^2 - 1 = (x - 1)(x + 1)$ is a product of two prime elements, so the gcd can only be $x - 1$ or $s^2 - 1$.

If n is odd, then $f(-1) = -2$, hence $x + 1$ does not divide n, and the claim is proved.

If n is even, then $f(-1) = 0$, hence in this case $(x - 1)(x + 1) = x^2 - 1$ divides f.

4.46. If $(N\alpha, N\beta) = 1$ in $\mathbb{Z}$, then there exist integers m and n with $mN\alpha + nN\beta = 1$. But then $m\alpha' \cdot \alpha + n\beta' \cdot \beta = 1$, hence $\gcd(\alpha, \beta) \sim 1$ in $\mathcal{O}_k$.

4.47. If a and m are coprime, then $ab + mn = 1$ for suitable integers b, n. But then $ab \equiv 1 \bmod m$, hence b represents the inverse of a modulo m.
Clearly $\frac{1}{2} \equiv \frac{1+21}{2} = 11 \bmod 21$. For computing the inverse of 5 mod 33, we use $1 = 2 \cdot 33 - 13 \cdot 5$, which shows that $\frac{1}{5} \equiv -13 \bmod 33$.

4.48. We write the equation in the form $x^3 = y^2 - 9 = (y - 3)(y + 3)$. The greatest common divisor of the factors on the right hand side divides their difference 6, hence there are four cases:

- $\gcd(y - 3, y + 3) = 1$. Then $y - 3 = a^3$ and $y + 3 = b^3$, hence $b^3 - a^3 = 6$. But his equation does not have any integral solution since $(b - a)(b^2 + ab + a^2) = 6$ is impossible in integers.
- $\gcd(y - 3, y + 3) = 2$. Changing the sign of y if necessary we may assume that $y - 3 = 2a^3$ and $y + 3 = 4b^3$, hence $4b^3 - 2a^3 = 6$ and therefore $2b^3 - a^3 = 3$.
  Solving this equation seems to require less elementary means. The solution $2 \cdot 4^3 - 5^3 = 3$ yields the solution $(x, y) = (40, \pm 253)$ of the original equation.
- $\gcd(y - 3, y + 3) = 3$. Here $y - 3 = 3a^3$ and $y + 3 = 9b^3$, which leads to $9b^3 - 3a^3 = 6$ and thus to $3b^3 - a^3 = 2$. The obvious solutions $a = b = 1$ gives the solutions $(x, y) = (3, \pm 6)$. Again, the solution of this cubic equation seems to be rather difficult.
- $\gcd(x - 3, x + 3) = 6$. Here either $x - 3 = 6a^3$ and $x + 3 = 36b^3$, or $x - 3 = 12a^3$ and $x + 3 = 18b^3$.
  In the first case we get $6b^3 - a^3 = 1$; in the second case we obtain $3b^3 - 2a^3 = 1$ with the obvious solution $a = b = 1$.

There seems to be no way of avoiding cubic equations of the form $ax^2 + by^3 = c$ in this approach. Such equations can be solved in a rather straightforward way (the keyword is Thue equations), but they require methods beyond the scope of this book. For avoiding these problems it might be a better idea to invoke cubic fields right from the start and factor the equation $y^2 = x^3 + 9$ in $\mathbb{Q}(\sqrt[3]{9})$.

4.49. As above, $\gcd(x - k, x + k) \mid 2k$. If the factors are coprime, then $x - k = a^3$ and $x + k = b^3$, hence $b^3 - a^3 = 2k$. If $k = 4m^3$, this equation does not have a nontrivial integral solution since the resulting equation is the cubic Fermat equation.

4.50. Assume that k is an odd prime number. From $y^2 + k^2 = x^3$ we deduce that y must be even; in fact if y is odd then the left hand side is divisible by 2, but not by 4, and such integers cannot be cubes. Write $(y + ki)(y - ki) = x^3$. Since $\gcd(y + ki, y - ki) \mid 2k$ and k is prime, we either have $\gcd(y + ki, y - ki) = 1$ or $\gcd(y + ki, y - ki) = k$. The last case is impossible for primes $k \equiv 3 \bmod 4$: If $y = km$, then $x^3 = y^2 + k^2 = k^2(m^2 + 1)$; since x^3 is a cube, k must divide $m^2 + 1$: Contradiction.

Assume now that $\gcd(y + ki, y - ki) = 1$; since all the units in $\mathbb{Z}[i]$ are cubes, we must have

$$y + ki = (a + bi)^3, \quad y - ki = (a - bi)^3.$$

Subtracting these equations we obtain $k = b(3a^2 - b^2)$. Since k is prime, we either have

- $b = \pm k$ and $3a^2 - b^2 = \pm 1$.
- $3a^2 - b = \pm k$ and $b = \pm 1$.
  In this case we find $k = 3a^2 - 1$ and $b = 1$. Examples:

a	2	4	6	8	12
k	11	47	107	191	407
x	5	17	37	65	145
y	2	52	198	488	1692

4.51. We compute the greatest common divisor of $1 + 8i$ and $5 + 4i$ using the Euclidean algorithm: Since $\frac{1+8i}{5+4i} = \frac{37}{41} + \frac{36}{41}i$ we take $1 + i$ as the quotient and obtain $1 + 8i - (5 + 4i)(1 + i) = -i$. This shows that $\gcd(1 + 8i, 5 + 4i) = 1$. Multiplying the equation through by i we obtain

$$(1 + 8i)i + (5 + 4i)(1 - i) = 1,$$

so we have obtained the special solution $x = i$, $y = 1 - i$.

The general solution is obtained by adding multiples of the homogeneous equation

$$(1 + 8i)(5 + 4i) + (5 + 4i)(-1 - 8i) = 0$$

to the special solution.

Chapter 5

5.1. We find that $1 - 2i$ is the nearest Gaussian integer to $\frac{26-29i}{13+4i} \approx 1.4 - 1.9i$; thus

$$26 - 29i - (1 - 2i)(13 + 4i) = 5 - 7i$$

is the first step in the Euclidean algorithm.
Next $\frac{13+4i}{5-7i} = 0.5 + 1.5i$, and so

$$13 + 4i - i(5 - 7i) = 6 - i.$$

Finally

$$5 - 7i = (1 - i)(6 - i).$$

Thus $\gcd(26 - 29i, 13 + 4i) \sim 6 - i \sim 1 + 6i$.
In fact, we have $(26-29i) = -i(1+6i)(5-4i)$ and $13+4i = (1+2i)(1+6i)$.
    This implies

$$6 - i = 13 + 4i - i(5 - 7i)$$
$$= 13 + 4i - i(26 - 29i - (1 - 2i)(13 + 4i))$$
$$= (3 + i)(13 + 4i) - i(26 - 29i).$$

5.2. Write $x^2 + 1 = kp$; then both $x + i$ and p are divisible by one of the primes above p, hence $\gcd(x + i, p) = (a + bi)$, where $a^2 + b^2 = p$.
    If $p \equiv 1, 3 \bmod 8$ and $x^2 \equiv -2 \bmod p$, then $\gcd(x - \sqrt{-2}, p) = c + d\sqrt{-2}$, where $c^2 + 2d^2 = p$.

5.3. We begin by observing that the elements of the second system are pairwise incongruent modulo $\pi = 1 + 2i$. Now $-1 \equiv 4$, $i \equiv 2$ and $-i \equiv 3 \bmod \pi$.

5.4. This is a trivial exercise: Just multiply $a + bi$ by the units ± 1 and $\pm i$.

5.5. Clearly $N(a + bi) = a^2 + b^2$ is odd if and only if $a \equiv b \bmod 2$. In this case, $\frac{a+bi}{1+i} = \frac{(a+bi)(1-i)}{2} = \frac{a+b}{2} - \frac{a-b}{2}i$. Conversely, $(1 + i)(c + di) = c - d + (c + d)i$ has even norm since $c - d \equiv c + d \bmod 2$.
    If $N(a + bi) = a^2 + b^2$ is odd, then a and b have different parity. If a is odd and b is even, then $q^2 \ equiv 1$ and $b^2 \equiv 0 \bmod 4$, hence $N(a + bi) = a^2 + b^2 \equiv 1 \bmod 4$.
    If $a + bi$ has odd norm and a is even, then $(a + bi)i = -b + ai$ has an odd real part. Thus every Gaussian integer with odd norm is associated with an element $a + bi \equiv 1 \bmod 2$. Observe that this congruent is equivalent to $a \equiv 1$ and $b \equiv 0 \bmod 2$.
    Finally observe that a complete system of coprime residue classes modulo $2 + 2i$ is $\{\pm 1, \pm i, \pm 1 + 2i, 2 \pm i\}$. If $a + bi \equiv 1 \bmod 2$, then $a + bi \equiv 1$

or $a + bi \equiv 1 + 2i \bmod 2 + 2i$. In the second case, $-a - bi \equiv -1 - 2i \equiv$ $-1 - 2i + 2 + 2i \equiv 1 \bmod 2 + 2i$. Thus every element $a + bi$ with odd norm has an associate congruent to $1 \bmod 2 + 2i$.

5.6. Since x and y have different parity, we may assume that y is even. We also assume that the solution is primitive, i.e., that $\gcd(x, y) = 1$. Since $x + yi$ and $x - yi$ are coprime, the equation $z^2 = (x + yi)(x - yi)$ implies that $x + yi = \varepsilon(a + bi)^2$ and $x - yi = \varepsilon'i(a - bi)^2$. Since y is even it follows that $\varepsilon = \pm 1$, and since $-1 = i^2$ we can subsume it into the square. Thus $x + yi = (a + bi)^2$, which implies $x = a^2 - b^2$, $y = 2ab$ and $z = a^2 + b^2$.

5.7. Let (x, y, z) be a nonzero solution in integers. Then x must be divisible by 3, say $x = 3x_1$. This implies $27x_1^3 + 3y^3 + 9z^3 = 0$, hence $9x_1^3 + y^3 + 3z^3 = 0$. Thus $y = 3y_1$, which leads to $3x_1^3 + 9y_1^3 + z^3 = 0$. Now $z = 3z_1$ yields $x_1^3 + 3y_1^3 + 9z_1^3 = 0$. If the equation $x^3 + 3y^3 + 9z^3 = 0$ has a nonzero solution (x, y, z), then (x_1, y_1, z_1) is a smaller nonzero solution. By infinite descent, this is impossible.

5.8. We have $1 + 2i \equiv 1 + 2i - (3 + 2i) \equiv -2 \bmod 3 + 2i$. Thus $[\frac{1+2i}{3+2i}] = [\frac{-2}{3+2i}] = (\frac{-2}{13}) = -1$, where we have used the fact that $[\frac{a}{\pi}] = (\frac{a}{p})$ for primes π with norm $N(\pi) = p$.
Next $1 + 4i \equiv 1 + 4i - 2(3 + 2i) \equiv -5 \bmod 3 + 2i$, hence $[\frac{1+4i}{3+2i}] = (\frac{-5}{13}) = -1$.
Finally $4i \equiv -1 \bmod (1 + 4i)$ implies $2i \equiv -9 \bmod (1 + 4i)$, hence $1 + 2i \equiv 1 - 9 \equiv -8 \bmod (1 + 4i)$. Thus $[\frac{1+2i}{1+4i}] = (\frac{-8}{13}) = -1$.

5.9. 1. We begin by showing that $[\frac{a}{\pi}] = (\frac{a}{p})$ for elements π with prime norm $N\pi = p$. By definition we have $[\frac{a}{\pi}] \equiv a^{\frac{p-1}{2}} \bmod \pi$. Since both sides are elements of $\mathbb{Z}$, the congruence even holds modulo p. But then $[\frac{a}{\pi}] \equiv a^{\frac{p-1}{2}} \equiv (\frac{a}{p}) \bmod p$. Since p is odd and both sides differ by ± 1, the congruence implies equality $[\frac{a}{\pi}] = (\frac{a}{p})$.

If $q \equiv 3 \bmod 4$ is prime, then $[\frac{a}{q}] \equiv a^{\frac{q^2-1}{2}} = (a^{q-1})^{\frac{q+1}{2}} \equiv 1 \bmod q$ by Fermat's Little theorem. Thus $[\frac{a}{q}] = 1$ by the same argument as above.

2. Next $[\frac{a}{a+bi}] = (\frac{a}{a^2+b^2}) = (\frac{a^2+b^2}{a}) = (\frac{b^2}{a}) = 1$, where we have used quadratic reciprocity.

3. Multiplying the trivial congruence $c + di \equiv 0 \bmod (c + di)$ through by i we find $ci \equiv d \bmod (c + di)$. Thus

$$\left[\frac{a+bi}{c+di}\right] = \left[\frac{a+bi}{c+di}\right]\left[\frac{c}{c+di}\right] = \left[\frac{ac+bci}{c+di}\right] = \left[\frac{ac+bd}{c+di}\right] = \left(\frac{ac+bd}{q}\right).$$

4. We have $ac + bd \equiv 1 \bmod 2$ and $pq \equiv 1 \bmod 4$; moreover

$$pq = (a^2 + b^2)(c^2 + d^2) = (ac + db)^2 + (ad - bc)^2.$$

Thus

$$\left(\frac{ac+bd}{pq}\right) = \left(\frac{pq}{ac+bd}\right) = \left(\frac{(ac+db)^2+(ad-bc)^2}{ac+bd}\right) = \left(\frac{ad-bc}{ac+bd}\right)^2 = 1.$$

5. Now

$$\left(\frac{ac+bd}{pq}\right) = \left[\frac{a+bi}{c+di}\right]\left[\frac{c+di}{a+bi}\right].$$

This implies the claim.

5.10. Each ideal class of $\mathbb{Q}(\sqrt{-19})$ contains an element with norm $\leq \sqrt{19/3} < 3$. Since $\left(\frac{-19}{2}\right) = 1$, $\mathcal{O}_k$ has unique factorization.
If $\Delta = -43$, then $\left(\frac{-43}{2}\right) = \left(\frac{-43}{3}\right) = -1$, so again there is nothing to check.
For $\Delta = -67$ we have $\left(\frac{\Delta}{p}\right) = -1$ for $p = 2$ and $p = 3$, and for $\Delta = -163$ we have $\left(\frac{\Delta}{p}\right) = -1$ for $p = 2, 3, 5$, and 7.

5.11. We claim that $N_u\left(\frac{1+\sqrt{-17}}{3} - \gamma\right) \geq 1$ for all $\gamma \in S$. Write $\gamma = \frac{a+b\sqrt{-17}}{2^n}$; since $N_u(2) = 1$, the inequality $N_u\left(\frac{1+\sqrt{-17}}{3} - \gamma\right) \geq 1$ is equivalent to $N_u\left(\frac{1+\sqrt{-17}}{3} \cdot 2^n - delta\right) \geq 1$ for $\delta = a + b\sqrt{-17} \in \mathbb{Z}[\sqrt{-17}]$. Finally $\frac{1+\sqrt{-17}}{3} \cdot 2^n \equiv \pm\frac{1+\sqrt{-17}}{3}$ mod 1 shows that for proving $N_u\left(\frac{1+\sqrt{-17}}{3} - \gamma\right) \geq 1$ we may assume that $\gamma \in \mathbb{Z}[\sqrt{-17}]$.
Write $\frac{1+\sqrt{-17}}{3} - \gamma = \frac{a+b\sqrt{-17}}{3}$ and observe that $a, b \neq 0$; if both a and b are even, the norm does not change if we divide through by the unit 2. Thus we may assume that a and b have different parity or are both odd. In the first case, $N_u(a + b\sqrt{-17}) = N(a + b\sqrt{-17}) \geq 18$, in the second case $a^2 + 17b^2 \equiv 2$ mod 4, hence $N_u(a + b\sqrt{-17}) = \frac{1}{2}N(a + b\sqrt{-17}) \geq 9$. This proves our claim.

5.12. The norm N_u is multiplicative, hence it is sufficient to find, for every $\xi \in K$, an element $\gamma \in R$ with $N_u(\xi - \gamma) < 1$. Write $\xi = \frac{a+b\sqrt{-5}}{2^j c}$ for ordinary integers a, b, c with c odd and $j \geq 0$. Since 2 is a unit, we may multiply through by 2^j and therefore may assume that $\xi = \frac{a+b\sqrt{-5}}{c}$ for some odd integer c; subtracting suitable integral multiples of 1 and $\sqrt{-5}$ we may assume that $|\frac{a}{b}| \leq \frac{1}{2}$ and $|\frac{b}{c}| \leq \frac{1}{2}$.
If $|\frac{b}{c}| \leq \frac{1}{3}$, then $N_u(\xi) = \frac{N_u(a+b\sqrt{-5})}{c^2} \leq \frac{a^2+5b^2}{c^2} \leq \frac{1}{4} + \frac{5}{9} < 1$.
Assume therefore that $\frac{1}{3} < |\frac{b}{c}| \leq \frac{1}{2}$. We now distinguish the following cases:

- $a \equiv b$ mod 2: Then $N_u(a + b\sqrt{-5}) \leq \frac{a^2+5b^2}{2}$, hence $N_u(\xi) \leq \frac{1}{2} \cdot \frac{1+5}{4} = \frac{3}{4} < 1$.
- $a \equiv 1, b \equiv 0$ mod 2: Replace a by $a \pm 1$ such that $|\frac{a}{b}| \leq 1$; then $N_u(\xi) \leq \frac{1}{4}\frac{a^2+5b^2}{c^2} \leq \frac{1}{4}(1 + \frac{5}{4}) < 1$.

- $a \equiv 0$, $b \equiv 1 \bmod 2$: Replace b by $b \pm 1$ such that $\frac{1}{2} < |\frac{b}{c}| < \frac{2}{3}$. Then $N_u(\xi) \leq \frac{1}{4}(\frac{1}{4} + \frac{20}{9}) < 1$.

Thus R is Euclidean with respect to N_u.

Clearly -1 and 2 are units in R. Now $a + b\sqrt{-5} \in R$ is a unit if and only if its inverse $\frac{a - b\sqrt{-5}}{a^2 + 5b^2} \in R$,—which is the case if and only if $a^2 + 5b^2 = 2^m$ is a power of 2. Writing $a = \frac{A}{2^n}$ and $b = \frac{B}{2^n}$ for ordinary integers A and B we find that $A^2 + 5B^2 = 2^{m+2n}$. If B is odd, then $2^{m+2n} \geq 5$, hence $2^{m+2n} \geq 8$, and reduction modulo 8 yields a contradiction. If B is even, then so is A, and we can cancel a common factor 4 and repeat the reasoning. This shows that $A^2 + 5B^2 = 2^{m+2n}$ is only possible if $B = 0$ and thus $b = 0$. Thus the unit group of R is generated by -1 and 2.

5.13. We have $7 = (3+\rho)(3+\rho^2)$, $13 = (4+\rho)(4+\rho^2)$ and $19 = (5+2\rho)(5+2\rho^2)$.

5.14. For a conceptual proof we simply observe that $\mathbb{Z}[\rho]/(2)$ has three elements, which implies $\alpha^3 \equiv 1 \bmod 2$ for all nonzero residue classes modulo 2.

For a computational proof write $\alpha = a + b\rho$. Then $(a + b\rho)^2 = a^3 + 3a^2b\rho + 3ab^2\rho^2 + b^3 = a^3 + 3a^2b\rho - 3ab^2(1+\rho) + b^3 = a^3 - 3ab^2 + b^3 + 3ab(a-b)\rho$. The claim now follows from the observation that $ab(a - b)$ is always even.

5.15. The integral solutions of the equation $y^2 = x^3 + 24$ are $(1, 5)$, $(-2, 4)$, $(10, 32)$ and $(8.158, 736.844)$. How close to this result can you come by factoring $y^2 - 24 = x^3$ in the quadratic number field $\mathbb{Q}(\sqrt{6})$?

Write $x^3 = (y - 2\sqrt{6})(y + 2\sqrt{6})$. Clearly the gcd of the factors on the right divides their difference $4\sqrt{6}$. Moreover, if y is even then it must be divisible by 4, which implies that the gcd of the two factors cannot have norm divisible exactly by 2. Similarly we cannot have $3 \mid y$ since in this case y^2 is divisible by 9, but $x^3 + 24$ is not. Thus we are left with the following possibilities:

1. $\gcd(y - 2\sqrt{6}, y + 2\sqrt{6}) \sim 1$,

   In this case $y - 2\sqrt{6} = \eta\alpha^3$, where η is a unit. Subsuming cubes of units into α^3 we may assume that $\eta \in \{1, \eta, \eta'\}$.

   If $\eta = 1$, then $y - 2\sqrt{6} = (a + b\sqrt{6})^3$ leads to $y = a^3 + 18ab^2$ and $2 = 3a^2b + 6b^3$. The second equation is impossible modulo 3.

   If $\eta = \varepsilon = 5 + 2\sqrt{6}$, then $y = 5a^3 + 36a^2b + 90ab^2 + 72b^3$ and $2 = 2a^3 + 15a^2b + 36ab^2 + 30b^3$. If a is even we obtain a contradiction after dividing through by 2 and reducing modulo 2; if $b = 2c$ then we obtain $1 = a^3 + 15a^2c + 72ac^2 + 120c^3$. The solution $a = 1$ and $c = 0$ yields $(x, y) = (1, 5)$. Without advanced techniques it does not seem possible to exclude other solutions.

2. $\gcd(y - 2\sqrt{6}, y + 2\sqrt{6}) \sim 4 + 2\sqrt{6}$; in this case $y \pm 2\sqrt{6}$ is divisible by $(2 + \sqrt{6})^3 = 44 + 18\sqrt{6}$, and we obtain the equations

$$y - 2\sqrt{6} = \varepsilon'(44 - 18\sqrt{6})(a - b\sqrt{6})^3,$$

$$y + 2\sqrt{6} = \varepsilon(44 + 18\sqrt{6})(a + b\sqrt{6})^3.$$

Subtracting these equations from each other we obtain, if $\varepsilon = 1$,

$$1 = 9a^3 + 66a^2b + 162ab^2 + 132b^3,$$

which is impossible since the right hand side is divisible by 3.
If $\varepsilon = 5 + 2\sqrt{6}$ we obtain

$$1 = 89a^3 + 654a^2b + 1602ab^2 + 1308b^3.$$

This equation has the following solutions:

a	b	x	y
-5	-2	-2	-4
$--7$	3	10	32
-211	90	8.158	736.844

Again, showing that there are no others seems to be very hard.

5.16. For $f(x) = x^2 + 19x - 19$ we can simply verify that

$$f(x^2 + 20x - 19) = f(x) \cdot f(x + 1).$$

The Taylor expansion of a polynomial is given by $f(x + h) = f(x) + hg(x)$ for a suitably chosen polynomial g with integral coefficients. Setting $h = f(x)$ we see that $f(x + f(x)) = f(x)(1 + g(x))$.

5.17. We factor the equation over $\mathbb{Z}[i]$ and get $(y+i)(y-i) = 2x^3$. Since y is odd, the factors on the left have greatest common divisor $1 + i$; since powers of i may be subsumed into the cube we get

$$y + i = (1 + i)(a + bi)^3, \quad y - i = (1 - i)(a - bi)^3.$$

Subtracting these equations from each other we find

$$1 = a^3 + 3a^2b - 3ab^2 - b^3 = (a - b)(a^2 + 4ab + b^2).$$

This implies $a - b = \pm 1$, hence

$$1 = 6b^2 \pm 6b + 1$$

and so $(a, b) = (\pm 1, 0)$ or $= (0, \pm 1)$. This shows $y = \pm 1$ and $x = 1$.

5.18. The identity can be verified by brute force. Since $\sqrt{5}$ divides the fifth power on the left, the expression on the right must be divisible by $\sqrt{5}^5$; thus either one of the factors a, b, or c is divisible by $\sqrt{5}$ or $\sqrt{5}$ divides the expression in the brackets; but this is impossible since squares are congruent to 0 or ± 1 mod $\sqrt{5}$.

5.19. This is a simple calculation:

$$
\begin{aligned}
\phi(x^2 - xy + y^2, x^2 - 2xy + y^2) &= (x^2 - xy + y^2)^2 \\
&\quad + (x^2 - xy + y^2)(x^2 - 2xy + y^2) - (x^2 - 2xy + y^2)^2 \\
&= x^4 - x^3 y + x^2 y^2 - xy^3 + y^4,
\end{aligned}
$$

and this implies the claim.

5.20. In the ring $\mathbb{Z}[\sqrt{m}\,]$ we have, for every prime number $p \nmid m$ and $\alpha = a + b\sqrt{m} \in \mathbb{Z}[\sqrt{m}\,]$,

$$
(a + b\sqrt{m}\,)^p \equiv a^p + b^p \sqrt{m}^{\,p} \equiv a + bm^{\frac{p-1}{2}} \sqrt{m}
$$

$$
\equiv a + \left(\frac{m}{p}\right) b\sqrt{m} \bmod p.
$$

Similar calculations work for elements $\alpha = \frac{a + b\sqrt{m}}{2}$ in the case where $m \equiv 1 \bmod 4$.

5.21. If $\left(\frac{5}{p}\right) = +1$, then, by Binet's formula,

$$
U_p = \frac{\omega^p - \omega'^{\,p}}{\omega - \omega'} \equiv \frac{\omega - \omega'}{\omega - \omega'} \equiv 1 \bmod p \qquad \text{and}
$$

$$
U_{p+1} = \frac{\omega^{p+1} - \omega'^{\,p+1}}{\omega - \omega'} \equiv \frac{\omega^2 - \omega'^2}{\omega - \omega'} = \omega + \omega' \equiv 1 \bmod p.
$$

If $\left(\frac{5}{p}\right) = -1$, on the other hand, then

$$
U_p = \frac{\omega^p - \omega'^{\,p}}{\omega - \omega'} \equiv \frac{\omega' - \omega}{\omega - \omega'} \equiv -1 \bmod p
$$

and

$$
U_{p+1} = \frac{\omega^{p+1} - \omega'^{\,p+1}}{\omega - \omega'} \equiv \frac{\omega\omega' - \omega'\omega}{\omega - \omega'} \equiv 0 \bmod p.
$$

The residue class of U_{p-1} mod p now follows from $U_{p-1} = U_{p+1} - U_p$.

Chapter 6

6.1. Write $a + b = 2c$; then $a + bi = a + (2c - a)i = a(1 - i) + 2ci$. Since 1-i $= -i(1+i)\,2 = (1 + i)(1 + i)$, these elements are multiples of $1 + i$.
Conversely, $(a+bi)(1+i) = a-b+(a+b)i$, and then $(a-b)+(a+b) = 2a$ is even.

6.2. If $(a, b) = d$, then there exist elements $r, s \in R$ with $d = ra + sb$. This implies that $\gcd(r, s)$ divides d.
Conversely, if d divides both a and b, then d divides $\gcd(a, b)$.

6.3. We find

$$\mathfrak{ab} = ((1 + \sqrt{-5})^2, 2(1 + \sqrt{-5}), 3(1 + \sqrt{-5}), 6)$$

$$= ((1 + \sqrt{-5})^2, 2(1 + \sqrt{-5}), 3(1 + \sqrt{-5}), (1 + \sqrt{-5})(1 - \sqrt{-5}))$$

$$= (1 + \sqrt{-5})(1 + \sqrt{-5}, 2, 3, (1 - \sqrt{-5})) = (1 + \sqrt{-5})$$

since the second ideal contains $3 - 2 = 1$. Similarly,

$$\mathfrak{ac} = (6, 2(1 - \sqrt{-5}), 3(1 + \sqrt{-5}), 6) = (1 - \sqrt{-5})(1 + \sqrt{-5}, 2, -2 + \sqrt{-5})$$

$$= (1 - \sqrt{-5})$$

since $3(1 + \sqrt{-5}) = (-2 + \sqrt{-5})(1 - \sqrt{-5})$. Finally

$$\mathfrak{bc} = (3)(2, 1 + \sqrt{-5}, 1 - \sqrt{-5}, 3) = (3).$$

6.4. If $\{n, a + \omega\}$ is a $\mathbb{Z}$-basis of $\mathcal{O}_k$, then so is $\{n, a + n + \omega\}$. Conversely, if $a + \omega \in M$ and $a + k + \omega \in M$, then $k \in M \cap \mathbb{Z}$, hence k is a multiple of n.

6.5. The equation $-2 + i = 5a + b(1 + 2i)$ implies $-2 + i = 5a + b + 2bi$; comparing real and imaginary parts shows $2b = 1$, which is impossible in integers. We have $(1 + 2i) = 5\mathbb{Z} + (-2 + i)\mathbb{Z}$.

6.6. Assume that $a\sqrt{m} \equiv b\sqrt{m} \mod \mathbb{Z}$; then $(a - b)\sqrt{m} \in \mathbb{Z}$, hence $a = b$ since $\sqrt{m}$ is irrational. Thus all residue classes $b\sqrt{m} + \mathbb{Z}$ ($b \in \mathbb{Z}$) in R/M are pairwise distinct, and $N(M) = \infty$.

6.7. Let $\mathfrak{a} = (7, 1+\sqrt{-5})$. Then $6 = (1+\sqrt{-5})(1+\sqrt{-5}) \in \mathfrak{a}$, hence $7-6 \in \mathfrak{a}$ and therefore $\mathfrak{a} = (1)$.
More generally, let $\mathfrak{a} = (a, \alpha)$ and assume that $\gcd(a, N\alpha) = 1$. Then $N\alpha = \alpha\alpha' \in \mathfrak{a}$, and since $(a, N\mathfrak{a}) = (1)$ in $\mathbb{Z}$ we have $1 \in \mathfrak{a}$ and therefore $\mathfrak{a} = (1)$.

6.8. Since $N(4 + \sqrt{-5}) = 21 = 3 \cdot 7$ the ideal $\mathfrak{a} = (4 + \sqrt{-5})$ is divisible by prime ideals above 3 and 7. Write $\mathfrak{p} = (3, 1+\sqrt{-5})$ and $\mathfrak{q} = (7, 3+\sqrt{-5})$; then $(3) = \mathfrak{pp}'$ and $(7) = \mathfrak{qq}'$. Since $4 + \sqrt{-5} = 3+1+\sqrt{-5} \in \mathfrak{p}$ we clearly have $\mathfrak{p} \mid \mathfrak{a}$. Next $4 + \sqrt{-5} = 7 - (3 - \sqrt{-5} \in \mathfrak{q}'$, hence $\mathfrak{q}' \mid \mathfrak{a}$ and thus $\mathfrak{a} = \mathfrak{pq}'$.

This can be verified computationally as follows: We have

$$(3, 1 + \sqrt{-5})(7, 3 - \sqrt{-5}) = (21, 9 - 3\sqrt{-5}, 7 + 7\sqrt{-5}, 8 + 2\sqrt{-5}).$$
$$= (4 + \sqrt{-5})(4 - \sqrt{-5}, 1 - \sqrt{-5}, 3 + \sqrt{-5}, 2)$$
$$= (4 + \sqrt{-5})$$

since the second ideal contains $1 = 4 - \sqrt{-5} - (1 - \sqrt{-5}) - 2$.

6.9. Since $N(8 + \sqrt{-14}) = 2 \cdot 3 \cdot 13$ and $N(4 - \sqrt{-14}) = 2 \cdot 3 \cdot 5$, the norm of a greatest common divisor must divide 6. Since the only elements of norm ≤ 6 are 1 and $4 = N(2)$, the greatest common divisor of $8 + \sqrt{-14}$ and $4 - \sqrt{-14}$ is 1.

6.10. Since $10 + \sqrt{-5} = 3 \cdot 3 + 1 + \sqrt{-5}$ we have $10 + \sqrt{-5} \in (3, 1 + \sqrt{-5})$. Similarly, $10 + \sqrt{-5} = 7 + 3 + \sqrt{-5}$, hence $10 + \sqrt{-5} \in (7, 3 + \sqrt{-5})$. This implies that $(21, 10 + \sqrt{-5}) = (3, 1 + \sqrt{-5}) \cdot (7, 3 + \sqrt{-5})$.

6.11. We have $(a, b + \sqrt{m})^2 = (a^2, a(b + \sqrt{m}), (b + \sqrt{m})^2)$. Since $-a^2 = b^2 - m = (b - \sqrt{m})(b + \sqrt{m})$ we find $(a, b + \sqrt{m})^2 = (b + \sqrt{m})(b - \sqrt{m}, a, b + \sqrt{m})$. The last ideal contains a and $2b = b - \sqrt{m} + b + \sqrt{m}$, hence it contains $\gcd(a, 2b) = 1$. This proves our claim.

6.12. We have $(2) = \mathfrak{p}_2^2$ for $\mathfrak{p}_2 = (2, \sqrt{-6})$ and $(3) == \mathfrak{p}_3^2$ for $\mathfrak{p}_3 = (3, \sqrt{-6})^2$. Then $(\sqrt{-6}) = \mathfrak{p}_2\mathfrak{p}_3$ and $(2) \cdot (3) = \mathfrak{p}_2^2 \cdot \mathfrak{p}_3^2$.

6.13. We have

$$\mathfrak{a}\mathfrak{a}' = (2, \tfrac{1+\sqrt{-23}}{2})(2, \tfrac{1-\sqrt{-23}}{2}) = (4, 1 + \sqrt{-23}, 1 - \sqrt{-23}, 6)$$
$$= (2)(2, \tfrac{1+\sqrt{-23}}{2}, \tfrac{1-\sqrt{-23}}{2}, 3) = (2)$$

Since an ideal containing 2 and 3 contains $3 - 2 = 1$ and thus is the unit ideal. Next $(\frac{3-\sqrt{-23}}{2})$ is contained in $\mathfrak{a}$ since $\frac{3-\sqrt{-23}}{2} = 2 - \frac{1+\sqrt{-23}}{2}$, and $(\frac{3-\sqrt{-23}}{2})$ is not contained in $\mathfrak{a}'$ since it is not divisible by $\mathfrak{a}\mathfrak{a}'$. Since $(\frac{3-\sqrt{-23}}{2})$ and $\mathfrak{a}^3$ both have norm 8, they must be equal.

This can also be proved by brute force: One shows that $\mathfrak{a}^2 = (4, \frac{3-\sqrt{-23}}{2})$ and then $\mathfrak{a}^3 = \mathfrak{a}^2\mathfrak{a} = (\frac{3-\sqrt{-23}}{2})$.

If $\mathfrak{a}^2$ is principal, then there must be an element with norm 4 in $\mathcal{O}_k$, which does not exist. Alternatively, if $\mathfrak{a}^2$ and $\mathfrak{a}^3$ are principal, then so is $\mathfrak{a}$ since $\mathfrak{a}^3 = \mathfrak{a} \cdot \mathfrak{a}^2$, but $\mathfrak{a}$ is not principal because there is no element with norm 2 in $\mathcal{O}_k$.

6.14. See Exercise 4.40.

6.15. We have

$$(2, 1 + \sqrt{-3})(2, 1 + \sqrt{-3}) = (4, 2 + 2\sqrt{-3}, 2 - 2\sqrt{-3}, 4) = (2)(2, 1 + \sqrt{-3}).$$

On the other hand, the ideals (2) and $(2, 1 + \sqrt{-3})$ are distinct since $1 + \sqrt{-3} \notin (2)$.

6.16. With $I = (2, 1 + \sqrt{m})$ we have $I^2 = (2)I$, yet clearly $I \neq (2)$.

6.17. With $\mathfrak{p} = (2, 1 + 3i)$ we easily check $I^2 = (2)$. The prime ideals (q) are inert in $\mathbb{Z}[i]$ and thus also in $\mathbb{Z}[i]$. If $p \equiv 1 \bmod 4$, write $p = a^2 + b^2$; then $\mathfrak{p}_1 = (p, 3a + 3bi)$ and $\mathfrak{p}_2 = (p, 3a - 3bi)$ satisfy $\mathfrak{p}_1\mathfrak{p}_2 = (p)$.

We have $1 - 3i = 2(3+6i) - 5 - 15i \in (5, 3+6i)$ and $3+6i = 5 - 2(1-3i) \in (5, 1 - 3i)$.

Clearly $(3) \supset (3+6i)$. If there was an ideal A in $\mathbb{Z}[3i]$ with $(3)A = (3+6i)$, then $3 + 6i = 3a$ for some $a \in A$; but $a = 1 + 2i$ is not even an element of $\mathbb{Z}[3i]$.

Finally we have

$$(45) = (3)^2(5, 3 + 6i)(5, 3 - 6i) = (3 + 6i)(3 - 6i),$$

and these ideals are irreducible.

6.18. Write $\mathfrak{A} = \prod \mathfrak{p}^{a_\mathfrak{p}}$ and $\mathfrak{B} = \prod \mathfrak{p}^{b_\mathfrak{p}}$. Then $\gcd(\mathfrak{A}, \mathfrak{B}) = (1)$ means that $a_\mathfrak{p} > 0$ implies $b_\mathfrak{p} = 0$. Since $a_\mathfrak{p} + b_\mathfrak{p}$ is a multiple of n, both $a_\mathfrak{p}$ and $b_\mathfrak{p}$ must be multiples of n. But this means that both ideals are n-th powers

6.19. Let $\alpha \in \mathfrak{a}$ and $\beta \in \mathfrak{b}$; then $\alpha\beta \in \mathfrak{a}$ and $\alpha\beta \in \mathfrak{b}$ since $\mathfrak{a}$ and $\mathfrak{b}$ are ideals, hence $\alpha\beta \in \mathfrak{a} \cap \mathfrak{b}$, and we have proved the claimed inclusion.

Now assume that $\mathfrak{a}$ and $\mathfrak{b}$ are coprime and that $\alpha \in \mathfrak{a} \cap \mathfrak{b}$. This implies $(\alpha) = \mathfrak{a}\mathfrak{c} = \mathfrak{b}\mathfrak{d}$ for ideals $\mathfrak{c}$ and $\mathfrak{d}$. Since $\mathfrak{a}$ and $\mathfrak{b}$ are coprime, we must have $\mathfrak{b} \mid \mathfrak{c}$, hence $\mathfrak{c} \subseteq \mathfrak{b}$ and thus $(\alpha) \subseteq \mathfrak{a}\mathfrak{b}$.

6.20. We check these claims one by one.

- $\mathfrak{a} \sim \mathfrak{a}$ is true since $1 \cdot \mathfrak{a} = 1 \cdot \mathfrak{a}$.
- $\mathfrak{a} \sim \mathfrak{b}$ implies $\mathfrak{b} \sim \mathfrak{a}$. In fact, $\mathfrak{a} \sim \mathfrak{b}$ implies $\alpha\mathfrak{a} = \beta\mathfrak{b}$; reading this equation from right to left proves the claim.
- $\mathfrak{a} \sim \mathfrak{b}$ and $\mathfrak{b} \sim \mathfrak{c}$ imply $\mathfrak{a} \sim \mathfrak{c}$. In fact, we have $\alpha\mathfrak{a} = \beta\mathfrak{b}$ and $\gamma\mathfrak{b} = \delta\mathfrak{c}$. But then $\alpha\gamma\mathfrak{a} = \beta\gamma\mathfrak{b} = \beta\delta\mathfrak{c}$.

6.21. If $m \equiv 2 \bmod 4$, then

$$(2, \sqrt{m})^2 = (4, 2\sqrt{m}, m) = (2)(2, \sqrt{m}, \tfrac{m}{2}).$$

Since $\frac{m}{2}$ is odd, the second ideal contains 1 and therefore is the unit ideal. Thus $(2) = (2, \sqrt{m})^2$ in this case.

If $m \equiv 3 \bmod 4$, then

$$(2, 1 + \sqrt{m})^2 = (4, 2 + 2\sqrt{m}, m + 1 + 2\sqrt{m}) = (4, 2 + 2\sqrt{m}, m - 1)$$

$$= (2)(2, 1 + \sqrt{m}, \tfrac{m-1}{2})$$

since $m + 1 + 2\sqrt{m} - (2 + 2\sqrt{m}) = n - 1$. The last ideal contains 2 and the odd integer $\frac{m-1}{2}$, hence is equal to (1). Thus $(2, 1 + \sqrt{m})^2 = (2)$ in this case. Now let $m \equiv 1 \bmod 8$; then

$$(2, \tfrac{1+\sqrt{m}}{2})(2, \tfrac{1-\sqrt{m}}{2}) = (2)(2, \tfrac{1+\sqrt{m}}{2}, \tfrac{1-\sqrt{m}}{2}, \tfrac{1-m}{4}).$$

The last ideal contains $\frac{1+\sqrt{m}}{2} + \frac{1-\sqrt{m}}{2} = 1$, hence it is equal to (1). Moreover, $(2, \frac{1+\sqrt{m}}{2}) \neq (2, \frac{1-\sqrt{m}}{2})$ since otherwise these ideals would contain $\frac{1+\sqrt{m}}{2} + \frac{1-\sqrt{m}}{2} = 1$.

Finally consider the case $m \equiv 5 \bmod 8$. If there exists a prime ideal with norm 2, then it must have basis $\{2, a + \frac{1+\sqrt{m}}{2})$ with $a = 0$ or $a = 1$. Since 2 must divide the norm of $a + \frac{1+\sqrt{m}}{2} = \frac{2a+1+\sqrt{m}}{2}$, we find $(2a + 1)^2 \equiv m \bmod 8$, which implies that $m \equiv 1 \bmod 8$. Thus if $m \equiv 5 \bmod 8$, then (2) is inert.

6.22. If $\Delta = -19, 21, 29, 37$, the Gauss bound says that each ideal class contains an ideal with norm $\leq \sqrt{|\Delta|/3}$, and this bound is < 3 for these discriminants. Thus each ideal class contains an ideal with norm 1, and this implies the claim.

If we demand in addition that $\Delta \equiv 2 \bmod 3$, then we get class number 1 if $\sqrt{|\Delta|/3} < 5$, i.e., if $|\Delta| < 75$. This gives $\Delta = -43$ and $\Delta = -67$ if $\Delta < 0$; for positive Δ, the Gauss bound $\sqrt{\Delta/5}$ implies that $h = 1$ for the discriminants $\Delta = 29, 53, 77$, and 101.

6.23. We have $(2, 1 + \sqrt{-m})^2 = (2)$; if the ideal $(2, 1 + \sqrt{-m}) = (\alpha)$ is principal, then $N\alpha = 2$, i.e., $a^2 + mb^2 = 2$. But this is impossible for $m > 1$. Thus the ideal class of $(2, 1 + \sqrt{-m})$ has order 2, hence the class number of $\mathbb{Q}(\sqrt{-m})$ is even.

6.24. This expression is equal to the class number h of the complex quadratic number field with discriminant Δ. If $\Delta = -23$, for example, we have

$$h = \frac{2}{23}(1 + 1 + 1 + 1 - 1 + 1 - 1 + 1 + 1 - 1 - 1) = 3.$$

6.25. In each case, the primes below the Gauss bound $\sqrt{|\Delta|/3}$ are inert.

6.26. If the prime ideals above (2) are principal, then there must be elements with norm 2, i.e., the equation $x^2 + my^2 = 8$ must have integral solutions. For $m \equiv 7 \bmod 8$, this implies $m = 7$.

6.27. Consider the equation $y^2 = x^3 - d$ for $d = 3t^2 - 1$ with $t = 3c^3$, that is, $y^2 = x^3 - 27c^6 + 1$. Clearly this equation has the solutions $(3t^2, \pm 1)$ not listed in Theorem 6.20. This implies that if $d = 27c^6 - 1$ is squarefree and $\not\equiv 7 \bmod 8$ (that is, if c is odd), then $\mathbb{Q}(\sqrt{-d})$ has class number divisible by 3. Computations suggest that this holds even in the case $d \equiv 7 \bmod 8$:

c	d	$h(d)$	c	d	$h(d)$
1	26	6	5	421, 874	900
2	1, 727	36	6	125, 9711	1608
3	19, 682	108	7	3, 176, 522	1512
4	110, 591	444	8	7, 077, 887	2088

6.28. We go through the statements one by one:

1. For each prime ideal $\mathfrak{p}$ of norm $p \neq 5$ either $\mathfrak{p}$ or $\mathfrak{p}\mathfrak{a}$ is principal. In fact, if $\mathfrak{p}$ lies in the principal class, then it is principal; if it lies in the class of $[\mathfrak{a}]$, then $\mathfrak{p} \sim \mathfrak{a}$, hence $\mathfrak{a}\mathfrak{p} \sim \mathfrak{a}^2 \sim [(1)]$ and $\mathfrak{a}\mathfrak{p}$ is principal.
2. If p is a prime with $(-5/p) = +1$, then p splits: $(p) = \mathfrak{p}\mathfrak{p}'$. If $\mathfrak{p}$ is principal, then $p = x^2 + 5y^2$; if $\mathfrak{a}\mathfrak{p}$ is principal, then $2p = x^2 + 5y^2$.
3. If $p = x^2 + 5y^2$, then x and y have different parity; this implies $p = x^2 + 5y^2 \equiv x^2 + y^2 \equiv 1 \bmod 4$ and thus $p \equiv 1, 9 \bmod 20$.
   If $2p = x^2 + 5y^2$, on the other hand, then x and y are both odd, hence $2p \equiv 1 + 5 \equiv 6 \bmod 8$ and therefore $p \equiv 3 \bmod 4$, which implies $p \equiv 3, 7 \bmod 20$.
4. It follows that primes $p \equiv 1, 9 \bmod 20$ are represented by the form $p = x^2 + 5y^2$, and primes $p \equiv 3, 7 \bmod 20$ by $2p = x^2 + 5y^2$.
5. Clearly $(a^2 + 5b^2)(c^2 + 5d^2) = (ac - 5bd)^2 + 5(ad + bc)^2$.
6. Assume that $2p = a^2 + 5b^2$; then a and b are both odd, and $4p^2 = (a^2 + 5b^2)^2 = (a^2 - 5b^2)^2 + 5(2ab)^2$; since $a^2 - 5b^2$ is divisible by 4, we can cancel 4 and obtain $p^2 = x^2 + 5y^2$.
7. If $2p = a^2 + 5b^2$ and $2q = c^2 + 5d^2$, then a, b, c and d are odd, hence the brackets in $4pq = (ac - 5bd)^2 + 5(ad + bc)^2$ are both even; canceling 4 then yields the claim that $pq = x^2 + 5y^2$.

6.29. The quadratic number field $K = \mathbb{Q}(\sqrt{-6})$ has class number 2; the nontrivial ideal class is generated by the ideal $\mathfrak{a} = (2, \sqrt{-6})$ above 2. This implies as above that primes p that split in K (those with $(\frac{-6}{p}) = +1$, i.e., with $p \equiv 1, 5, 7, 11 \bmod 24$) either are represented by the quadratic form $Q_0(x, y) = x^2 + 6y^2$, or $2p$ is represented by Q_0. In the latter case, $2p = X^2 + 6y^2$ implies that $X = 2x$ is even, hence $p = 2x^2 + 3y^2$.
Now $p = x^2 + 6y^2 \equiv 1, 7 \bmod 8$ if p and therefore x is odd, and $p = 2x^2 + 3y^2 \equiv 3, 5 \bmod 8$ if p and therefore y is odd.
Thus the primes $p \equiv 1, 7 \bmod 24$ are represented by Q_0, and the primes $p \equiv 5, 11 \bmod 24$ are represented by $Q_1(x, y) = 2x^2 + 3y^2$.
The field $\mathbb{Q}(\sqrt{-10})$ also has class number 2, and the nonprincipal ideal class is generated by the prime ideal $\mathfrak{a} = (2, \sqrt{-10})$ above 2. The primes $p \equiv 1, 7, 9, 11, 13, 19, 23, 37 \bmod 40$ split in K, and either $p = x^2 + 10y^2$ (if the prime ideals $\mathfrak{p}$ and $\mathfrak{p}'$ above p are principal) or $p = 2x^2 + 5y^2$ (if $\mathfrak{p}$ and $\mathfrak{p}'$ lie in the same class as $\mathfrak{a}$). Since $x^2 + 10y^2 \equiv \pm 1 \bmod 8$, this form represents the primes $p \equiv 1, 7, 9, 23 \bmod 40$, and the form $2x^2 + 5y^2$ represents the primes $p \equiv 11, 13, 19, 37 \bmod 40$.

6.30. The smallest primes $p \nmid 23$ for which $f(x) = x^3 - x + 1$ splits into three linear factors modulo p are the following:

p	$f(x) \bmod p$	(x, y)
59	$(x + 4)(x + 13)(x + 42)$	$(5, 2)$
101	$(x + 20)(x + 89)(x + 93)$	$(1, 4)$
167	$(x + 73)(x + 127)(x + 134)$	$(11, 2)$
173	$(x + 97)(x + 110)(x + 139)$	$(7, 4)$
211	$(x + 97)(x + 120)(x + 205)$	$(-1, 6)$
223	$(x + 33)(x + 63)(x + 127)$	$(1, 6)$

The last column contains the values of x and y for which $p = x^2 + xy + 6y^2$; observe that $4p = (2x + 1)^2 + 23y^2$. Since x is odd and y is even, each such prime is actually represented by the form $X^2 + 23Y^2$.

6.31. The ideal (2) is irreducible since $R/(2) \simeq (\mathbb{Z}/2\mathbb{Z})[x]$ is the polynomial ring over $\mathbb{Z}/2\mathbb{Z}$ and thus a domain.

6.32. This is a standard exercise in the construction of the reals from the rational numbers. The set of null sequences forms a subring of $\mathcal{C}$ since sums, differences, and products of null sequences are null sequences (the verification is straightforward). The same holds for the other sets of sequences.

The subring $\mathcal{N}$ is an ideal in $\mathcal{D}$ since the product of a null sequence with a converging sequence is again a null sequence. In fact, $\mathcal{N}$ is also an ideal in $\mathcal{C}$ and $\mathcal{B}$.

The subring $\mathcal{D}$ of converging sequences is not an ideal in $\mathcal{C}$ since the product of a sequence converging to a rational number and of a Cauchy sequence need not converge to a rational number.

Similarly the product of a Cauchy sequence not converging to 0 with the bounded sequence $a_n = (-1)^n$ is not Cauchy, so $\mathcal{C}$ is not an ideal in $\mathcal{B}$.

The sequences $(1, 0, \frac{1}{3}, 0, \frac{1}{5}, 0, \ldots)$ and $(0, \frac{1}{2}, 0, \frac{1}{4}, 0, \ldots)$ are nonzero sequences whose product is the zero sequence $(0, 0, 0, 0, \ldots)$. Thus each ring contains zero divisors.

Finally we claim that $\mathcal{N}$ is a maximal ideal in $\mathcal{C}$. Assume that I is an ideal in $\mathcal{C}$ with $\mathcal{N} \subseteq I \subseteq \mathcal{C}$. If the first inclusion is strict, then there is a Cauchy sequence $(a_n) \in I$ that is not a null sequence. Take an arbitrary Cauchy sequence $(b_n) \in \mathcal{C}$. Then the sequence (c_n) defined by $c_n = \frac{b_n}{a_n}$ if $a_n \neq 0$ and $c_n = 0$ if $a_n = 0$ is a Cauchy sequence, and the product $(a_n)(c_n) = (b_n)$ up to a null sequence, which implies that $\mathcal{C} = I$. Thus $\mathcal{N}$ is a maximal ideal in $\mathcal{C}$.

6.33. If $y^2 = x^3 - 4f$ with $f \equiv 3 \bmod 8$ and if $x = 2x_1$ and $y = 2y_1$ are even, then $y_1^2 + f = 2x_1^3$. Since $f \equiv 3 \bmod 8$ the integer y_1 is odd; but then $y_1^2 + f \equiv 4 \bmod 8$ is divisible by 4, but not by 8.

Thus x and y are odd. We find $x^3 = y^2+4f = (y+2\sqrt{-f})(y-2\sqrt{-f})$. The gcd of the factors on the right must be an ideal with odd norm dividing $\sqrt{-f}$. Assume that $\mathfrak{p}$ is a prime ideal dividing both factors; since $\mathfrak{p}$ is ramified, we have $\mathfrak{p}^2 = p$ for some prime $p \mid f$; but $p \mid y$ and $p \mid x$ then imply $p^2 \mid 4f$, which contradicts our assumption that f is squarefree.

By unique factorization into prime ideals we conclude that $(y + 2\sqrt{-f}) = \mathfrak{a}^3$. Since we have assumed that the class number of $\mathbb{Q}(\sqrt{-f})$ is not divisible by 3, the ideal $\mathfrak{a}$ must be principal, say $\mathfrak{a} = (\alpha)$

Now write $\alpha = \frac{r+s\sqrt{-f}}{2}$; then, up to sign,

$$y + 2\sqrt{-f} = \alpha^3 = \frac{r^3 - 3frs^2 + s(3r^2 - fs^2)\sqrt{-f}}{8},$$

and comparing coefficients of $\sqrt{-f}$ we obtain

$$16 = s(3r^2 - fs^2).$$

Now there are the following cases:

- $s = 1$, $f = 3r^2 - 16$; this implies $y = \frac{r^3-3fr}{8} = r(r^2 - 6)$ and finally $x = r^2 - 4$.
- $s = -1$, $f = 3r^2 + 16$; this implies $y = \frac{r^3-3fr}{8} = r(r^2 + 6)$ and finally $x = r^2 + 4$.
- $s = 2$; then $r = 2t$ is even and we find $f = 3t^2 - 4$, but this contradicts $f \equiv 3 \bmod 8$.
- $s = 4$: Then $r = 2r_1$ is even, and we find $4f = 3r_1^2 - 1$, which is impossible in integers.
- $s = -4$: Then $r = 2r_1$ is even, and we find $4f = 3r_1^2 + 1$; setting $r_1 = 2t + 1$ we obtain $f = 3t^2 + 3t + 1$. In this case $x = 16t^2 + 16t + 5$ and $y = \frac{r^3-48fr}{8} = (2t + 1)(32t^2 + 32t + 11)$.
- $s = \pm 8$ or $s = \pm 16$ does not lead to any solutions.

The value of f for which there are two essentially distinct solutions have the form $f = 3r^2 \pm 16$ and $f = 3t^2 + 3t + 1$. The equation $3r^2 + 16 = 3t^2 + 3t + 1$ leads to $r^2 = t^2 + t - 5$, which is easily seen to have the only integral solutions $(r, t) = (1, 2)$ and $(r, t) = 1, -3)$ leading to $f = 19$, and $(r, t) = (5, 5)$ leading to $f = 91$. Since the equation $3r^2 - 16 = 3t^2 + 3t + 1$ is not solvable in integers, only the equations $y^2 = x^3 - 4f$ with the values $f = 19$ and $f = 91$ possess a pair of integral solutions.

f	h	f	integral points
11	1	$3 \cdot 3^2 - 16$	$(5, 9)$
19	1	$3 \cdot 1^2 + 16$	$(5, 7), (101, 1015)$
43	1	$3 \cdot 3^2 + 16$	$(13, 45)$
59	3	$3 \cdot 5^2 - 16$	$(21, 95)$
91	2	$3 \cdot 5^2 + 16$	$(29, 155), (485, 10681)$
131	5	$3 \cdot 7^2 - 16$	$(45, 301)$
163	1	$3 \cdot 7^2 - 16$	$(53, 385)$

The equation $y^2 = x^3 - 339$ has two solutions $(13, 291)$ and $(61, 475)$ not predicted by this result. We conclude that the class number of $\mathbb{Q}(\sqrt{-339})$ must be divisible by 3. In fact, the class number is $h = 6$.

6.34. We claim that $\mathfrak{a}_1 = (\frac{11+\sqrt{85}}{2})$. Since $\mathfrak{a}_1^2 = (2 + \sqrt{85})$ has norm 81, the ideal $\mathfrak{a}$ has norm 9. Moreover, $9 = \frac{11+\sqrt{85}}{2} \cdot \frac{11-\sqrt{85}}{2}$ and $2 + \sqrt{85} = 2 \cdot \frac{11+\sqrt{85}}{2} - 9$ are both contained in $(\frac{11+\sqrt{85}}{2})$, hence $\mathfrak{a}$ divides this ideal, and since they have the same norm, they must be equal.

If the second ideal is principal, then there exists an element $\beta = \frac{x+y\sqrt{85}}{2}$ with norm ± 7. But the equation $x^2 - 85y^2 = \pm 4 \cdot 7$ is not solvable modulo 5.

6.35. We have $m = 9^2 + 32^2 = 23^2 + 24^2 = 31^2 + 12^2 = 33^2 + 4^2$. Let $\mathfrak{a}_1 = (32 + \sqrt{m}, 9)$, $\mathfrak{a}_2 = (24 + \sqrt{m}, 23)$, $\mathfrak{a}_3 = (12 + \sqrt{m}, 31)$ and $\mathfrak{a}_4 = (4 + \sqrt{m}, 33)$. Since K has class group $\simeq (2, 2)$, each square of an ideal is principal. In particular, $\mathfrak{a}$ must be principal. The elements $\alpha = \frac{31+\sqrt{m}}{2}$ and $\beta = \frac{33+\sqrt{m}}{2}$ have norms $N\alpha = -36$ and $N\beta = -4$, hence $\gamma = -\frac{31+\sqrt{m}}{33-\sqrt{m}} = 133 + 4\sqrt{m}$ has norm $N\gamma = 9$.
$N(1795 + 54\sqrt{m}) = -5 \cdot 31$.

6.36. We first show that x and y must be odd. If both are even, set $x = 2X$ and $y = 2Y$; then $8X^3 + 4 = 4pY^2$, i.e., $2X^3 + 1 = pY^2$. Then Y must be odd, hence $2X^3 + 1 \equiv p \equiv 5 \bmod 8$; but this implies $2X^3 \equiv 4 \bmod 8$, which is impossible.
Thus x and y are odd. Then

$$(-a)^3 = 4 - py^2 = (2 - y\sqrt{p})(2 + y\sqrt{p}).$$

Since the factors are coprime, we must have $(2 - y\sqrt{p}) = \mathfrak{a}^3$. Since 3 does not divide the class number h of $K = \mathbb{Q}(\sqrt{p})$, $\mathfrak{a}$ must be principal, hence

$$2 - y\sqrt{p} = \eta\alpha^3$$

for some unit η and an element $\alpha \in \mathcal{O}_K$.
Subsuming cubes into α^3 we may assume that $\eta = 1$, $\eta = \varepsilon$ or $\eta = \varepsilon'$. Now 2 is inert in K, hence $\alpha^3 \equiv 1 \bmod 2$ by Fermat's Little Theorem (observe that $N\alpha$ is odd); thus $\eta \equiv y\sqrt{p} \equiv 1 \bmod 2$ and $\eta' \equiv 1 \bmod 2$. But the

fundamental unit $\varepsilon \equiv \frac{\pm 1 \pm \sqrt{p}}{2}$ mod 2, and this implies that $\eta = 1$. Thus

$$2 + y\sqrt{p} = \left(\frac{c + d\sqrt{p}}{2}\right)^3,$$

and this implies

$$16 = c(c^2 + 3pd^2) \quad \text{and} \quad b = \frac{d(3c^2 + pd^2)}{8}.$$

If $c = 1$, then $3pd^2 = 15$, hence $p = 5$ and $d = 1$. If c is even, we get a contradiction.

Primes of the form $p = x^3 + 4$ have an obvious integral point $(x, 1)$, hence the class number of $\mathbb{Q}(\sqrt{p})$ must be divisible by 3 for these primes. Examples are $p = 9^3 + 4 = 733$ and $p = 25^3 + 4 = 15629$.

6.37. If y is odd, then $x = 2x_1$ is even, and we have

$$2x_1^3 = \left(\frac{y + \sqrt{k}}{2}\right)\left(\frac{y - \sqrt{k}}{2}\right)$$

in $\mathcal{O}_K$. If q is a prime number dividing both factors on the right, then q divides their sum y and their difference k. But then $q \mid x$ and $q^2 \mid (x^3 - y^2) = k$, which contradicts the assumption that k is squarefree.

Thus the factors on the right are coprime; we choose the sign of y in such a way that $y \equiv 1 \bmod 4$. This implies

$$\left(\frac{y + \sqrt{k}}{2}\right) = \mathfrak{p}a^3 \quad \text{and} \quad \left(\frac{y - \sqrt{k}}{2}\right) = \mathfrak{q}b^3,$$

where $\mathfrak{p}\mathfrak{q} = (2)$ and $\mathfrak{a}\mathfrak{b} = (x_1)$. But then $[\mathfrak{p}]$ lies in the cube of an ideal class, which contradicts our assumption.

6.38. The prime ideal $\mathfrak{p} = (2, \frac{1 + \sqrt{-31}}{2})$ generates the class group of $\mathbb{Q}(\sqrt{-31})$, which has order 3. By the preceding exercise, the equation $y^2 = x^3 - 31$ does not have an integral solution with y odd. By Exercise 1.23 there is no solution with y even since $31 = 3^3 + 2^2$.

Chapter 7

7.1. Observe that $\alpha = \sqrt{\eta} > 0$. Since $N\alpha = \alpha\alpha' < 0$ we have $\sqrt{\eta'} = -\alpha'$ and therefore $\text{Tr}\,\alpha = \alpha + \alpha' = \sqrt{\eta} - \sqrt{\eta'}$.

7.2. If $m = n^2$, then

$$1 = x^2 - my^2 = m^2 - n^2y^2 = (x - ny)(x + ny),$$

hence $x - ny = \pm 1$ and $x + ny = \pm 1$. Adding these equations yields $x = \pm 1$ and $y = 0$.

7.3. The real numbers $\{ax\} = ax - \lfloor ax \rfloor$ for $0 \le a \le q$ lie in the interval $[0, 1)$. By Dirichlet's pigeonhole principle there must be two such numbers a and b for which $\{ax\}$ and $\{bx\}$ differ at most by $\frac{1}{q}$. Since

$$\{ax\} - \{bx\} = ax - bx + \lfloor bx \rfloor - \lfloor ax \rfloor = \lfloor ax - bx \rfloor + \{ax - bx\} + \lfloor bx \rfloor - \lfloor ax \rfloor$$

we find with $q = a - b$ and $p = \lfloor ax - bx \rfloor + \lfloor bx \rfloor - \lfloor ax \rfloor$ that $-\frac{1}{q} < qx - p < \frac{1}{q}$; dividing through by q proves the claim.

7.4. If $m = t^2 - 1$, then $t + \sqrt{m}$ is a unit, and the elements with smallest nontrivial norm are $N(t - 1 + \sqrt{m}) = -2t + 2$ and $N(t + 1 + \sqrt{m}) = 2t + 2$.

We will prove that the only norms n with $|n| - 2t + 2$ are $N(a) = a^2$ for integers a and $N(t - 1 + \sqrt{m}) = -2t + 2$ along the lines of the proof of Prop. 7.9.

Set $\xi = x + y\sqrt{m}$; we will show that if $|N\xi| = n$ is not a square, then $|n| \ge 2t + 2$. Assume therefore that $|n| < 2t + 2$; since $\varepsilon = t + \sqrt{m} > 1$ is a unit in $\mathbb{Z}[\sqrt{m}]$, we can find a power η of ε for which $\xi \eta = a + b\sqrt{m}$ has coefficients a and b that satisfy the bounds from Theorem 7.8. Because of $2t < \varepsilon < 2\sqrt{m}$ we find

$$|b| \le \frac{\sqrt{n}}{2\sqrt{m}} \left(\sqrt{\varepsilon} + \frac{1}{\sqrt{\varepsilon}} \right) < 2.$$

Thus $|b| \le 1$. If $b = 0$ then $|N\xi| = a^2$ is a square; thus $b = \pm 1$, and this shows that $\alpha = \xi \eta = a \pm \sqrt{m}$. Now $|N\xi| = |N\alpha| = |a^2 - m|$ is minimal for values of a close to $\sqrt{m}$, and we find

$$|a^2 - m| = \begin{cases} -2t + 2 & \text{if } a = t - 1, \\ 1 & \text{if } a = t, \\ 2t + 2 & \text{if } a = t + 1. \end{cases}$$

If t is composite and divisible by an odd prime number q, this prime splits in $\mathbb{Q}(\sqrt{m})$; since it cannot be a norm, the class number must be > 1. This also follows from the ambiguous class number formula (Chap. 9) and the observation that $m = t^2 - 1 = (t - 1)(t + 1)$ is composite and cannot be of the form $m = pq$ for primes $p \equiv q \equiv 3 \bmod 4$.

Now consider the case $m = t^2 + 4$ for odd values of t. Then $\varepsilon = \frac{t + \sqrt{m}}{2}$ is a unit with norm -1, and the elements $\alpha = \frac{t \pm 2 + \sqrt{m}}{2}$ have norms $N(\alpha) = \pm t$. We claim that all elements norms have absolute value $< t$ are integers.

For a proof, set $\xi = \frac{a + b\sqrt{m}}{2}$. We will show that if $|N\xi| = n$ is not a square, then $|n| \ge t$. Assume therefore that $|n| < t$; since $\varepsilon = \frac{t + \sqrt{m}}{4} > 1$ is a unit in

$\mathbb{Z}[\sqrt{m}]$, we can find a power η of ε for which $\xi\eta = a+b\sqrt{m}$ has coefficients a and b that satisfy the bounds from Theorem 7.8. Because of $t < \varepsilon < \sqrt{m}$ we find

$$|b| \le \frac{\sqrt{n}}{\sqrt{m}}\left(\sqrt{\varepsilon} + \frac{1}{\sqrt{\varepsilon}}\right) < 1.$$

This implies that $b = 0$ (and then ξ is an integer) or $b = \pm\frac{1}{2}$. In this case, the elements with minimal norm are $\frac{t\pm 2+\sqrt{m}}{2}$ as claimed.

7.5. Let $\varepsilon = \frac{t+u\sqrt{m}}{2}$ be the fundamental unit of $\mathbb{Q}(\sqrt{m})$, and let n denote the smallest natural number for which $x^2 - my^2 = \pm 4n$ is solvable in nonzero integers. Among all solutions, choose one for which $|y|$ is minimal. If we multiply $\alpha = \frac{x+y\sqrt{m}}{2}$ by ε we get

$$\alpha\varepsilon = \frac{\frac{tx-muy}{2} + \frac{ux-ty}{2}\sqrt{m}}{2}.$$

Since y was chosen minimal we must have

$$\left|\frac{ux-ty}{2}\right| \ge y.$$

Now there are two cases:

- $ux \ge (t+2)y$. Then the norm equation $N(\varepsilon\alpha) = \pm m$ implies

$$n \ge \frac{\left(\frac{t+2}{u}\right) - m}{4} \cdot y^2 \ge \frac{t^2 + 4t + 4 - mu^2}{4u^2} = \begin{cases} \frac{t+2}{v^2} & \text{if } N\varepsilon = +1, \\ \frac{t}{v^2} & \text{if } N\varepsilon = -1. \end{cases}$$

- $ux \le (t-2)y$; then we get, similarly as above,

$$-n \le \frac{\left(\frac{t-2}{u}\right) - m}{4} \cdot y^2 \le \frac{t^2 - 4t + 4 - mu^2}{4u^2} = \begin{cases} -\frac{t-2}{v^2} & \text{if } N\varepsilon = +1, \\ -\frac{t}{v^2} & \text{if } N\varepsilon = -1. \end{cases}$$

This proves the claim.

7.6. Let $t^2 - 2pu^2 = 1$ be the minimal positive solution of the Pell equation. Then t is odd and $2pu^2 = (t-1)(t+1)$, hence we are in one of the following cases:

$$t - 1 = 4a^2, \qquad\qquad t + 1 = 2pb^2,$$
$$t - 1 = 2a^2, \qquad\qquad t + 1 = 4pb^2,$$
$$t - 1 = 4pa^2, \qquad\qquad t + 1 = 2b^2,$$
$$t - 1 = 2pa^2, \qquad\qquad t + 1 = 4b^2.$$

In each of these cases we obtain

$$1 = pb^2 - 2a^2; \quad 1 = 2pb^2 - a^2; \quad 1 = b^2 - 2pa^2; \quad 1 = 2b^2 - pa^2.$$

The first and the last equation are impossible modulo p, the third contradicts the minimality of u. Thus the second equation $a^2 - 2pb^2 = -1$ must be solvable in integers.

7.7. Assume that $t^2 - 2pu^2 = 1$ for minimal $u \geq 1$. Then $(t-1)(t+1) = 2pu^2$, and since the factors on the left have greatest common divisor 2 we have one of the following equations:

$$\begin{array}{ll}
t - 1 = 4a^2, & t + 1 = 2pb^2, \\
t - 1 = 2a^2, & t + 1 = 4pb^2, \\
t - 1 = 4pa^2, & t + 1 = 2b^2, \\
t - 1 = 2pa^2, & t + 1 = 4b^2.
\end{array}$$

In each of these cases we obtain

$$1 = pb^2 - 2a^2; \quad 1 = 2pb^2 - a^2; \quad 1 = b^2 - 2pa^2; \quad 1 = 2b^2 - pa^2.$$

Since $p \equiv 3 \bmod 4$, the second equation is impossible, and the third contradicts the minimality of u. Thus $2a^2 - pb^2 = 1$ or $2b^2 - pb^2 = -1$. Multiplying these equations through by 2 proves the claims.

7.8. For $m = 3$ we have $\varepsilon = 2 + \sqrt{3}$ since this corresponds to the smallest positive solution of the Pell equation $t^2 - 3u^2 = 1$.
$m = 19$. The element $\alpha = 4 + \sqrt{19}$ generates a prime ideal with norm 3. Since $N(5 + \sqrt{19}) = 6$ we find that

$$\alpha = \frac{5 + \sqrt{19}}{4 - \sqrt{19}} = -(13 + 3\sqrt{19})$$

generates the prime ideal above 2, hence

$$\varepsilon = \frac{1}{2}\alpha^2 = 170 + 39\sqrt{19}$$

is a unit. Since the coefficient of $\sqrt{19}$ for squares is even, ε cannot be a square. Since $(1 + \sqrt{19})^5 > \varepsilon$, the unit ε is either fundamental or a cube. But $\varepsilon = \alpha^3$ is easily shown to be impossible.
$m = 43$: We set $\mathfrak{a} = (2, 1 + \sqrt{43})$, $\mathfrak{p} = (3, 1 + \sqrt{43})$ and $\mathfrak{q} = (7, 1 + \sqrt{43})$.

a	$N(a + \sqrt{43})$	$(a + \sqrt{43})$
5	$-2 \cdot 3^2$	$\mathfrak{a}\mathfrak{p}'^2$
6	-7	$\mathfrak{q}'$
7	$2 \cdot 3$	$\mathfrak{a}\mathfrak{p}$
8	$3 \cdot 7$	$\mathfrak{p}'\mathfrak{q}$

The ideal $\mathfrak{a}$ is generated by

$$\alpha = \frac{2(5 + \sqrt{43})}{(7 - \sqrt{43})^2} = 59 + 9\sqrt{43},$$

hence $\varepsilon = \frac{1}{2}\alpha^2 = 3482 + 531\sqrt{43}$ is a unit in $\mathbb{Z}[\sqrt{43}]$. Clearly ε is not a square; since $\varepsilon \equiv 4 \bmod \mathfrak{p}$, it is not a cube. This implies that ε is fundamental.
$m = 67$: We set $\mathfrak{a} = (2, 1 + \sqrt{67})$, $\mathfrak{p} = (3, 1 + \sqrt{67})$ and $\mathfrak{q} = (7, 2 + \sqrt{67})$. Here $\mathfrak{p} = (8 - \sqrt{67})$ and $(7 + \sqrt{67}) = \mathfrak{a}\mathfrak{p}^2$. Thus $\mathfrak{a} = (\alpha)$ for

$$\alpha = \frac{7 + \sqrt{67}}{(8 - \sqrt{67})^2} = 221 + 27\sqrt{67},$$

and $\varepsilon = \frac{1}{2}\alpha^2 = 48842 + 5967\sqrt{67}$ is a unit. We have to show that ε is not a k-th power for $1 < k \le 5$. Clearly ε is not a square; $\varepsilon \equiv 4 \bmod \mathfrak{p}$ shows that ε is not a cube. Next $6 + \sqrt{67}$ generates a prime ideal $\mathfrak{r}$ of norm 31, and $\varepsilon \equiv 20 \bmod \mathfrak{r}$ shows that ε is no fifth power.
$m = 131$: Here we need $\mathfrak{a} = (2, 1 + \sqrt{131})$ and $\mathfrak{p} = (5, 1 + \sqrt{131})$. Using $(11 + \sqrt{131}) = \mathfrak{a}\mathfrak{p}$ and $(16 + \sqrt{131}) = \mathfrak{p}^3$ we obtain the generator

$$\alpha = \frac{11 + \sqrt{131}}{2(16 + \sqrt{131})} = 103 + 9\sqrt{131}$$

of $\mathfrak{a}$ and the unit $\varepsilon = \frac{1}{2}\alpha^2 = 10610 + 927\sqrt{131}$. Here it is sufficient to show that ε is not a square (this is obvious) and not a cube. But $(12 + \sqrt{131}$ generates a prime ideal $\mathfrak{r}$ of norm 13, and $\varepsilon \equiv 6 \bmod 13$ shows that ε is not congruent to a cube modulo $\mathfrak{r}$.
$m = 159$: Here we need $\mathfrak{a} = (1, 1 + \sqrt{159})$, $\mathfrak{p} = (3, \sqrt{159})$, $\mathfrak{q} = (5, 2 + \sqrt{159})$. From $(12 + \sqrt{159}) = \mathfrak{p}\mathfrak{q}$ we obtain $\mathfrak{q}^2 = (101 + 8\sqrt{159})$. Next $(13 + \sqrt{159}) = \mathfrak{a}\mathfrak{p}$ then gives us $(\mathfrak{q}')^2 = (164 + 13\sqrt{159})$. Thus

$$\varepsilon = \frac{164 + 13\sqrt{159}}{101 - 8\sqrt{159}} = 1324 + 105\sqrt{159}$$

is a unit that is easily shown to be fundamental.

$m = 199$: Here $14 + \sqrt{199}$ generates a prime ideal $\mathfrak{p}$ of norm 3 and $19 + \sqrt{199}$ has norm $2 \cdot 3^4$. Thus

$$\alpha = \frac{19 + \sqrt{199}}{(14 - \sqrt{199})^4} = 127539 + 9041\sqrt{199}$$

generates the prime ideal above 2, and

$$\varepsilon = \frac{1}{2}\alpha^2 = 16266196520 + 1153080099\sqrt{199}$$

is a unit, which we can show to be fundamental by checking that ε is not a k-th power for $k = 2, 3, 5,$ and 7.

7.9. From $\pm 4 = t^2 - mu^2 \equiv t^2 - u^2 \bmod 8$ we immediately deduce that $t \equiv u \equiv 0 \bmod 2$.

7.10. Clearly $N\varepsilon = n^2 - m = 1$, so ε is a unit. Since $(1 + \sqrt{m})^2 = m + 1 + 2\sqrt{m} > \varepsilon$, the unit must be fundamental.

If $m = n^2 + 1$, the element $\varepsilon = n + \sqrt{m}$ is a unit with norm -1. If $m = n^2 \pm 4$ is squarefree, then n is odd, and $\frac{n + \sqrt{m}}{2}$ is a unit.

7.11. We have $(2) = \mathfrak{a}^2$ for $\mathfrak{a} = (2, \sqrt{478})^2$, $(3) = \mathfrak{p}\mathfrak{p}'$ for $\mathfrak{p} = (3, 1 + \sqrt{478})$, and $(7) = \mathfrak{q}\mathfrak{q}'$ for $\mathfrak{q} = (7, 3 + \sqrt{478})$. We find

a	$N(a + \sqrt{478})$	$(a + \sqrt{478})$
10	$2 \cdot 3^3 \cdot 7$	$\mathfrak{a}\mathfrak{p}^3\mathfrak{q}$
17	$3^3 \cdot 7$	$\mathfrak{p}'^3\mathfrak{q}$
22	$2 \cdot 3$	$\mathfrak{a}\mathfrak{p}$
24	$2 \cdot 7^2$	$\mathfrak{a}\mathfrak{q}^2$
25	$3 \cdot 7^2$	$\mathfrak{p}\mathfrak{q}'^2$

Next $(10 + \sqrt{478})(17 + \sqrt{478}) = (27)\mathfrak{a}\mathfrak{q}$, hence $\mathfrak{a}\mathfrak{q}^2 = (24 + \sqrt{478})$, but we already knew that. But

$$\frac{2(10 + \sqrt{478})}{((22 + \sqrt{478}))^3} = -4635 + 212\sqrt{478}$$

generates $\mathfrak{q}$, and therefore

$$\alpha = \frac{24 + \sqrt{478}}{(4635 - 212\sqrt{478})^2}$$

generates $\mathfrak{a}$. This implies that

$$\varepsilon = \frac{1}{2}\alpha^2 = 1617319577991743 + 73974475657896\sqrt{478}$$

is a unit, and in fact the fundamental unit. The last claim requires showing that ε is no p-th power for all primes $p \le 11$.

Since $\mathfrak{a}$ is principal, so is $\mathfrak{p}$, and this implies that $\mathfrak{q}$ is principal. For the class number to be 1 we need to show that the prime ideals with norm less than the Minkowski bound (thus with norm ≤ 19) are principal. This is now a matter of a few simple calculations involving the elements $18 + \sqrt{478}$, $19 + \sqrt{478}$, $23 + \sqrt{478}$: It follows that the prime ideals above 11, 13, and 17 are also principal.

7.12. The equation $2x^2 - 5y^2 = \pm 1$ is impossible in integers since it is not solvable modulo 5. Thus $|2x^2 - 5y^2| \le 1$ implies $2x^2 - 5y^2 = 0$, which is only possible for $x = y = 0$ since $\sqrt{10}$ is irrational.

7.13. Let x denote the continued fraction expansion

$$x = \cfrac{1}{2 + \cfrac{1}{1 + \cfrac{1}{2 + \dots}}}.$$

Then $x = \cfrac{1}{1 + \cfrac{1}{1 + x}}$ yields, after simplifying the fraction, the quadratic

equation $2x^2 + 2x - 1 = 0$, whose unique positive solution is $x = \sqrt{3} - 1$.

The partial convergents in this case are

$$\frac{5}{3}, \frac{7}{4}, \frac{19}{11}, \frac{26}{15}, \dots.$$

7.14. Euclidean division shows

$$\sqrt{m} = t + \sqrt{m} - t$$

$$\frac{1}{\sqrt{m} - t} = \frac{\sqrt{m} + t}{2} = t + \frac{\sqrt{m} - t}{2}$$

$$\frac{2}{\sqrt{m} - t} = \sqrt{m} + t = 2t + \sqrt{m} - t.$$

This implies that

$$\sqrt{m} = t + \cfrac{1}{t + \cfrac{1}{2t + \sqrt{m} - t}},$$

which implies the claim.

7.15. Clearly

$$\left(\frac{a-1}{2}\right)^3 + \left(\frac{a+1}{2}\right)^3 = \frac{a^3 + 3a}{4} = \left(\frac{x}{2}\right)^2.$$

For solving $y^2 = x^3 + 3x$ we observe that $x \geq 0$; excluding the point $(0, 0)$ we may assume that $x > 0$. Now there are two cases:

1. $x = a^2$, $x^2 + 3 = b^2$. Since the only squares differing by 3 are 1 and 4 we obtain $a = 1$, $b = 2$ and $x = 1$, $y = \pm 2$.
2. $x = 3a^2$, $x^2 + 3 = 3b^2$. Here $9a^4 + 3 = 3b^2$ yields $3a^4 + 1 = b^2$. If b is odd, then we may assume that either

   - $b + 1 = 8r^4$ and $b - 1 = 6s^4$; then $4r^4 - 3s^4 = 1$. In the equation $3s^4 = 4r^4 - 1 = (2r^2 - 1)(2r^2 + 1)$ we cannot have $3 \mid 2r^2 - 1$ since $\left(\frac{2}{3}\right) = -1$; thus $2r^2 + 1 = 3t^4$ and $2r^2 - 1 = u^4$, hence $3t^4 - u^4 = 2$. The unique integral solution $t^2 = u^2 = 1$ gives $b = 7$ and finally $x = 12$.
   - $b + 1 = 2r^4$ and $b - 1 = 24s^4$. Then $r^4 - 12s^4 = 1$, hence $12s^4 = r^4 - 1 = (r^2 - 1)(r^2 + 1)$. Since $r^2 + 1 \equiv 2 \bmod 4$ we must have $r^2 + 1 = 2t^4$ and $r^2 - 1 = 6u^4$, which implies $t^4 - 3u^4 = 1$. This equation does not have an integral solution by Exercise 1.20.

   If b is even, then $b + 1 = \pm r^4$ and $b - 1 = \pm 3s^4$, hence $r^4 - 3s^4 = \pm 2$; clearly the minus sign must hold, hence $r^4 - 3s^4 = -2$. The only integral solutions are given by $r^2 = s^2 = 1$, hence $b = -2$ ($b = 2$ is impossible since $b + 1 = 3 \neq \pm r^4$) and $x = 3$.

7.16. Write $t^2 - 1 = pu^2$; if t is odd, then $\gcd(t - 1, t + 1) = 2$, and there are two cases:

   - $t - 1 = 2a^2$ and $t + 1 = 2pb^2$; then $a^2 - pb^2 = -1$, which contradicts the fact that $p \equiv 3 \bmod 4$.
   - $t - 1 = 2pa^2$ and $t + 1 = 2b^2$; then $a^2 - pb^2 = +1$, and this contradicts the minimality of t (recall that $t + u\sqrt{p}$ is fundamental).

   Thus t must be even. Clearly $t^2 - pu^2 = 1$ implies $t^2 = 1 + pu^2 \equiv 4 \bmod 8$ if $p \equiv 3 \bmod 8$, and $t^2 = 1 + pu^2 \equiv 0 \bmod 8$ if $p \equiv 7 \bmod 8$.

In fact we can show that t is divisible by 8 in the latter case. Since $\gcd(t - 1, t + 1) = 1$ we have the two cases

- $t - 1 = a^2, t + 1 = pb^2$; then $a^2 - pb^2 = -2$, which contradicts the fact that $p \equiv 7 \bmod 8$.
- $t + 1 = a^2, t - 1 = pb^2$. Since a is odd, the first equation implies $t \equiv 0 \bmod 8$.

7.17. It follows immediately from $f(a, b) = f(a', b')$ that $(a - a')\xi_1 + (b - b')\xi_2 = 0$. Thus $\xi_1 / \xi_2 = \frac{b - b'}{a - a'} \in \mathbb{Q}$ if $a - a' \neq 0$. This contradiction shows that $a - a' = 0$ and therefore $b - b' = 0$, hence $(a, b) = (a', b')$ and finally the injectivity of f.

7.18. Since $\varepsilon + \varepsilon' = 4$ and $\varepsilon\varepsilon' = 1$ we have

$$V_{n+1} = (\varepsilon^n + \varepsilon'^n)(\varepsilon + \varepsilon') - \varepsilon\varepsilon'^n - \varepsilon'\varepsilon^n = 4V_n - V_{n-1}.$$

Similarly,

$$V_{2n} = \varepsilon^{2n} + \varepsilon'^{2n} = (\varepsilon^n + \varepsilon'^n)^2 - 2 = V_n^2 - 2.$$

Chapter 8

8.1. (See Mignotte [93]) We already know that in this case $y + 1 = qa^2$ and $\frac{y^q + 1}{y + 1} = qb^2$, where $ab = x$.

Assume that $q = 8k + r$ for $r = 5, 7$. Then $y = qa^2 - 1 \equiv a - 1 \bmod 8$. Moreover,

$$x^2 = y^q + 1 = (y^2 - 1 + 1)^{4k}y^a + 1 \equiv y^a + 1 \bmod (y^2 - 1).$$

Thus $y^a + 1$ is a quadratic residue modulo each prime divisor of $y^2 - 1$, and, in particular, we have

$$1 = \left(\frac{y^a + 1}{y - 1}\right) = \left(\frac{2}{y - 1}\right),$$

which contradicts the fact that $y - 1 \equiv a - 2 \equiv \pm 3 \bmod 8$.

Now assume that $q = 8k + 3 = 24h + a$ for $a \in \{11, 19\}$ (here we are using $q \neq 3$); then $y \equiv 2 \bmod 8$ and

$$x^2 = y^q + 1 = (y^3 - 1 + 1)^{8h}x^a + 1 \equiv x^a + 1 \bmod (x^3 - 1).$$

If $a = 11$, then $x^{11} - x^2 = x^2(x^9 - 1)$, hence

$$1 = \left(\frac{x^{11}+1}{x^3-1}\right) = \left(\frac{x^2+1}{x^3-1}\right) = \left(\frac{x^3-1}{x^2+1}\right) = \left(\frac{-x-1}{x^2+1}\right) = \left(\frac{x^2+1}{x+1}\right) = \left(\frac{2}{x+1}\right) = -1,$$

and this is a contradiction since $x + 1 \equiv 3 \bmod 8$.
If $a = 19$, then similarly

$$1 = \left(\frac{x^{19}+1}{x^3-1}\right) = \left(\frac{x+1}{x^3-1}\right) = \left(\frac{x^3-1}{x+1}\right) = \left(\frac{2}{x+1}\right) = -1.$$

8.2. The congruence

$$(-1)^m \equiv (1 + 2m - m^2)S_0 \bmod 3$$

implies

$$S_0 \equiv \frac{(-1)^m}{1 + 2m - m^2} \bmod 3.$$

The residue class modulo 3 of the numerator depends on $m \bmod 2$, that of the denominator on $m \bmod 3$. Thus the residue class of $S_0 \bmod 3$ only depends on $m \bmod 6$, and the claim follows by verifying it for all integers m with $1 \le m \le 6$.

8.3. We have

$$(1+1)^m = \binom{m}{0} + \binom{m}{1} + \binom{m}{2} + \binom{m}{3} + \dots,$$

$$(1-1)^m = \binom{m}{0} - \binom{m}{1} + \binom{m}{2} - \binom{m}{3} + \dots, \quad \text{hence}$$

$$2^m = 2\binom{m}{0} + 2\binom{m}{2} + 2\binom{m}{4} + \dots$$

$$= 2\binom{m}{1} + 2\binom{m}{3} + 2\binom{m}{5} + \dots.$$

8.4. Let ρ denote a primitive cube root of unity. Then for $k = 0, 1, 2$ we have

$$(1 + \rho^k)^m = \binom{m}{0} + \rho^k \binom{m}{1} + \rho^{2k} \binom{m}{2} + \binom{m}{3} + \rho^k \binom{m}{4} + \dots.$$

Adding these equations for $k = 0, 1, 2$ yields, since $1 + \rho = -\rho^2$

$$2^m + (-\rho^2)^m + (-\rho)^m = 3\left[\binom{m}{0} + \binom{m}{3} + \binom{m}{6} + \dots\right].$$

- If $m \equiv 0 \bmod 3$, then

$$\binom{m}{0} + \binom{m}{3} + \binom{m}{6} + \ldots = \frac{2^m + 2(-1)^m}{3}.$$

- If $m \equiv 1, 2 \bmod 3$, then the left hand side becomes $2^m + (-1)^m(\rho + \rho^2) = 2^m - (-1)^m$, and the claim follows.

8.5. As above we find

$$(1 + i^k)^m = \binom{m}{0} + i^k \binom{m}{1} + (-1)^k \binom{m}{2} + i^{3k} \binom{m}{3} + \binom{m}{4} + \ldots$$

for $k = 0, 1, 2, 3$. Adding these equations yields

$$4\left[\binom{m}{0} + \binom{m}{4} + \binom{m}{8} + \ldots\right] = 2^m + (1 + i)^m + (1 - i)^m,$$

hence

$$\binom{m}{0} + \binom{m}{4} + \binom{m}{8} + \ldots = \begin{cases} 2^{m-2} + (-1)^{\frac{m}{4}} 2^{\frac{m-2}{2}} & \text{if } m \equiv 0 \bmod 4, \\ 2^{m-2} + (-1)^{\frac{m-1}{4}} 2^{\frac{m-3}{2}} & \text{if } m \equiv 1 \bmod 4, \\ 2^{m-2} & \text{if } m \equiv 2 \bmod 4, \\ 2^{m-2} + (-1)^{\frac{m+1}{4}} 2^{\frac{m-3}{2}} & \text{if } m \equiv 3 \bmod 4 \end{cases}$$

as claimed.

8.6. If y and $y + 1$ are S-smooth, then $2 \in S$ since either y or $y + 1$ is even. Thus $4y(y + 1) = (2y + 1)^2 - 1$ is S-smooth. We claim that there are at most 3^n integers $a > 0$ for which $a^2 - 1$ is S-smooth.
   Assume that $a^2 - 1 = p_1^{e_1} \cdots p_n^{e_n}$. Write $e_j = 2f_j + g_j$ with $g_j \in \{0, 1, 2\}$, where $g_j = 0$ if $e_j = 0$ and $g_j = 2$ otherwise, and set $d = \prod p_j^{g_j}$ and $b = \prod p_j^{f_j}$; then $a^2 - 1 = db^2$, or $a^2 - db^2 = 1$, and each prime dividing b also divides d. Størmer's Theorem 8.4 then implies that, for a fixed value of d, the equation $a^2 - 1 = db^2$ has at most one positive solution with the property that primes dividing b also divide d. Since $d = \prod p_j^{g_j}$ and $0 \le g_j \le 2$ there are at most 3^n choices for d, and this proves our claim.
   For improvements of this procedure see [75].

8.7. Let x and y be natural numbers satisfying $x^2 + x + 1 = 3y^2$. Then $x \equiv 1 \bmod 3$, hence we can write $x = 3z + 1$ for some natural number z. Then $9z^2 + 3 = 12y^2$ implies $(2y)^2 - 3z^2 = 1$, hence $2y + z\sqrt{3}$ is a unit in $\mathbb{Z}[\sqrt{3}\,]$, and we can write $2y + z\sqrt{3} = (2 + \sqrt{3}\,)^h$ for some exponent $h \ge 1$. If h is even, then so is z, which is impossible since $(2y)^2 - 3z^2 = 1$; this shows that $h = 2n + 1$ must be odd. Thus

$$2y + z\sqrt{3} = (2 + \sqrt{3})^{2n+1} \quad \text{and} \quad 2y - z\sqrt{3} = (2 - \sqrt{3})^{2n+1}.$$

These equations quickly imply the claims since $x = 3z + 1$.

Chapter 9

9.1. The equality $[\mathfrak{a}] = [\mathfrak{b}]$ of ideal classes is by definition equivalent to the existence of an element $\alpha \in k^\times$ with $\mathfrak{a} = (\alpha)\mathfrak{b}$. Applying σ to this equation shows that $\mathfrak{a}^\sigma = (\alpha^\sigma)\mathfrak{b}\sigma$, which in turn implies that $[\mathfrak{a}^\sigma] = [\mathfrak{b}^\sigma]$. Since $\sigma^2 = \text{id}$, applying σ to the last equation now proves the converse.

9.2. The norm of the fundamental unit $\varepsilon = 3 + \sqrt{10}$ of $\mathbb{Q}(\sqrt{10})$ has norm -1, hence an ideal above a prime number p is principal if and only if the equation $x^2 - 10y^2 = p$ has integral solutions. Clearly $x^2 - 10y^2 = 2$ and $x^2 - 10y^2 = 5$ are impossible modulo 5 and modulo 8, respectively. Moreover, $(2-\sqrt{10}) = (2, \sqrt{10})(3, 1+\sqrt{10})$, hence $(3, 1+\sqrt{10})$ cannot be principal either.

The ideals $(2, \sqrt{10})$ and $(5, 1+\sqrt{10})$ are generated by ramified prime ideals, hence are ambiguous, whereas $(3, 1 + \sqrt{10})$ is not ambiguous.

9.3. We know that the fundamental unit ε has norm -1, so ideals above primes q are principal if and only if $q = x^2 - 2py^2$ has integral solutions. The equation $2 = x^2 - 2py^2$ is not solvable modulo p since 2 is a nonsquare modulo primes $p \equiv 5 \bmod 8$.

9.4. Assume that $(2, \sqrt{2p}) = (\alpha)$ is principal. Then $\varepsilon = \frac{1}{2}\alpha^2$ is a unit, and ε cannot be a square since 2 is no square in K. But $N\varepsilon = \frac{1}{4}(N\alpha)^2 = +1$, hence the fundamental unit must have norm $+1$.

Conversely, assume that $N\varepsilon = +1$. By Hilbert's Theorem 90 there is an element $\alpha \in \mathcal{O}_K$ with $\alpha^{1-\sigma} = \varepsilon$. Then (α) is ambiguous, and getting rid of rational prime we find that (α) is one of the ideals (1), $(2, \sqrt{2p})$, $(p, \sqrt{2p})$ or $(\sqrt{2p})$.

If $(\alpha) = (1)$, then $\alpha = \eta$ for some unit η, but then $\varepsilon = \eta^{1-\sigma}$ cannot be fundamental.

Similarly, if $(\alpha) = (\sqrt{2p})$, then $\alpha = \eta\sqrt{2p}$ for some unit η, and again $\alpha^{1-\sigma} = -\eta^{1-\sigma}$ contradicts the fact that ε is fundamental.

Thus (α) must be one of the ideals $(2, \sqrt{2p})$ or $(p, \sqrt{2p})$; since their product is $(\sqrt{2p})$, they must both be principal.

Finally assume that $2p = a^2 + b^2$ with $a > b > 0$ and set $\mathfrak{a} = (a, b + \sqrt{m}\,)$. Clearly

$$\mathfrak{a}^2 = (a^2, a(b + \sqrt{2p}\,), (b + \sqrt{2p}\,)^2) = (2p - b^2, a(b + \sqrt{2p}\,), (b + \sqrt{2p}\,)^2)$$
$$= (b + \sqrt{2p}\,)(b - \sqrt{2p}, a, b + \sqrt{2p}\,) = (b + \sqrt{2p}\,).$$

Note that $\mathfrak{a}\mathfrak{a}^\sigma = (a) \neq \mathfrak{a}^2$, so $\mathfrak{a}$ is not an ambiguous ideal. But $\mathfrak{a}\mathfrak{a}^\sigma \sim \mathfrak{a}^2$ clearly implies that $\mathfrak{a}^\sigma \sim \mathfrak{a}$, hence the ideal class of $\mathfrak{a}$ is ambiguous.

9.5.   Let c be the nontrivial ideal class. Since c^σ must also be nontrivial, we have $c = c^\sigma$, and so the nontrivial ideal class is ambiguous.

9.6.   Let c be an ambiguous ideal class. Then $c = c^\sigma$, hence $c^2 = c \cdot c^\sigma = 1$ since the norm of an ideal is principal. Thus ambiguous ideal classes have order 1 or 2. Since the class group has odd order, every ambiguous ideal class must have order 1, and this shows that $\mathrm{Am}(k) = 1$.

9.7.   Consider the natural map $\pi : A \longrightarrow AB/B$ defined by sending an element $a \in A$ to the coset $aB + B$. Its kernel consists of all elements a with $aB \in B$; this is equivalent to $a \in B$. Thus $\ker \pi = A \cap B$. On the other hand, π is clearly onto, and this implies that $A/A \cup B = A/\ker \pi \simeq \mathrm{im}\,\pi = AB/B$.

9.8.   Let H denote the group of nonzero principal ideals, H^G its subgroup fixed by the Galois group G, A the group of nonzero ambiguous ideals and P the group of all fractional ideals in $\mathbb{Q}^\times$. The kernel of the map $\iota : H^G/P \longrightarrow A/P$ sending a coset $(\alpha)P$ with $(\alpha') = (\alpha)$ to the same coset in A/P consists of all cosets $(\alpha)P$ satisfying $(\alpha) \in P$, hence of the coset P. Thus ι is injective. The map $\pi : A/P \longrightarrow \mathrm{Am}_{\mathrm{st}}(k)$, which sends a coset $\mathfrak{a}P$ with $\mathfrak{a}^\sigma = \mathfrak{a}$ to the ideal class $[\mathfrak{a}]$, is well defined since changing $\mathfrak{a}$ by a principal ideal does not change its ideal class. Since every strongly ambiguous ideal class is generated by an ambiguous ideal, the map π is surjective.
For showing that $\ker \pi = \mathrm{im}\,\iota$ assume that $\pi(\mathfrak{a}P) = [(1)]$; then $\mathfrak{a} = (\alpha)$ is principal and therefore in the image of ι. Since the image of ι is obviously contained in $\ker \pi$, we have proved the claim.

9.9.   We have

$$\alpha = \frac{m + ni}{m - ni} = \frac{(m + ni)^2}{(m - ni)(m + ni)} = \frac{m^2 - n^2}{m^2 + n^2} + \frac{2mn}{m^2 + n^2} \cdot i.$$

Clearing denominators we find

$$x = m^2 - n^2, \quad y = 2mn \quad \text{and} \quad z = m^2 + n^2$$

as desired.

The generalization to arbitrary quadratic number fields is straightforward; elements of norm 1 in $\mathbb{Q}(\sqrt{m})$ have the form

$$\frac{a + b\sqrt{m}}{a - b\sqrt{m}} = \frac{(a + b\sqrt{m})^2}{(a + b\sqrt{m})(a - b\sqrt{m})} = \frac{a^2 + mb^2}{a^2 - mb^2} + \frac{2ab}{a^2 - mb^2} \cdot \sqrt{m},$$

which implies that

$$x = a^2 + mb^2, \quad y = 2ab, \quad \text{and} \quad z = a^2 - mb^2$$

is a solution of $x^2 - my^2 = z^2$.

9.10. An element $\alpha \in k^\times$ belongs to H^G if and only if $(\alpha) = (\alpha^\sigma)$. This is equivalent to the existence of a unit ε with $\alpha = \varepsilon\alpha^\sigma$, i.e., to $\varepsilon = \alpha^{1-\sigma}$. Clearly $\varepsilon \in E[N]$ since $N\varepsilon = 1$, and the kernel of the homomorphism $\lambda((\alpha)) = \varepsilon E^{1-\sigma}$ consists of all principal ideals (α) with $\alpha^{1-\sigma} = \eta^{1-\sigma}$ for some unit η. But then $\beta = \alpha/\eta$ has the property that $(\alpha) = (\beta)$ and $\beta^\sigma = \beta$, which shows that $\beta \in \mathbb{Q}^\times$. Thus $\ker \lambda = P$.

9.11. The ideal $(3, 1 + \sqrt{10})$ has norm 3 and is not principal, whereas $(1, 3 + \sqrt{10}) = (1)$ lies in the principal class.

9.12. If $\mathfrak{p} = (\pi)$, then $(p) = \mathfrak{p}^2 = (\pi^2)$. Thus $\pi^2 = \varepsilon p$ for some unit ε. If $\pm\varepsilon$ is a square, then so is $\pm p$, and we must have $m = \pm p$, which we have excluded. This works more generally for products n of disjoint ramified primes. If $m = 30$, the elements $\alpha = 6 + \sqrt{30}$ and $\beta = 5 + \sqrt{30}$ generate ramified ideals, and we have $\frac{\alpha^2}{6} = \frac{\beta^2}{5} = 11 + 2\sqrt{30}$.

9.13. If $\left(\frac{q}{p}\right) = +1$, then $\left(\frac{q^*}{p}\right) = +1$ for $q^* = \left(\frac{-1}{q}\right)q$. Thus p splits in $k = \mathbb{Q}(\sqrt{q^*})$, and we have $\pm 4p^h = x^2 + q^*y^2$, where h is the odd class number of k. Reduction modulo q then implies $\left(\frac{p}{q}\right) = +1$.

9.14. If $p \equiv 3 \bmod 4$, then the equation $u^2 - pu^2 = -1$ is impossible modulo p (as well as modulo 4). Assume therefore that $p \equiv 1 \bmod 4$ is prime, and let (t, u) be the smallest positive solution of the Pell equation $t^2 - pu^2 = 1$. Then t must be odd, and in $pu^2 = (t - 1)(t + 1)$ we have $\gcd(t + 1, t - 1) = 2$. Thus either $t - 1 = 2a^2$ and $t + 1 = 2pb^2$ or $t - 1 = 2pa^2$ and $t + 1 = 2b^2$. In the second case we obtain $b^2 - pa^2 = 1$ for a smaller pair (a, b) contradicting our assumptions. Thus $pb^2 - a^2 = 1$ and therefore $a^2 - pb^2 = -1$, hence the fundamental unit of $\mathbb{Q}(\sqrt{p})$ has negative norm if $p \equiv 1 \bmod 4$ is prime.

9.15. Since $x^2 - qy^2$ is even, x and y must have the same parity. If x and y are odd, then $x^2 - qy^2 \equiv x^2 + y^2 \equiv 2 \bmod 4$, which contradicts the fact that $x^2 - qy^2$ is divisible by 4. Thus $x = 2A$ and $y = 2Y$ are even, and we find $\pm p^h \equiv X^2 - qY^2$.

Again we have $X^2 - qY^2 \equiv X^2 + Y^2 \bmod 4$, and we find that X and Y must be odd and that $\pm p^h \equiv 1 \bmod 4$. Since $p \equiv 1 \bmod 4$, the plus sign must hold.

9.16. If $N\varepsilon_m = +1$, then $\varepsilon_m = \alpha^{\sigma-1}$ by Hilbert's Theorem 90. Set $\mathfrak{a} = (\alpha)$; then $\mathfrak{a}^\sigma = (\alpha^\sigma) = (\alpha\varepsilon_m) = (\alpha)$, so $\mathfrak{a}$ is an ambiguous principal ideal. If $\mathfrak{a} = (1)$ then $\varepsilon_m = 1$, which is nonsense; similarly, $\mathfrak{a} = (\sqrt{m})$ leads to $\varepsilon_m = -1$.

9.17. Assume that $p \equiv 1 \bmod 4$ is a prime number, and let ε denote the fundamental unit of $k = \mathbb{Q}(\sqrt{p})$. If $N\varepsilon = +1$, then there is an ambiguous principal ideal $(\alpha) \neq (1), (\sqrt{p})$. But ambiguous principal ideals are generated by ramified primes, and the only ramified prime in $\mathbb{Q}(\sqrt{p})$ is p (here we have used $p \equiv 1 \bmod 4$, so the discriminant of k is $\Delta = p$).

9.18. The fundamental unit $\varepsilon = 170 + 39\sqrt{19}$ of $\mathbb{Z}[\sqrt{19}]$ is $\equiv 1 \bmod 13$, so reduction modulo the prime ideals above 13 only yields the trivial residue classes ± 1.

The fundamental unit $\varepsilon = \frac{1+\sqrt{5}}{2}$, on the other hand, is a primitive root modulo the prime ideals above 11, so in this case the image of the reduction homomorphism is the whole coprime residue class group modulo $\pi = 4 + \sqrt{5}$.

9.19. It suffices to prove the result for coprime values of x and y since rational primes p have the form $p = p + 0\sqrt{10}$. Now write the norm $n = x^2 - 10y^2$ of $\alpha = x + y\sqrt{10}$ as a product of primes p satisfying $(\frac{2}{p}) = (\frac{5}{p}) = +1$ and primes q satisfying $(\frac{2}{p}) = (\frac{5}{p}) = -1$. If $p \mid n$, then $p = \pi\pi'$, and either $\pi \mid \alpha$ or $\pi' \mid \alpha$. Thus it remains to prove the result for elements α whose norm is a product of primes q.

Let $(\alpha) = \mathfrak{q}_1 \cdots \mathfrak{q}_t$ denote its prime ideal factorization; observe that $t = 2s$ must be even since each ideal $\mathfrak{q}_j$ has order 2 in the class group. Now write $\mathfrak{q}_j(2, \sqrt{10}) = \prod(2a_j + b_j\sqrt{10})$. Then $(2, \sqrt{10})^t(\alpha) = \prod(2a_j + b_j\sqrt{10})$, where $(2, \sqrt{10})^t = 2^s$. Dividing each factor on the right by $\sqrt{2}$ we obtain $(\alpha) = \prod(a_j\sqrt{2} + b_j\sqrt{5})$, and this implies our claim. Analogous results hold for other fields with class number 2.

9.20. By the ambiguous class number formula, the class number is odd since there are two ramified primes (2 and q) and since the fundamental unit has norm $+1$ since the equation $t^2 - 2pu^2 = -1$ does not have a solution modulo p by the first supplementary law.

Since the prime ideal $\mathfrak{a} = (2, \sqrt{pq})$ satisfies $\mathfrak{a}^2 = (2)$, it must be principal (there is no class of even order since the class number is odd); thus there must be an element with norm ± 2: $X^2 - 2py^2 = \pm 2$. Clearly $X = 2x$ must be even, and we deduce $2x^2 - py^2 = \pm 1$. The equation $2x^2 - qy^2 = 1$ is impossible modulo 8; thus $2x^2 - qy^2 = -1$ must be solvable. Reducing this equation modulo q implies $2x^2 \equiv -1 \bmod q$, hence $(\frac{2}{q}) = (\frac{-1}{q}) = -1$.

Chapter 10

10.1. This is a purely formal exercise. If 1_A and 1_B denote the neutral elements of A and B, then $(1_A, 1_B)$ is the neutral element of $A \oplus B$. The inverse element of (a, b) is (a^{-1}, b^{-1}), and associativity is directly inherited from A and B.

10.2. For a subgroup U of $A \oplus B$ define the subgroup A_1 of A as the set of all $a \in A$ for which there exists an element $(a, b) \in U$, and define B_1 similarly. Clearly $A_1 \oplus B_1$ is a subgroup of U. Conversely, given an element $(a, b) \in U$, we have $a \in A_1$ and $b \in B_1$ by definition, hence $(a, b) \in A_1 \oplus B_1$.

10.3. This is easy: Changing a by a multiple of $N = n_1 N_2$ does not change the residue classes $a + N_1 \mathbb{Z}$ and $a + N_2 \mathbb{Z}$.

10.4. Let $d = \gcd(N_1, N_2)$ and write $d = mN_1 + nN_2$. Then

$$
\begin{aligned}
\chi(a + d + N\mathbb{Z}) &= \chi(a + mN_1 + nN_2 + N\mathbb{Z}) \\
&= \chi(a + nN_2 + N\mathbb{Z}) && \chi \text{ defined modulo } N_1 \\
&= \chi(a + N\mathbb{Z}) && \chi \text{ defined modulo } N_2
\end{aligned}
$$

This shows that χ is defined modulo d.

10.5. This is a special case of Lemma 10.1 since, by the Chinese Remainder Theorem, $(\mathbb{Z}/N\mathbb{Z})^{\times} \simeq (\mathbb{Z}/N_1)^{\times} \times (\mathbb{Z}/N_2)^{\times}$ because N_1 and N_2 are coprime.

10.6. Since 2 is a primitive root modulo 5, each residue class coprime to 5 can be represented in the form $a \equiv 2^j \bmod p$, and therefore $\chi(a) = \chi(2)^j$. Since $\chi(2) \in \{\pm 1, \pm i\}$ there exist exactly four nontrivial Dirichlet character defined modulo 5.

10.7. The kernel of the projection map $\pi : (\mathbb{Z}/N\mathbb{Z})^{\times} \longrightarrow (\mathbb{Z}/n\mathbb{Z})^{\times}$ consists of all residue classes $a \bmod N$ for which $a \equiv 1 \bmod n$. Clearly χ is trivial on this kernel if and only if χ is defined modulo N/n. This implies the claim.

10.8. For $\Delta = -3, -4, 5$, and 8, we have computed the Pell forms in Sect. 10.2. The only missing cases are the following:

$$
\Delta = -7: \quad f(q) = \frac{q + q^2 - q^3 + q^4 - q^5 - q^6}{1 - q^7} = \frac{q + 2q^2 + q^3 + 2q^4 + q^5}{1 + q + q^2 + q^3 + q^4 + q^5 + q^6}
$$

$$
\Delta = -8: \quad f(q) = \frac{q + q^3 - q^5 - q^7}{1 - q^8} = -\frac{q + q^3}{1 + q^4}
$$

$$
\Delta = -11: \quad f(q) = \frac{q + q^3 + 2q^4 + 3q^5 + 2q^6 + q^7 + q^9}{1 + q + q^2 + \ldots + q^{10}}
$$

$$
\Delta = 12: \quad f(q) = \frac{q - q^5 - q^7 + q^{11}}{1 - q^{12}} = \frac{q - q^3}{1 - q^2 + q^4}.
$$

10.9. Since

$$f_\chi(q) = \frac{\text{Fek}_\chi(q)}{1 - q^N}$$

we have

$$f_\chi\left(\frac{1}{q}\right) = \frac{\text{Fek}_\chi(\frac{1}{q})}{1 - \frac{1}{q}^N} = -\frac{q^N \text{Fek}_\chi(\frac{1}{q})}{1 - q^N}.$$

If $N = \Delta > 0$, then $(\frac{N}{N-1}) = +1$, hence $(\frac{N}{a}) = (\frac{N}{N-a})$. This shows that

$$q^N \text{Fek}_\chi\left(\frac{1}{q}\right) = q^N\left(\frac{1}{q} + \left(\frac{N}{2}\right)\frac{1}{q^2} + \ldots + \left(\frac{N}{N-1}\right)\frac{1}{q^{N-1}}\right)$$

$$= \left(\frac{N}{N-1}\right)q^{N-1} + \left(\frac{N}{N-2}\right)\frac{1}{q^{N-2}} + \ldots + \left(\frac{N}{1}\right)q$$

$$= \text{Fek}_\chi(q).$$

If $N = \Delta > 0$, then $(\frac{N}{N-1}) = -1$, and the same calculation as above shows

$$q^N \text{Fek}_\chi\left(\frac{1}{q}\right) = -\text{Fek}_\chi(q).$$

This proves the functional equation.

10.10. From

$$f_\chi(q) = \frac{q}{1 + q^2} = \frac{A}{1 - qi} + \frac{B}{1 + qi}$$

we obtain

$$q = A(1 + qi) + B(1 - qi) = A + B + (A - B)qi.$$

This implies $A + B = 0$ and $A - B = i$, hence $A = \frac{1}{2i}$ and $B = -\frac{1}{2i}$.

10.11. Here we find

$$f_\chi(q) = \frac{q - q^3}{1 + q^4} = \frac{1}{2\sqrt{2}}\left(\frac{1}{1 - \zeta q} - \frac{1}{1 - \zeta^3 q} - \frac{1}{1 - \zeta^5 q} + \frac{1}{1 - \zeta^7 q}\right).$$

10.12. The partial fraction decomposition of

$$\frac{\frac{\text{Fek}_\chi(q)}{q}}{1 - q^N} = \sum_{k=0}^{N-1} \frac{a_k}{q - \zeta^k}$$

is given by Euler's formulas:

$$a_k = \frac{\frac{\text{Fek}_\chi(\zeta^k)}{\zeta^k}}{-N\zeta^{k(N-1)}} = -\frac{\text{Fek}_\chi(\zeta^k)}{N}.$$

10.13. We begin by observing that $\left(\frac{n}{p}\right) \equiv n^m \bmod p$ implies

$$\text{Fek}_p(x) = \sum_{n=1}^{p-1} \left(\frac{n}{p}\right) x^n \equiv \sum_{n=1}^{p-1} n^m x^n \bmod p.$$

Therefore $\text{Fek}_p(1) \equiv \sum_{n=1}^{p-1} n^m \equiv 0 \bmod p$ by Gauss's congruence (3.12).

10.14. We form the derivatives of the polynomials in the preceding exercise:

$$\text{Fek}'_p(x) \equiv \sum n^m \cdot n x^{n-1} \bmod p,$$

$$\text{Fek}''_p(x) \equiv \sum n^m \cdot n(n-1) x^{n-2} \bmod p,$$

$$\cdots \qquad \cdots$$

$$\text{Fek}_p^{(k)} \equiv \sum n^m \cdot n(n-1) \cdots (n-k+1) x^{n-k} \bmod p.$$

After plugging in $x = 1$, the sum in $\text{Fek}_p^{(k)}(1)$ is over polynomials in n whose degree is equal to $m + k$. According to (3.12), we have $\text{Fek}_p^{(k)}(1) \equiv 0 \bmod p$ for $0 \le k < m$. For $k = m$, on the other hand, we get $\text{Fek}_p^{(m)}(1) \equiv -1 \bmod p$.

10.15. Since $\Phi_p(x) = \frac{x^p - 1}{x-1}$ we have

$$\Phi_p(x+1) = \frac{(x+1)^p - 1}{x} = x^{p-1} + \binom{p}{1} x^{p-2} + \ldots + \binom{p}{p-1}.$$

This polynomial is Eisenstein since $p \mid \binom{p}{k}$ for $1 \le k \le p-1$ and since $p^2 \nmid \binom{p}{p-1} = p$. Therefore $\Phi_p(x+1)$ is irreducible.

10.16. We find the following values for N_p:

p	5	7	11	13	17	19	23	29	31	37
N_p	5	3	11	15	17	27	23	29	27	27

Clearly $N_p = p$ for primes $p \equiv 2 \bmod 3$. If $p \equiv 1 \bmod 3$, write $p = a^2 + 3b^2$ with $a \equiv 1 \bmod 3$. Then $N_p = p + 2a$.

10.17. We have

$$\text{Fek}_p(x)^n = \sum_{t_1,\dots,t_n} \binom{t_1}{p}\binom{t_2}{p}\cdots\binom{t_n}{p}x^{t_1+t_2+\dots+t_n}.$$

If x is a p-th root of unity, then x^m only depends on the residue class of $m \bmod p$. Thus if $t_1 + \dots + t_n \equiv a \bmod p$, then

$$\text{Fek}_p(x)^n = \sum_{a=0}^{p-1} J_n(a)x^a$$

as claimed.

Bibliography

1. M. Aigner, *Markov's Theorem and 100 Years of the Uniqueness Conjecture* (Springer, Cham, 2013)
2. N.C. Ankeny, S. Chowla, H. Hasse, On the class-number of the maximal real subfield of a cyclotomic field. J. Reine Angew. Math. **217**, 217–220 (1965)
3. G. Arendt, *Éléments de la théorie des nombres complexes de la forme $a + b\sqrt{-1}$*. Programme Collège Royal Français, September 1863
4. A. Ash, R. Gross, *Fearless Symmetry. Exposing the Hidden Patterns of Numbers* (Princeton University Press, Princeton, 2006)
5. A. Ash, R. Gross, *Elliptic Tales: Curves, Counting, and Number Theory* (Princeton University Press, Princeton, 2012)
6. R. Ayoub, On L-functions. Monatsh. Math. **71**, 193–202 (1967)
7. R. Ayoub, S. Chowla, On Euler's polynomial. J. Numb. Theory **13**, 443–445 (1981)
8. E.J. Barbeau, *Pell's Equation* (Springer, New York, 2003)
9. E. Benjamin, C. Snyder, Elements of order four in the narrow class group of real quadratic fields. J. Aust. Math. Soc. **100**, 21–32 (2016)
10. C. Bergmann, *Über Eulers Beweis des großen Fermatschen Satzes für den Exponenten 3*. Math. Ann. **164**, 159–175 (1966)
11. D. Bernoulli, Observationes de seriebus quae formantur ex additione vel subtractione quacuncque terminorum se mutuo consequentium. Commentarii Acad. Sci. Imp. Petropol. III (1728), 85–100
12. Y.F. Bilu, Y. Bugeaud, M. Mignotte, *The Problem of Catalan* (Springer, New York, 2014)
13. J.H. Bruinier, G. van der Geer, G. Harder, D. Zagier, *The 1-2-3 of Modular Forms* (Springer, New York, 2008)
14. J.W.S. Cassels, *Lectures on Elliptic Curves* (Cambridge University Press, Cambridge, 1991)
15. J.W.S. Cassels, *Local Fields* (Cambridge University Press, Cambridge, 1986)
16. W. Castryck, A shortened classical proof of the quadratic reciprocity law. Am. Math. Monthly **115**, 550–551 (2008)
17. H.H. Chan, L. Long, Y. Yang, A cubic analogue of the Jacobsthal identity. Am. Math. Monthly **118**, 316–326 (2011)
18. P. Chebyshev, Sur les formes quadratiques. J. Math. Pures Appl. **16**, 257–282 (1851)
19. K. Chemla, S. Guo, *Les neuf chapitres. Le Classique mathématique de la Chine ancienne et ses commentaires* (Dunod, Paris, 2004)
20. H. Cohen, *A Course in Computational Algebraic Number Theory* (Springer, Berlin, Heidelberg, 1993)

© The Author(s), under exclusive license to Springer Nature Switzerland AG 2021
F. Lemmermeyer, *Quadratic Number Fields*, Springer Undergraduate
Mathematics Series, https://doi.org/10.1007/978-3-030-78652-6

21. J.H.E. Cohn, Eight Diophantine equations, Proc. Lond. Math. Soc. (3) **16**, 153–166 (1966); Corr. ibid. (3) **17**, 381 (1967)

22. H. Cohn, *Advanced Number Theory* (Dover Publications, New York, 1980); Original: *A Second Course in Number Theory* (Wiley, New York, 1962)

23. H. Cohn, *A Classical Invitation to Algebraic Numbers and Class Fields*, 2nd edn. Universitext (Springer, New York, 1988)

24. J.B. Cosgrave, K. Dilcher, *An Introduction to Gauss Factorials*, Am. Math. Monthly **118**, 812–829 (2011)

25. D. Cox, *Primes of the Form $x^2 + ny^2$* (Wiley, New York, 1989)

26. D. Cox, *Why Eisenstein proved the Eisenstein criterion and why Schönemann discovered it first*. Am. Math. Monthly **118**, 3–21 (2011)

27. H. Davenport, *The Higher Arithmetic*, 8th edn. (Cambridge University Press, Cambridge, 2008)

28. R. Dedekind, *Gesammelte mathematische Werke* (Friedrich Vieweg & Sohn, Braunschweig, 1932)

29. P.G.L. Dirichlet, Mémoire sur l'impossibilité de quelques équations indéterminées du cinquième degré. Acad. Sci. Royale France 1825; Werke I, 1–46

30. P.G.L. Dirichlet, Einige Resultate von Untersuchungen über eine Classe homogener Functionen des dritten und der höheren Grade, Ber. Verh. Königl. Preuß. Akad. Wiss. **1841**, 280–285 (1841); Werke I, 625–632

31. P.G.L. Dirichlet, *Vorlesungen über Zahlentheorie*, 2nd edn., ed. by R. Dedekind (Brunswick 1871); English translation *Lectures on Number Theory* (American Mathematical Society and London Mathematical Society, London , 1999)

32. F.W. Dodd, Number theory in the integral domain $\mathbb{Z}[\frac{1}{2} + \frac{1}{2}\sqrt{5}]$, Dissertation Univ. Northern Colorado, 1981; published as *Number Theory in the Quadratic Field with Golden Section Unit* (Polygonal Publishing House, Passaic, NJ, 1983)

33. D.S.L. Eelkema, Integer factorisation using conics, Bachelor thesis, Groningen (2020)

34. R.B. Eggleton, C.B. Lacampagne, J.L. Selfridge, Euclidean quadratic fields. Am. Math. Monthly **99**, 829–837 (1992)

35. G. Eisenstein, Neuer und elementarer Beweis des Legendre'schen Reciprocitäts-Gesetzes. J. Reine Angew. Math. **27**, 322–329 (1844); Math. Werke I, 100–107

36. L. Euler, *Theoremata circa divisores numerorum in hac forma paa $\pm$ qbb contentorum*. Commun. Acad. Sci. Petropol. **14**, 151–158 (1751); Opera Omnia **I - 2**, 194–222

37. L. Euler, De numeris, qui sunt aggregata duorum quadratorum. Nova Comm. Acad. Sci. Petropol. **4**(1752/3), 1758, 3–40; Opera Omnia **I - 2**, 295–327

38. L. Euler, *Vollständige Anleitung zur Algebra* (Birkhäuser, Basel, 1770); Leipzig (1883)

39. L. Euler, *Opera Postuma. Fragmenta arithmetica ex adversariis mathematicis deprompta*, vol. 1, (1862), pp. 231–232; S. 157

40. T. Evink, A. Helminck, Tribonacci numbers and primes of the form $p = x^2 + 11y^2$. Math. Slovaca **69**, 521–532 (2019)

41. F.G. Frobenius, Über das quadratische Reziprozitätsgesetz I, Sitzungsberichte Berliner Akad. 335–349 (1914). Ges. Abhandl. 628–642

42. F.G. Frobenius (unter Benutzung einer Mitteilung des Herrn Dr. R. Remak), *Über quadratische Formen, die viele Primzahlen darstellen*. Sitz. Kön. Preuß. Akad. Wiss. Berlin (1912), 966–980; Ges. Abh. III, 573–587

43. C.F. Gauß, *Disquisitiones Arithmeticae*, 1801; deutsche Übersetzung Maser 1889; Neuauflage (K. Reich, Hrsg.), Georg Olms Verlag 2015

44. C.F. Gauß, *Theorematis fundamentalis in doctrina de residuis quadraticis demonstrationes et amplicationes novae*, 1818; Werke II, 47–64

45. C.F. Gauß, *Theorie der biquadratischen Reste. Zweite Abhandlung* (Göttingen, 1832); deutsche Übersetzung Maser 1889

46. K. Girstmair, Kroneckers Lösung der Pellschen Gleichung auf dem Computer. Math. Semesterber. **53**, 45–64 (2006)

47. M. Hall, *Some equations* $y^2 = x^3 - k$ *without integer solutions*. J. Lond. Math. Soc. **28**, 379–383 (1953)
48. F. Halter-Koch, Quadratische Ordnungen mit großer Klassenzahl. J. Numb. Theory **34**, 82–94 (1990)
49. K. Halupczok, *Euklidische Zahlkörper*. Diplomarbeit (Hartung-Gorre Verlag, Konstanz, 1997)
50. S. Hambleton, Generalized Lucas-Lehmer tests using Pell conics. Proc. Am. Math. Soc. **140**, 2653–2661 (2012)
51. S. Hambleton, F. Lemmermeyer, Arithmetic of Pell surfaces. Acta Arith. **146**, 1–12 (2011)
52. S. Hambleton, V. Scharaschkin, Pell conics and quadratic reciprocity. Rocky Mt. J. Math. **42**, 91–96 (2012)
53. G.H. Hardy, E.M. Wright, *Einführung in die Zahlentheorie* (R. Oldenbourg Verlag, München, 1958)
54. M. Harper, A proof that $\mathbb{Z}[\sqrt{14}]$ *is Euclidean*, Ph.D. thesis, McGill University, 2000
55. M. Harper, $\mathbb{Z}[\sqrt{14}]$ is Euclidean. Can. J. Math. **56**, 55–70 (2004)
56. K. Hashimoto, L. Long, Y. Yang, Jacobsthal identity for $\mathbb{Q}(\sqrt{-2})$, Forum Math. **24**, 1225–1238 (2012)
57. H. Hasse, Über eindeutige Zerlegung in Primelemente oder in Primhauptideale in Integritätsbereichen. J. Reine Angew. Math. **159**, 3–12 (1928)
58. H. Hasse, Über mehrklassige, aber eingeschlechtige reellquadratische Zahlkörper. Elem. Math. **20**, 49–59 (1965)
59. T.L. Heath, *Diophantus of Alexandria. A Study in the History of Greek Algebra* (Cambridge University Press, Cambridge, 1910)
60. E. Hecke, *Lectures on the Theory of Algebraic Numbers* (Springer, Berlin, 1981)
61. D. Hilbert, *Die Theorie der Algebraischen Zahlkörper*, Jahresber. DMV **4**, 175–546 (1897); Engl. Transl. I. Adamson, *The Theory of Algebraic Number Fields* (Springer, New York, 1998)
62. F. Hirzebruch, D. Zagier, *The Atiyah-Singer Theorem and Elementary Number Theory* (Publish or Perish, Boston, 1974)
63. J. Høyrup, *Algebra in Cuneiform* (Max-Planck-Gesellschaft zur Förderung der Wissenschaften, Berlin, 2017)
64. A. Hurwitz, Über eine Aufgabe der unbestimmten Analysis. Archiv. Math. Phys. **3**, 185–196 (1907); Mathematische Werke **2**, 410–421
65. K. Ireland, K. Rosen, *A Classical Introduction to Modern Number Theory* (Springer, New York, 1990)
66. M. J. Jacobson, H. C. Williams, *Solving the Pell Equation* (CMS, New York, 2009)
67. E. Jacobsthal, *Anwendungen einer Formel aus der Theorie der quadratischen Reste*, Diss., Berlin, 1906
68. E. Jacobsthal, Über die Darstellung der Primzahlen der Form $4n + 1$ als Summe zweier Quadrate. J. Reine Angew. Math. **132**, 238–246 (1907)
69. H.W.E. Jung, *Einführung in die Theorie der quadratischen Zahlkörper* (Jänicke, Leipzig, 1936)
70. L. Kronecker, *Ueber die Potenzreste gewisser complexer Zahlen* (Monatsber, Berlin, 1880), pp. 404–407; Werke II, 95–101
71. E.E. Kummer, Zur Theorie der complexen Zahlen. J. Reine Angew. Math. **35**, 319–326 (1847)
72. R.C. Laubenbacher, D. Pengelley, Eisenstein's misunderstood geometric proof of the quadratic reciprocity theorem. College Math. J. **25**, 29–34 (1994)
73. V.A. Lebesgue, Recherches sur les nombres. J. Math. Pures Appl. **3**, 113–144 (1838)
74. V.A. Lebesgue, Sur l'impossibilité, en nombres entiers, de l'équation $x^m = y^2 + 1$. Nouv. Ann. Math. (1) **9**, 178–181 (1850)
75. D.H. Lehmer, On a problem of Störmer. Illinois J. Math. **8**, 57–79 (1964)
76. F. Lemmermeyer, The Euclidean algorithm in algebraic number fields. Expositiones Math. **13**, 385–416 (1995)

77. F. Lemmermeyer, *Reciprocity Laws* (Springer, Berlin, 2000)
78. F. Lemmermeyer, Higher Descent on Pell Conics, I. From Legendre to Selmer, arXiv:math/0311309; II. Two Centuries of Missed Opportunities, math/0311296; III. The First 2-Descent, math/0311310 (2003)
79. F. Lemmermeyer, Zur Zahlentheorie der Griechen. I: Euklids Fundamentalsatz der Arithmetik. Math. Semesterber. **55**, 181–195 (2008)
80. F. Lemmermeyer, Zur Zahlentheorie der Griechen. II: Gaußsche Lemmas und Rieszsche Ringe. Math. Sem.ber. **56**, 39–51 (2009)
81. F. Lemmermeyer, Jacobi and Kummer's ideal numbers. Abh. Math. Sem. Hamburg **79**, 165–187 (2009)
82. F. Lemmermeyer, Relations in the 2-class group of quadratic number fields, J. Austr. Math. Soc. **93**, 115–120 (2012)
83. F. Lemmermeyer, Parametrization of algebraic curves from a number theorist's point of view. Am. Math. Monthly **119**, 573–583 (2012)
84. F. Lemmermeyer, Binomial squares in pure cubic number fields. J. Théor. Nombres Bordeaux **24**, 691–704 (2012)
85. F. Lemmermeyer, *Mathematik à la Carte. Elementargeometrie an Quadratwurzeln mit einigen geschichtlichen Bemerkungen* (Springer-Spektrum, 2015)
86. F. Lemmermeyer, *Mathematik à la Carte. Quadratische Gleichungen mit Schnitten von Kegeln* (Springer-Spektrum, Berlin, 2016)
87. F. Lemmermeyer, Composite values of irreducible polynomials. Elemente d. Math. **74**, 36–37 (2019)
88. F. Lemmermeyer, *4000 Jahre Zahlentheorie* (Springer-Verlag, to appear)
89. F. Lemmermeyer, M. Mattmüller (Hrsg.), *Leonhardi Euler Opera Omnia (IV) 4. Correspondence of Leonhard Euler with Christian Goldbach* (Birkhäuser, Basel, 2015)
90. F. Lemmermeyer, P. Roquette (Hrsg.), *Helmut Hasse und Emmy Noether – Die Korrespondenz 1925 – 1935* (Univ.-Verlag, Göttingen, 2006)
91. H.W. Lenstra, Solving the Pell equation. Notices Am. Math. Soc. **49**, 182–192 (2002)
92. W.J. Leveque, *Topics in Number Theory*, vol. II (Addison-Wesley, Reading, MA, 1961)
93. M. Mignotte, A new proof of Ko Chao's Theorem, Math. Notes **76**, 358–367 (2004)
94. P. Mihailescu, Primary cyclotomic units and a proof of Catalan's conjecture. J. Reine Angew. Math. **572**, 167–195 (2004)
95. R. Mollin, On the divisor function and class numbers of real quadratic fields. I. Proc. Jpn. Acad. **66**, 109–111 (1990)
96. M.G. Monzingo, An elementary evaluation of the Jacobsthal sum. J. Number Theory **22**, 21–25 (1986)
97. L.J. Mordell, *Diophantine Equations* (Academic, London, 1969)
98. T. Motzkin, The Euclidean algorithm. Bull. Am. Math. Soc. **55**, 1142–1146 (1949)
99. T. Nagell, Solution complète de quelques équations cubiques à deux indéterminées. J. Math. Pures Appl. **4**, 209–270 (1925)
100. M. Nyberg, Culminating and almost culminating continued fractions (Norwegian). Norsk Mat. Tidskr. **31**, 95–99 (1949)
101. A. Oppenheim, Quadratic fields with and without Euclid's algorithm. Math. Ann. **109**, 349–352 (1934)
102. D. Pengelley, F. Richman, Did Euclid need the Euclidean algorithm to prove unique factorization?. Am. Math. Monthly **113**, 196–205 (2006)
103. J. Plemelj, Die Unlösbarkeit von $x^5 + y^5 + z^5 = 0$ im Körper $k\sqrt{5}$. Monatsh. Math. Phys. **23**, 305–308 (1912)
104. K. Plofker, *Mathematics in India* (Princeton University Press, Princeton, 2009)
105. G. Rabinovitch, Eindeutigkeit der Zerlegung in Primzahlfaktoren im quadratischen Zahlkörper. Proc. Int. Congr. Math. **1912**, 418–421 (1912)
106. L. Rédei, Über die quadratischen Zahlkörper mit Primzerlegung. Acta Sci. Math. (Szeged) **21**, 1–3 (1960)

107. L.W. Reid, *The Elements of the Theory of Algebraic Numbers* (The Macmillan Co., New York, 1910)

108. P. Ribenboim, *Catalan's Conjecture (are* 8 *and* 9 *the only Consecutive Primes?* (Academic, Boston, 1994)

109. P. Ribenboim, *Meine Zahlen, meine Freunde. Glanzlichter der Zahlentheorie* (Springer, Berlin, 2009)

110. P. Roquette, *The Riemann Hypothesis in Characteristic p in Historical Perspective* (Springer, Cham, 2018)

111. W. Scharlau, H. Opolka, *Von Fermat bis Minkowski. Eine Vorlesung über Zahlentheorie und ihre Entwicklung* (Springer, Berlin, 1980)

112. A. Scholz, *Einführung in die Zahlentheorie* (de Gruyter, Berlin, 1939)

113. R. Schoof, *Catalan's Conjecture* (Springer, London, 2008)

114. C.-O. Selenius, Rationale of the Chakravala Process of Jayadeva and Bhaskara II. Hist. Math. **2**, 167–184 (1975)

115. D. Shanks, On Gauss's class number problems. Math. Comp. **23**, 151–163 (1969)

116. H. Siebeck, Die recurrenten Reihen, vom Standpuncte der Zahlentheorie aus betrachtet. J. Reine Angew. Math. **33**, 71–77 (1846)

117. J. Silverman, J. Tate, *Rational Points on Elliptic Curves* (Springer, New York, 1992)

118. J. Sommer, *Vorlesungen über Zahlentheorie. Einführung in die Theorie der algebraischen Zahlkörper* (Teubner, Leipzig, 1907)

119. C. Størmer, *Solution d'un problème curieux qu'on rencontre dans la théorie élémentaire des logarithmes*, Nyt tidsskrift for matematik **19**, 1–7 (1908)

120. G. Szekeres, On the number of divisors of $x^2 + x + A$, J. Number Theory **6**, 434–442 (1974)

121. G. Terjanian, Sur l'équation $x^{2p} + y^{2p} = z^{2p}$. C. R. Acad. Sci. Paris **285**, 973–975 (1977)

122. A. Thue, Über Annäherungswerte algebraischer Zahlen. J. Reine Angew. Math. **135**, 284–305 (1909)

123. E. Trost, Eine Bemerkung zur diophantischen Analysis. Elem. Math. **26**, 60–61 (1971)

124. E. Trost, Solution of Problem E 2332. Am. Math. Monthly **87**, p. 77 (1972)

125. L. Tschakaloff, Unmöglichkeitsbeweis der Gleichung $\alpha^5 + \beta^5 = \eta\gamma^5$ im quadratischen Körper $K(\sqrt{5})$. Tôhoku Math. J. **27**, 189–194 (1926)

126. K. Vogel, *Vorgriechische Mathematik II. Die Mathematik der Babylonier* (Schroedel, Hannover, 1959)

127. K. Vogel (Hrsg.), *Neun Bücher arithmetischer Technik* (Ostwalds Klassiker der exakten Naturwissenschaften, Braunschweig, 1968)

128. F. von Schafgotsch, *Abhandlung über einige Eigenschaften der Prim- und zusammengesetzten Zahlen*, Abhandlung der Böhmischen Gesellschaft der Wissenschaften in Prag (1786), pp. 123–159

129. L. von Schrutka, *Ein Beweis für die Zerlegbarkeit der Primzahlen von der Form* $6n + 1$ *in ein einfaches und ein dreifaches Quadrat*. J. Reine Angew. Math. **140**, 252–265 (1911)

130. S. Wagstaff, *The Joy of Factoring* (AMS, Providence, 2013)

131. A. Wakulicz, On the equation $x^3 + y^3 = 2z^3$. Colloq. Math. **5**, 11–15 (1957)

132. A. Weil, *Number Theory: An Approach Through History from Hammurapi to Legendre* (Birkhäuser, New York, 1984)

133. A. Widmer, Über die Anzahl der Lösungen gewisser Kongruenzen nach einem Primzahlmodul, Diss. ETH Zurich, 1919

134. D. Zagier, *Zetafunktionen und quadratische Körper* (Springer, Berlin, Heidelberg, 1981)

135. Ch. Zeller, *Beweis des Reciprocitätsgesetzes für die quadratischen Reste* (Monatsber, Berlin, 1872), pp. 846–847

136. G. Zolotareff, Nouvelle démonstration de la loi de réciprocité de Legendre. Nouv. Ann. Math. (2) **11**, 354–362 (1872)

Name Index

© The Author(s), under exclusive license to Springer Nature Switzerland AG 2021   337
F. Lemmermeyer, *Quadratic Number Fields*, Springer Undergraduate
Mathematics Series, https://doi.org/10.1007/978-3-030-78652-6

Subject Index

Printed in the United States
by Baker & Taylor Publisher Services